OECD Environmental Performance Reviews

CHINA

OECD

ORGANISATION FOR ECONOMIC CO-OPERATION AND DEVELOPMENT

ORGANISATION FOR ECONOMIC CO-OPERATION AND DEVELOPMENT

The OECD is a unique forum where the governments of 30 democracies work together to address the economic, social and environmental challenges of globalisation. The OECD is also at the forefront of efforts to understand and to help governments respond to new developments and concerns, such as corporate governance, the information economy and the challenges of an ageing population. The Organisation provides a setting where governments can compare policy experiences, seek answers to common problems, identify good practice and work to co-ordinate domestic and international policies.

The OECD member countries are: Australia, Austria, Belgium, Canada, the Czech Republic, Denmark, Finland, France, Germany, Greece, Hungary, Iceland, Ireland, Italy, Japan, Korea, Luxembourg, Mexico, the Netherlands, New Zealand, Norway, Poland, Portugal, the Slovak Republic, Spain, Sweden, Switzerland, Turkey, the United Kingdom and the United States. The Commission of the European Communities takes part in the work of the OECD.

OECD Publishing disseminates widely the results of the Organisation's statistics gathering and research on economic, social and environmental issues, as well as the conventions, guidelines and standards agreed by its members.

This work is published on the responsibility of the Secretary-General of the OECD. The opinions expressed and arguments employed herein do not necessarily reflect the official views of the Organisation or of the governments of its member countries.

Also available in French under the title:
Examens environnementaux de l'OCDE : CHINE

and in Chinese under the title:
OECD 中国环境绩效评估报告

FOREWORD

A healthy economy needs a healthy environment. In line with its mission to promote sustainable economic growth and rising living standards, the OECD promotes better integration of environmental concerns into economic and sectoral policies. In this context, the OECD's Environmental Performance Reviews – conducted since 1992 – provide a systematic analysis of countries' actions to reach their environmental goals and specific recommendations to improve their outcomes.

We have learned much from past exercises of OECD member countries. But we have also continued to extend our reach and effectiveness with the work with non member countries. This co-operation has proven fruitful. In the case of China, the present review builds on over a decade of OECD-China environmental collaboration and is part of a wider strategy that relies on studies of several other areas such as: the Economic Surveys and the reviews of Agricultural Policies and Governance Reform.

The OECD Environmental Performance Review of China was based on the same methodology used for environmental reviews of all OECD countries. This review evaluates China's performance in reducing pollution, improving the management of natural resources, implementing economically efficient and environmentally effective policies, and strengthening international co-operation. It includes Conclusions and 51 Recommendations which were presented to China and all OECD countries in a peer review process.

The review confirms that rapid economic development, industrialisation and urbanisation have generated severe and growing pressures on the environment, resulting in significant damage to human health and depletion of natural resources. For example, air pollution levels in some cities are among the worst in the world, one-third of water courses are severely polluted, and illnesses and injuries are associated with poor environmental and occupational conditions. In terms of resource use, energy consumption per unit of GDP is about 20% higher than the OECD average, and large-scale transfers of water from southern to northern China are needed to keep up with growing demand.

The review supports recent declarations by the Chinese authorities to encourage more balanced patterns of development by promoting concepts such as "harmonious society" and "scientific development". It also found that China has embraced "polluter pays" and "user pays" principles, approaches promoted by OECD. This is reflected in China's environmental legislation and the introduction of emission charges and water pricing.

However, the implementation of these environmental policies lacks environmental effectiveness and economic efficiency. Energy, transport, industry and urban development policies need to integrate environmental concerns institutionally and through market-based mechanisms. Energy, water and other natural resources still remain under-priced. Further, international environmental co-operation needs to be strengthened in the mutual interest of China and the international community. Collective actions are necessary to protect the ozone layer, to combat climate change, and to tackle regional environmental issues such as acid rain, desertification and protection of the marine ecosystems.

I urge China to strengthen its efforts, to move towards a more environmentally "harmonious society", and to step up international environmental co-operation. It is in China's own interest to address the negative impacts of environmental degradation and natural resources depletion on the health and well-being of its people and on its economic development. It is also in the interest of making globalisation and environment compatible and mutually supportive. As China enjoys a continued high rate of economic growth, further environmental progress will reward China with major economic and social benefits.

Angel Gurría
Secretary-General

TABLE OF CONTENTS

Part II
SUSTAINABLE DEVELOPMENT

Part III
INTERNATIONAL CO-OPERATION

LIST OF FIGURES, TABLES AND BOXES

Figures

Tables

Boxes

Signs

The following signs are used in Figures and Tables:

.. : not available

− : nil or negligible

. : decimal point

n.a. : not applicable

* : all countries are not included.

Country Aggregates

OECD Europe: All European member countries of the OECD (Austria, Belgium, Czech Republic, Denmark, Finland, France, Germany, Greece, Hungary, Iceland, Ireland, Italy, Luxembourg, Netherlands, Norway, Poland, Portugal, Slovak Republic, Spain, Sweden, Switzerland, Turkey and United Kingdom).

OECD: The countries of OECD Europe plus Australia, Canada, Japan, the Republic of Korea, Mexico, New Zealand and the United States.

Country aggregates may include Secretariat estimates.

Currency

Monetary unit: CNY (yuan)

In 2006, CNY 7.95 = USD 1.

Cut-off Date

This report is based on information and data available up to November 2006.

LIST OF TEAM MEMBERS

Mr. Leif Benergård   Expert from reviewing country: Sweden
Mr. Alan Colodey   Expert from reviewing country: Canada
Mr. Andrea De Angelis   Expert from reviewing country: Italy
Mr. Ryo Fujikura   Expert from reviewing country: Japan
Ms. Gørild Merethe Heggelund   Expert from reviewing country: Norway
Mr. Mark Kasman   Expert from reviewing country: United States
Ms. Kim Kwang-Yim   Expert from reviewing country: Korea
Mr. Helmut Schnurer   Expert from reviewing country: Germany
Mr. Simon Spooner   Expert from reviewing country: United Kingdom

Mr. Christian Avérous   OECD Secretariat
Mr. Gérard Bonnis   OECD Secretariat
Ms. Sylvie Dénaux   OECD Secretariat
Mr. Brendan Gillespie   OECD Secretariat
Ms. Emmanuelle Guidetti   OECD Secretariat
Mr. Krzysztof Michalak   OECD Secretariat
Mr. Eduard Goldberg   OECD Secretariat (Consultant)
Mr. Karl Hallding   OECD Secretariat (Consultant)
Mr. Bill Long   OECD Secretariat (Consultant)
Ms. Chen Mo   OECD Secretariat (Consultant)

The peer review meeting of the OECD Working Party on Environmental Performance took place in Beijing on 8-9 November 2006. It was opened by *Mr. PAN Yue* (Vice Minister, State Environmental Protection Administration, China) and *Mr. Kiyo AKASAKA* (Deputy Secretary General, OECD). The Delegation of China and the Delegations of OECD member countries examined the environmental performance of China. The meeting was closed by *Mr. ZHOU Jian* (Vice Minister, SEPA, China) and *Mr. Lorents LORENTSEN* (Director, Environment Directorate, OECD).

The *environmental data* in the report are based on official environmental statistics from Chinese and intergovernmental organisations sources. In particular, the *OECD and the UN Statistical Division* collected jointly environmental data from China, by use of internationally recognised methodology.

1

CONCLUSIONS AND RECOMMENDATIONS *

Since the launch of the "open door" policy in 1978, economic growth has been remarkable. Over the last 15 years, the average rate of *economic growth has been 10.1% per year*. China now has the *fourth largest economy in the world*. Large foreign direct investment and the increased role of market forces have facilitated the *country's integration into the global economy*. In the process, large numbers of people have escaped extreme poverty. However, while China has about 20% of the world's population, *GDP per capita is still low* (USD 6 000 for China compared to USD 25 000 for OECD countries at PPP 2000) and is *unevenly distributed* across the country, with wealthier coastal provinces and less-developed western provinces. Large migrations have contributed to rapid urbanisation (now 43% of the population), and *income disparity* between urban and rural areas has increased. *Poverty remains a serious challenge* in rural China.

This rapid economic growth, industrialisation and urbanisation have generated high pressures on the environment, with consequent damage to health and natural resources. Chinese authorities, aware of the deterioration of the environment, are promoting more balanced patterns of development, using concepts such as *"harmonious society"* and "scientific development". Their responses have included planning for national economic and social development (covering environmental management efforts), modern environmental legislation, strengthened environmental institutions, and higher priority to environmental and natural resources management. Nevertheless, air pollution in some Chinese cities reaches levels that are among the worst in the world, energy intensity is about 20% higher than the OECD average, and about a third of the water courses are severely polluted. Challenges with waste management, desertification, and nature and biodiversity protection remain. To achieve a *new*

---

* These Conclusions and Recommendations have been approved by all the Delegations of the OECD Working Party on Environmental Performance, including the Chinese Delegation, at its meeting (Beijing, 8-9 November 2006).

economic and social development model (a resource saving and environmentally friendly society according to the 11th FYP), China will need to i) strengthen the effectiveness and efficiency of the implementation of its environmental policies; and ii) enhance the integration of environmental concerns into economic decisions (e.g. fiscal, energy, agriculture, transport and land-use decisions).

Environmental issues in China have often a strong *international dimension*, reflecting regional environmental interdependencies (e.g. transboundary air and water issues, regional seas pollution, desertification) and *global economic and environmental interdependencies*. The environmental pressures and demand for energy and other resources associated with China's rapid economic development dramatically underlines questions about the environmental sustainability of current production and consumption patterns globally. Trade as well as financing of development (e.g. official development assistance, foreign direct investment) have important environmental dimensions. China therefore has a *shared interest* with OECD and other countries to address related challenges, and has significantly enhanced its engagement in international environmental co-operation in recent years.

This report examines progress made by China since 1990 and evaluates the extent to which the country's *domestic objectives and international commitments* are being met. Some 51 recommendations are made that could help strengthen China's environmental performance in the context of sustainable development.

1. Environmental Management

Implementing environmental policies more effectively and efficiently

China's comprehensive and modern set of environmental laws, together with its successive Five-Year Plans for National Economic and Social Development (FYPs) and Five-Year Environmental Plans (FYEPs), provide a high-quality framework for pursuing sustainable development and environmental progress. In December 2005, the State Council issued a decision for better implementing environmental policies. In April 2006, the Chinese Premier announced, in the sixth national environmental protection meeting, three new policy directions, including: integrating environmental protection and economic decision-making on an equal footing, further decoupling pollutant emissions from economic growth, applying a mix of instruments to resolve environmental problems. The proposed directions and measures are being implemented and go a long way towards addressing the environmental policy implementation gap. Within their mandates, *departments under the State Council have worked hard to support environmental policy implementation.* A range of regulatory and economic instruments (e.g. pollution charges, user charges, emissions trading) and policy approaches that harness markets and public interest in the environment have been developed. Campaigns and award schemes to support implementation at the local level have been organised; work with non-governmental organisations (NGOs) to develop procedures for public participation in *environmental impact assessment* (EIA) is an important recent example. There is evidence that *local leaders* in some of the richer provinces are responding to demands from the public for better environmental conditions, and are recognising the benefits to the economy and the society. More than 8 000 companies are registered under *ISO 14000.* In 2004, pollution abatement and control (PAC) investment expenditure was 1.2% of GDP.

However, these efforts have not been sufficient to keep pace with the environmental pressures and challenges generated by the very rapid growth of China's developing economy nor to capture the potential economic benefits to be obtained from improved pollution abatement and nature protection. Overall, environmental efforts have lacked effectiveness and efficiency, largely as a result of an *implementation gap.* The weaknesses in the present system are demonstrated by the failure to achieve some of the key objectives of the 10th FYP, and the severity of environmental problems in many parts of China. National environmental legislation and regulations should be *compiled* in an environmental code, to make them more consistent and user-friendly. Environmental policy priorities should be *focused on human health and key natural resources.* Consistent nationwide implementation of environmental regulations for products and industrial/energy facilities should be enhanced and given priority. The biggest obstacles to environmental policy implementation are at the *local level.* The performance objectives of local leaders, the pressures to raise revenues locally to finance un-funded mandates, and the limited accountability to local populations have generally meant that *economic priorities have over-ridden environmental concerns.* There is a need for much stronger *monitoring, inspection and enforcement capabilities* to establish a better mix of incentives and sanctions. In addition, *environmental expenditure* needs to be made more efficiently, and environmental policy instruments need to be made more effective. Implementation of the polluter pays and user pays principles should be strengthened. Special provisions are needed to integrate environment into the development strategies of the less-developed regions and to ensure the affordability of environmental services for the poor. There is an increase in damages associated with disasters of climatic and industrial origin, requiring improved prevention and mitigation measures.

Recommendations:

- *implement environmental law and regulations nationwide* for products and industrial/energy facilities; strengthen *monitoring, inspection and enforcement capabilities* throughout the country, including through the independence of the enforcement functions of Environmental Protection Bureaus (EPBs);

- consider establishing *SEPA as a ministry*; strengthen SEPA's supervisory capacity of EPBs in local government;

- continue efforts to make *local leaders more accountable* to the higher level government and to local populations for their environmental performance;

- strengthen the *integrated permitting system* and establish it as a more central instrument for pollution prevention and control; strengthen the integration of environmental protection in land-use planning and regulations, as well as in other relevant plans and regulations;

- extend the use of *pollution charges, user charges, emissions trading and other market-based instruments* and their incentive functions, taking social factors into account.

Air

During the review period, China achieved improvements in *ambient air quality* (e.g. lowering the concentration of SO_2 in urban areas and designated control zones) and in decoupling emissions of SO_2, NO_2 and CO_2 from economic growth. The overall emission reduction targets for SO_2, soot and dust from stationary sources set out in the 9th FYP (1996-2000) were met and surpassed; those for *stationary source emissions* of soot (–10%) and industrial dust (–20%) stipulated in the 10th FYP (2001-05) are also likely to have been met. The *legislative and regulatory framework* was updated with the tightening of some emission limits, the introduction of total emission control, and the designation of special control zones (covering 39% of the population). The rate of *emission charges* was trebled. A start was made with flue gas desulphurisation at large emission sources. A nationwide *air quality monitoring* network was put in place. *Energy policy and institutions* (including a renewable energy law) were strengthened, and efforts to diversify energy sources had some success. In the domestic sector, the dependence on coal was reduced from 69% to 30% during 1990-2004. Concerning *transport*, environment-related efforts included the adoption of fuel-efficiency standards for light-duty passenger vehicles in 2004, the adoption of the various EURO standards for vehicle emissions at set dates, and the development of bus rapid transit systems in some cities.

Despite these efforts, air quality in some Chinese cities remains among the worst in the world. About 60% of cities above county level are likely to have met the grade II *ambient air quality* standard by 2005. The SO_2 concentration in urban air, after dropping steadily since the early 1990s, began to climb again in 2002. Nationwide SO_2 emissions from stationary sources increased by 13% during 2000-04, and therefore are not likely to have reached the 10% reduction target of the FYP. In the special air pollution control zones, SO_2 emissions fell by 2% instead of the targeted 20%. Likewise, the proportion of cities suffering from highly *acid rain* (i.e. pH under 4.5) rose again to 10% in 2004, after a low of 2% in 2000. Current emission reduction targets are not sufficient to meet ambient air quality standards. To date, insufficient attention has been paid to *VOCs and toxic air pollutants*. Air pollution regulations and permit conditions are not well enforced. China's *energy intensity* per unit of GDP is about 20%

Recommendations:

- translate the *energy intensity* improvement target into more ambitious *energy efficiency* targets in all sectors; use a mix of instruments to achieve them, including pricing policies, demand management, introduction of cleaner technologies, and energy-efficient buildings, houses and appliances;
- bolster the adoption of *cleaner fuels* (including cleaner coal technology, coal washing and flue gas desulphurisation) and cleaner fuels for vehicles, as well as cleaner cars;
- implement more ambitious *air emission reduction targets* capable of achieving ambient quality objectives already adopted; manage a *wider range of air pollutants*, including VOCs and toxic substances;
- further improve the quality of *monitoring* data needed for effective air quality management and widen their scope (e.g. sources and pollutants);
- develop and implement a *national transportation strategy* that recognises the environmental externalities of transport and takes an integrated approach to private and public transport; streamline the institutional framework for developing sustainable transport systems; use a mix of regulation and economic instruments (e.g. taxes) to give citizens incentives for rational transport decisions;
- strengthen *mass transport in urban areas*, and take measures to encourage the urban use of cleaner transport modes (e.g. bicycles).

However, China's water situation is of high concern. First, many *water courses, lakes and coastal waters are severely polluted* as a result of agricultural, industrial and domestic discharges. The pollution has severely degraded aquatic ecosystems, is a major threat to human health, and may limit economic growth. The use of untreated water affects development especially in the poorer, more disadvantaged regions. Large investment in water services will continue to be necessary: i) in urban areas to address the investment backlog and meet the needs of the large influx of rural migrants; ii) in rural areas, taking into account affordability issues; and iii) in the least developed areas in the form of development assistance and transfers. Secondly, China has *very low water resources per capita* (one quarter of the world average), and they are unevenly distributed (e.g. one tenth in northern and western areas). Among the 600 larger cities, 400 suffer from water shortages. Thirdly, with surface water polluted and scarce, demand for *groundwater* far exceeds the rate of replenishment in many areas in both rural and urban areas. It will be impossible to maintain the high (and inefficient) levels of urban and agricultural water consumption. The country is undertaking a *major project* to transfer more than 40 billion m^3 a year from the southern Yangtze basin to the North China plain by 2020. However, this will still not meet the needs for economic growth and ecological recovery, without determined demand management and sustainable use by urban, industrial and agricultural users. Finally, around 70% of water withdrawal in China is for *agriculture*, with 40% of farmland being irrigated. Agriculture and the rural communities (that lack sewer systems) are also major sources of pollution. To make water management more sustainable, the demand for water by agriculture must be reduced and diffuse pollution must be identified and prevented.

Waste

During the review period, China significantly decoupled the generation of municipal and, to a lesser extent, industrial waste from economic growth. Concerning *industrial solid waste*, the country met and surpassed the targets set out in the 9th and 10th FYPs with respect to recovery, reuse of waste material and safe disposal in landfills. China also stepped up its efforts to put in place an adequate *legal framework for modern waste management* by adopting a cleaner production law in 2003 and updating its 1995 waste law in 2004. A series of more specific regulations and standards for various types of waste, such as medical waste, were adopted. A national programme ("Construction Programme of Hazardous Waste and Medical Waste Disposal Facilities") was put in place in 2003 to significantly increase capacity for treating *hazardous and medical waste*, and good progress was made consequently with the treatment of medical waste. Considerable amounts of materials are recycled through informal activities (e.g. by freelancers). The opening of the market to foreign waste management technology is a positive signal for further improvement. Chinese authorities wish to curb the generation of all types of waste by fostering a high quality, *low material intensity economic growth model*. Indeed, given the rapid growth of its economy and of its imports, China's drive to reduce its material intensity parallels the drive to reduce its energy intensity. The *concepts of the "3Rs" (reduce, reuse, recycle) and of the "circular economy"* are part of the 11th FYP.

Recommendations:

- foster the move *towards a circular economy* by focusing on waste reduction, reuse of waste material and waste recycling, and related targets; require provincial and local governments to adopt and implement comprehensive *waste management plans* (including accurate verification of volumes of waste – municipal, industrial and hazardous – generated and treated) covering elements of the waste hierarchy;

- accelerate the pace of extending *waste treatment capacity* by building treatment infrastructure and establishing systems for the collection, reuse and recycling of waste (e.g. separate collection of household waste), including in rural areas;

- formulate *enforcement plans* for different sectors (e.g. households, large industry, small and medium-sized enterprises) and types of waste;

- streamline the allocation of *responsibility for the management* of different types of waste; ensure that waste facilities operate efficiently and comply with standards; further develop workable regulations and policy instruments for waste management; improve the collection of *waste data* and develop tools to evaluate the effectiveness of waste management policies at national and provincial levels;

- establish *financing mechanisms* with a mix of public and private financing, and move to charging for waste services more progressively in less developed areas; improve the collection rate of waste charges and set them at a level consistent with the government's aim to achieve a circular economy;

- provide the *informal sector* (freelancers) with equipment, organisational assistance and training to continue collection and recycling under improved hygienic and environmental conditions, as part of waste management plans;

- *raise awareness* of waste management and efficient resource use among the public, small and medium-sized enterprises, and industry.

Nevertheless, the amounts of municipal waste, industrial waste and hazardous waste far exceed what can safely be treated and disposed of. Some of those waste are stored waiting for treatment (e.g. close to 50% of municipal waste) or *are dumped in an uncontrolled fashion*. Human health and the environment are put at risk through a proliferation of uncontrolled dumps surrounding the cities. The 10th FYP target of increasing the capacity of *municipal landfills* to 150 kt/day was not achieved. Total *waste generation* increased by as much as 80% during the review period. Waste management is still the "poor cousin", compared to air and water management, in its share of national government *funding* of investments. Local bodies have trouble collecting waste charges, which remain too low to cover the operational cost of waste management. Overall, the emphasis remains *too heavily on landfilling* (the destination of 44% of municipal waste), and few local governments implement separate collection and recycling. Incineration and composting account only for 3% and 5%, respectively of municipal waste treatment. *Responsibility for waste management* is fragmented across too many agencies. *Enforcement* is inadequate and does not distinguish sufficiently between large industries and small and medium-sized enterprises.

Nature

China has established a comprehensive *legal framework* for managing nature and biodiversity, which includes wildlife and marine protection as well as terrestrial and marine protected areas. China actively reports on its international commitments and also publishes annual state of the environment reports related to its internal goals and targets. *Protected areas* at the national, provincial, prefecture and county levels have been dramatically increased over the review period, and China has received international recognition for its wetlands, biosphere reserves, and natural and cultural heritage preservation programmes. Outside of protected areas, ecological considerations have led to *afforestation* of large areas. New forestry initiatives have been taken to further develop shelter forests in arid, mountainous and coastal areas, to streamline forest management (e.g. more stringent harvest quotas) and to promote farm forestry on land sensitive to soil erosion (e.g. grain for green policy). Various environmental protection programmes within the country have begun to recognise the value of environmental outreach (alien species, endangered wildlife). China has been proactive in developing bilateral and regional co-operation in the area of nature conservation. There has been a regular increase in the number of world heritage sites and Ramsar wetlands.

However, there is a need for more *institutional co-ordination* and integration of efforts to assess and protect nature and biodiversity inside and outside of protected areas, given the number of agencies and stakeholders involved. There is insufficient monitoring to assess trends and evaluate the *protection status of nature reserves*. The main targets for species and habitat protection are in terms of percentage of land area. There is a need to ensure that the key natural habitat types and ecosystems are adequately protected and that they support species recovery plans. Although China has a relatively high percentage of total area classified as protected, *marine habitats and species* are not sufficiently represented and are subject to land-based sources of pollution and habitat alteration, in addition to exploitation pressures. Management level of reserves needs to be improved and attention should be paid to integrated habitats

---

Recommendations:

- modernise and implement *legislation on nature protection*, in particular adopt a law on the protection and management of Nature Reserves, notably favouring an increase of marine protected areas and of protected areas with higher protection status; consider ratification of the Bonn Convention;

- enhance the *capacity of national, provincial, prefecture and county level agencies* to manage biodiversity protection of existing reserves and integrate nature conservation within economic and social development projects outside protected areas;

- increase the *financial and human resources* for nature and biodiversity protection and further involve local residents in patrolling, monitoring and habitat enhancement, in the context of poverty alleviation; diversify the sources of financing of nature conservation;

- develop the use of *economic instruments related to nature and biodiversity protection*, not only as income supporting measures, but to reward the provision of environmental services;

- integrate long-term plans for rehabilitating and maintaining species and protected areas (including managing alien species) with *land*-use and river basin management plans, and any subordinate provincial, prefecture and country plans;

- integrate the economic and social values of protecting habitats and species (e.g. ecological services, tourism development) within *development decision-making*, in particular as part of EIAs;

- promote *sustainable forest management* through issuance of forest management plans, certification of foresting practices, and labelling of forest products in China; expand co-operation with supplying countries in the *forestry sector*, to ensure that imported wood and wood products are sourced from forests that are managed on a sound, sustainable basis.

---

protection to minimise fragmentation and to *enhance habitats continuity through biodiversity corridors*. There is a need to integrate nature protection concerns into *development plans* especially in impoverished central and western regions with abundant biodiversity. Little has been done to promote biodiversity protection on forestland and to tailor payments to forest owners to the provision of forest ecosystem services. China has not yet ratified the Bonn Convention on migratory species, although it is active in regional co-operation on migratory waterbirds.

2.   Towards "Harmonious Society" and Environmentally Sustainable Development

Integration of environmental concerns in economic decisions

China's two digit average economic growth was accompanied by some *decoupling of pollution from economic growth* in the period 1990-2005. This was the case, in particular, for SO_2 and recently NO_x emissions. Energy intensity has improved by about a half since 1990, though the decrease has levelled off. Water withdrawal and municipal waste have also been significantly decoupled from the economic growth. Successive *Five-Year Plans for National Economic and Social Development* (FYPs) have provided an important means for identifying and addressing priority environmental problems: they are underpinned by solid analysis, they establish quantitative targets, and they frame investment programming and budgeting. The Chinese leadership has announced its intention to place environmental protection in a more strategic position. In this perspective, the 11th FYP advocates a *new economic model in which growth is guided by resource conservation* rather than by continued expansion of resource use. Improved energy intensity and the concept of the "circular economy" are recognised as key to help reduce the pollution and resources intensity of the Chinese economy. Various measures have been taken to better *integrate environmental and economic decision-making*: provision has been made in the 2003 EIA law to assess the potential environmental impacts of sectoral programmes. Some energy prices have been deregulated (e.g. some coal prices). The use of environment-related taxes has expanded, but accounts only for about 3% of total tax revenues.

However, the pollution, energy and material intensities of the Chinese economy remain high, as well as its water use intensity, and *pollution remains very serious in many locations*. China generates more pollution and consumes more resources per unit of GDP than OECD averages. There is a high rate of environmentally significant accidents, and resource degradation is constraining economic development. *Health costs and ecological damages of present development are high.* The target of quadrupling GDP between 2000 and 2020 requires *commensurate strengthening of environmental management and finance*, so that economic growth is environmentally sustainable. It is not sure that present policies, although going in the right direction, are *sufficiently ambitious* to meet the strategic environmental objectives identified by Chinese leaders. The *under-pricing of energy, water and other resources* needs to be addressed. More effective arrangements at the level of the State Council are needed to better integrate environment into economic and sectoral decision-making, including a *strengthened role for SEPA.*

Recommendations:

- review *price levels* for energy, water and other natural resources so as to better reflect their scarcity value and internalise externalities; consider mechanisms to compensate or mitigate their impact on poorer sections of the population and regions that would be adversely affected by such price increases;

- consider establishing an inter-ministerial group to examine how *environment-related taxes* might be restructured to help better achieve environmental policy objectives;

- *increase and diversify the sources of environmental finance* by fuller implementation of the polluter pays and user pays principles, and increase the effectiveness and efficiency of allocating public environmental expenditure;

- strengthen the institutional mechanisms for *better integrating environment into economic and sectoral policies*, possibly by establishing a Leading Group on environment or on sustainable development; fully implement the provisions in the EIAs law for assessing the potential environmental impacts of sectoral programmes;

- continue to establish *national targets* to achieve key environmental objectives, taking into account scientific, economic and social analysis.

Integration of environmental and social decisions

China's economic growth has helped raise living standards and has contributed to significantly reduce poverty. In recent years, government policies have emphasised *economic growth with due attention to social and environmental concerns*: environmental issues associated with rapid urbanisation and development of coastal regions, with poverty, and with development challenges in less-advanced western parts of the country are being addressed. Considerable progress has been made since the mid-1990s in the development of *environmental information*, access to this information, and participation on environmental issues. China produces each year comprehensive environmental statistics and environmental reports. The media and the rise of committed and outspoken environmental NGOs reinforce the demand for environmental progress. Progress can also be seen in environmental education and awareness-raising through primary education.

Recommendations:

- further improve health and living standards, particularly in less developed areas, by reducing the share of people without *access to sound environmental services* (safe water, basic sanitation, electricity); taking account of affordability constraints, give higher priority to water infrastructure in development strategies (e.g. for the poorer central and western China);
- consolidate and strengthen information on health and the environment and develop a *national health*-environment plan of action; implement the most cost-effective measures; promote pollution release and transfer reporting by enterprises; build capacity to report on exposures of specific population groups to environmental health risks (e.g. occupational health, health impacts near polluting facilities, children's health);
- continue to improve environmental information by developing and using *indicators of environmental performance*, environment-related *economic information* and analysis, and environmental accounting tools such as *material flows accounts*; expand the coverage of environmental information (e.g. to diffuse pollution, toxic substances, hazardous waste); continue to improve consumer protection and public *access to environmental information*;
- further expand *environmental education* and awareness, particularly among young people;
- continue efforts to work with *NGOs and the public* to achieve environmental policy goals; strengthen co-operation and partnerships with *enterprises* and corporate social responsibility.

However, the rapid economic growth has led to *very wide and increasing disparities* between the rich and the poor, urban and rural communities, and coastal and inland provinces. While some aspects of the urban environment have improved in China's mega and large cities, additional *demands for environmental services* (e.g. water supply, water sanitation, solid waste management) are resulting from the large population migration from western and central China to coastal China. At the same time, the needs for environmental services of the expanding towns and townships and of the rural poor, particularly in the central and western regions, are also growing. To reduce industry relocation and environment-related distortions to competitiveness and trade within China, *national environmental standards* (i.e. product, emission and quality standards) should be implemented by all provinces effectively and efficiently, minimising transition periods when transitions are necessary. Concerning *health*, pollution is contributing to an increase in *respiratory diseases*, cancer and birth defects. Environmental and health information should be strengthened to support priority setting and to generate related economic and health benefits. Concerning *environmental information*, improvements could be made with respect to indicators of environmental performance, environment-related economic information, environmental and material flows accounts, the coverage of environmental information, and monitoring. *Environmental education* should be further strengthened (e.g. at university level) and expanded, particularly for young people. Environmental awareness should be increased in Chinese enterprises.

3. International Co-operation

The last decade has seen a *dramatic increase in China's engagement with other countries* in addressing environmental challenges. This reflects a growing recognition across the spectrum of Chinese institutions of the important economic, social and ecological stakes that China has in meeting these challenges, and also of its shared interests with the international community. China is now an active, constructive participant in a broad array of regional and global environmental conventions, institutions and programmes, and is *drawing heavily on international financial institutions and special mechanisms* (e.g. the Montreal Protocol's Multilateral Fund) to augment its own resources and ensure that China's international commitments are met. Since 1995, it has reduced its production and consumption of ozone-depleting substances more than any other country; established comprehensive and ambitious policies and legal regimes in the areas of marine pollution and fisheries management; provided international leadership in efforts to control transboundary movement of hazardous waste; recognised and taken initial steps to confront its emissions of greenhouse gases; and undertaken a detailed examination of how its *trade and investment policies* can work to support environmental management goals.

Recommendations:

- continue China's *active engagement in international environmental co-operation,* seeking to improve the effective and efficient use of i) domestic resources, and ii) international support mechanisms (e.g. the World Bank's Clean Development Fund, the Multilateral Fund under the Montreal Protocol, and the Global Environment Facility);

- strengthen *monitoring, inspection and enforcement capabilities* in support of the implementation of international commitments (e.g. on trade in endangered species, in forest products, in hazardous waste and in ozone-depleting substances, as well as on sound chemicals management, ocean dumping and fisheries management);

- improve governmental oversight and environmental performance in the *overseas operations of Chinese corporations* (in the spirit of the OECD guidelines for multinational enterprises);

- develop *partnerships with foreign enterprises* to contribute to environmental progress through provision of training, technical support and cleaner technology; ensure environmental requirements are not relaxed to attract *foreign direct investments;*

- continue to assign high priority to domestic and regional *anti-desertification efforts;*

- intensify domestic and international co-operation to reduce *transboundary air pollution* in Northeast Asia by, inter alia , introducing cleaner coal technology, improving energy efficiency and fuel switching;

- ensure that the interim and final targets for the phase-out of *ozone-depleting substances* under the Montreal Protocol continue to be achieved on schedule;

- prepare a *coherent national plan on climate change* which draws together the array of climate-related activities currently underway and planned to improve their collective efficiency and impact;

- strengthen efforts to protect and improve *water quality in coastal waters and adjacent regional seas* from land-based pollution sources, and upgrade environmental management regulations and government oversight in the aquaculture industry;

- integrate environmental considerations systematically into China's growing *development co-operation* programme.

China, however, remains the *second largest contributor of greenhouse gases,* and is still the world's largest producer and consumer of ozone-depleting substances. Its largely coal-fired economy is a major source of acid rain and other transboundary air pollutants in Northeast Asia, and is a significant contributor to global-scale air pollution, including mercury. Its *coastal waters and regional seas* are suffering from an increasing burden of land-based pollution in many areas; and the environmental management and food-sanitation regimes for China's rapidly expanding marine aquaculture industry need strengthening. A lack of strong *monitoring, inspection and enforcement capabilities* and associated penalties are limiting the effectiveness of otherwise sound policies, laws and regulations established to further China's domestic objectives and international commitments in the areas of marine fisheries, coastal water quality, hazardous waste transport, and the control of *illegal trade in endangered species, forest products and ozone-depleting chemicals.* Stronger efforts are needed by the government to ensure that *Chinese corporations operating overseas*, particularly in such environmentally-sensitive industries as forest products and mining, are positive contributors to China's stated goal of building an international reputation for sound environmental management and sustainable development. *Funding limitations* and *inadequate institutional co-ordination* are constraining the pace at which China is able to carry out an ambitious international environmental agenda that includes a range of difficult challenges (e.g. desertification control, greenhouse gas reduction, marine management). To achieve success, increased financial efforts from China as well as major technical support and targeted financial assistance to China from *OECD countries and international financial institutions* will be required.

THE CONTEXT

Features

- Economic context
- Social context
- Energy: state and outlook
- Institutional environmental context

---

1. Physical Context

With a land area of *9.6 million km*² China is one of the largest countries in the world (covering about 7% of the earth's land surface). It stretches over 5 200 kilometres from east to west, and over 5 500 kilometres from north to south. It shares borders with 14 countries (Afghanistan, Bhutan, India, Kazakhstan, Democratic People's Republic of Korea, Kyrgyzstan, Laos, Mongolia, Myanmar, Nepal, Pakistan, Russia, Tajikistan and Vietnam). It has around 18 000 kilometres of coastline on its east side. About 5 000 islands plot its territorial seas.

China has a complex *topography*. A terraced terrain descends step by step from the western Qinghai-Tibet Plateau (known as the "roof of the world") to the eastern coastal area. Nearly 60% of China's total land area is made up of mountains, plateaus and hills more than 1 000 metres high. Several of the world's highest summits are located in China (including Mount Qomolangma, or Everest, 8 848 metres). The eastern part of the country is characterised by low mountains, hills and alluvial plains. Arable and permanent crop land covers 18% of the national territory (134 million hectares), mainly concentrated on the North and Northeast China Plain, the Middle-Lower Yangtze Plain, the Pearl river delta and the Sichuan basin; grassland covers 40% (288 million hectares) and forests and wooded land cover around 32% (235 million hectares) and are concentrated in the northeast and the southwest (Figure 2.1).

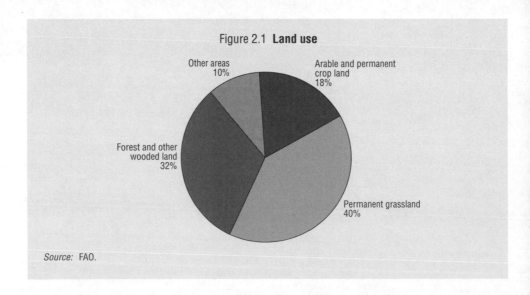

Figure 2.1 **Land use**

Other areas 10%
Arable and permanent crop land 18%
Forest and other wooded land 32%
Permanent grassland 40%

Source: FAO.

Owing to its vast territory and complex topography, China's *climate* is extremely diverse. Precipitation decreases progressively from southeast to northwest, and is mainly concentrated during the summer season. Climate varies from tropical in the south to sub-arctic in the north. China can be divided into six main thermal areas (cold-moderate, warm-moderate, moderate, subtropical, tropical and equatorial) and four rainfall zones (humid, semi-humid, semi-arid, arid), leading to 18 different climatic regions.

Freshwater resources amount to 2.8 trillion cubic metres and are unevenly distributed, both temporally and spatially. The Yangtze river (6 300 km) originates in the Qinghai province and ends up in the East China sea near Shanghai; it accounts for 39% of China's total water flow and is the country's major inland waterway. The Yellow river (5 464 km) accounts for 3% of total water flow and carries such a large amount of silt that its bed is constantly rising; floods occur downstream and droughts upstream. The Pearl river accounts for 14% of the country's total water flow. Lakes are numerous. Canals have been built or are planned to move water from south to north. China has the largest hydropower potential in the world. Presently, only a small part is used, mostly in the southwest of the country, the Yangtze river accounting for around 40% of the national total. Built recently in the middle section of the Yangtze river, the Three Gorges Dam has an installed capacity of 10 GW (to increase to 20 GW by 2009). Another major dam is to be built in the Jinsha river (upstream tributary to the Yangtze river), which is expected to add another 40 GW capacity (Chapter 4).

Total proved *coal reserves*[1] (114 billion tonnes or 13% of the world's total) are mainly located in the north and the northwest (IEA, 2005). Proved oil reserves[1] (17 billion barrels or 1.4% of the world's total) are concentrated in the far-west, the northeast and off-shore. Concerning *mineral resources*, ferrous metallic mineral reserves are located in provinces around Beijing, while all non-ferrous metallic minerals can be found in China.

Due to the variety of climatic conditions, China is known as the "*kingdom of plants and animals*". More than 30 000 species of vascular plants and some 6 300 species of vertebrates can be found across the territory. The three southwestern provinces (Yunnan, Guizhou, Guangxi) are home to over 3 000 species of medicinal plants. Species endemic to China include the Giant Panda, Red Ibis, Golden Monkey and Metasequoia (Chapter 6).

China is subject to *disasters of climatic or geological origins* including floods, droughts, typhoons and earthquakes (Table 2.1). The need to improve prevention of and response to floods has increased in recent years (Chapter 4). Economic development has been accompanied by industrial or mining accidents and severe ecological damage such as deforestation, desertification and soil erosion.

Table 2.1 **Selected natural disasters in China**, 1906-2005

	Number of events	Deaths	Total population affected	Total damage (000 USD)
Flood	142	6 592 917	1 553 294 054	75 715 374
Ave. per event		*46 429*	*10 938 691*	*533 207*
Drought	35	3 501 400	239 116 000	1 845 832
Ave. per event		*100 040*	*6 831 886*	*52 738*
Epidemic	10	1 561 484	9 823	0
Ave. per event		*156 148*	*982*	*0*
Earthquake	100	784 725	20 267 450	7 775 200
Ave. per event		*7 847*	*202 675*	*77 752*
Windstorm[a]	163	170 256	323 416 375	20 819 139
Ave. per event		*1 045*	*1 984 150*	*127 725*
Extreme temperature	9	243	36 880	3 000 000
Ave. per event		*27*	*4 098*	*333 333*
Landslide	38	2 726	89 502	62 400
Ave. per event		*72*	*2 355*	*1 642*
Wildfire	5	243	56 613	110 000
Ave. per event		*49*	*11 323*	*22 000*
Wave/surge	3	126	0	0
Ave. per event		*42*	*0*	*0*

a) Includes tornadoes, typhoons, storms.
Source: EM-DAT OFDA/CRED International Disaster database.

2.   Economic Context

2.1   GDP and related indicators

China has one of the world's fastest growing economies, with an average annual growth rate of 10.8% over almost 15 years. Growth was 10.1% in 2004 and 9.9% in 2005 and is expected to be 10.5% in 2006. In 2005, China's GDP^2 reached CNY^3 18 232 billion (current USD 2 259 billion or USD 8 536 billion using PPP-2000[4]) (NBSC, 2006a). Accordingly, the Chinese economy ranked sixth in 2004 (or fourth including Hong Kong). In 2006, it will *rank fourth in size*.

Since 1978, a range of reforms have contributed to open up China's *economy* and to make it more market-oriented. The inflation rate is low, but it increased in 2004 to reach 3.9%. In 2005 it was 1.8%. The private sector has grown to represent more than 50% of GDP (OECD, 2005a).

The *GDP per capita* in 2004 (CNY 12 299, USD 1 486 or USD 5 975 using PPP-2000) was more than three times its 1990s level, but it still remains *very low* compared to average GDP per capita in OECD countries (USD 25 300 using PPP-2000). GDP is *unevenly distributed* among the provinces (Table 2.2). Coastal provinces are wealthy compared to inland provinces, which leads to large workforce migrations. For instance, Shanghai has a per capita GDP ten times greater than Guizhou province.

Table 2.2 **Provinces, autonomous regions and centrally-administered municipalities,** 2004

	Administrative division[a]	Seat of government	Area (1 000 km²)	Population[b] (million persons)	GDP per capita (USD/capita)[c]	GDP per capita (USD/capita)[c] (China = 1)
Beijing	M	Beijing	16.8	14.9	3 466	2.7
Tianjin	M	Tianjin	11.3	10.2	3 459	2.7
Hebei	P	Shijiazhuang	190.0	68.1	1 556	1.2
Shanxi	P	Taiyuan	150.0	33.4	1 102	0.9
Inner Mongolia	AR	Hohhot	1 197.5	23.8	1 375	1.1
Liaoning	P	Shenyang	150.0	42.2	1 969	1.5
Jilin	P	Changchun	187.4	27.1	1 319	1.0
Heilongjiang	P	Harbin	460.0	38.2	1 679	1.3
Shanghai	M	Shanghai	6.3	17.4	5 167	4.1
Jiangsu	P	Nanjing	102.6	74.3	2 504	2.0
Zhejiang	P	Hangzhou	101.8	47.2	2 878	2.3
Anhui	P	Hefei	139.6	64.6	900	0.7
Fujian	P	Fuzhou	121.4	35.1	2 083	1.6
Jiangxi	P	Nanchang	166.9	42.8	986	0.8
Shandong	P	Jinan	156.7	91.8	2 039	1.6
Henan	P	Zhengzhou	167.0	97.2	1 096	0.9
Hubei	P	Wuhan	185.9	60.2	1 267	1.0
Hunan	P	Changsha	211.8	67.0	1 012	0.8
Guangdong	P	Guangzhou	179.6	83.0	2 334	1.8
Guangxi	AR	Nanning	236.7	48.9	821	0.6
Hainan	P	Haikou	33.9	8.2	1 136	0.9
Chongqing	M	Chongqing	82.4	31.2	1 032	0.8
Sichuan	P	Chengdu	485.0	87.3	908	0.7
Guizhou	P	Guiyang	170.0	39.0	493	0.4
Yunnan	P	Kunming	394.0	44.2	810	0.6
Tibet	AR	Lhasa	1 274.9	2.7	933	0.7
Shaanxi	P	Xi'an	205.6	37.1	940	0.7
Gansu	P	Lanzhou	390.0	26.2	719	0.6
Qinghai	P	Xining	720	5.4	1 044	0.8
Ningxia	AR	Yinchuan	62.8	5.9	946	0.7
Xinjiang	AR	Urumqi	1 655.8	19.6	1 354	1.1

a) M: centrally administered municipality, P: province, AR: autonomous region.
b) Excludes military personnel.
c) Converted using the exchange rate (CNY/USD) 8.2765.
Source: "Statistical Yearbook, 2005", National Bureau of Statistics of China; China Information Internet Center.

2.2 Structure of the Chinese economy

During 1990-2004, the structure of China's economy changed significantly. The share of primary activities dropped from 27% to 13 %, secondary activities remained more or less stable around 42-46% and tertiary activities increased from 31% to 41%.

Concerning *agriculture*, China is the world's largest producer of rice, wheat, cotton and tobacco. It is also one of the main producers of silk, tea and sugarcane. Agriculture is mainly concentrated in the eastern part of the country.

Concerning *industry*, growth is driven by branches such as communication equipment, computers, transport equipment, non-ferrous metals, the manufacture of raw chemical products, crude steel, coal, cement, fertilisers and textiles. Between 2003 and 2004, industrial output (CNY 18 722 billion) increased by 32%. Of this around 67% comes from heavy industry. Private enterprise accounts for 17% of industrial output.

Concerning *energy*, owing to its size and natural endowment, China looms large in global statistics. The country's total primary energy demand is the world's second largest, after that of the United States[5] (IEA, 2006a, 2006b). In 2004, China, with more than 38% of global output, was the world's largest coal producer, the largest producer of hydroelectricity, the second-largest producer of electricity (after the United States), and the sixth largest producer of crude oil (IEA, 2006a). Due to its dependency on coal, China accounted in 2004 for 18% of world *carbon emissions*, ranking second behind the United States. A quadrupling of its GDP by 2020 (compared to 2000) is expected to be accompanied by a doubling of its TPES (Box 2.1, Table 2.3).

2.3 Trade and FDI

In 2004 China became the third destination and origin of world *trade*. Its 2005 exports and imports reached USD 762 billion and USD 660 billion respectively, an increase of 28% and 18% compared to 2004. In 2005, China exported particularly to the United States (21% of total exports), the European Union (19%), and Japan (14%). Chinese imports came particularly from Japan (15%), the Republic of Korea (12%) and the European Union (11%). In December 2001, China joined the World Trade Organisation.

Foreign direct investment (FDI) in China has increased very rapidly since 1990. Beginning in 1993, China emerged as the largest recipient of FDI among developing countries. In 2005 FDI in China reached USD 60.3 billion, with 62% from Asia. FDI is mainly concentrated on manufacturing (70%) and real estate (9%). FDI is unevenly distributed among the regions: 65% of the total is in the five coastal provinces/ municipalities of Guangdong, Jiangsu, Shandong, Shanghai and Zhejiang.

Box 2.1 **The Chinese energy sector at a glance**

State

China's *total primary energy supply* (TPES) amounted to 1 609 Mtoe in 2004. The energy mix is dominated by coal, which accounted for 61.7% of TPES in 2004. The share of oil was 19.3%, of renewable energies 15.6%, of nuclear 0.8%, and of natural gas 2.6%. The total installed capacity for electricity generation amounted to 440 GW in 2004. In the same year, coal accounted for 78%, hydro 16%, and nuclear 2.3% of electricity generation; gross electricity generation reached 2 200 TWh. The low thermal efficiency of China's power stations contributes to its low TFC/TPES ratio of 0.65.

Total final consumption (TFC) amounted to 1 038 Mtoe in 2004. Consumption from the residential sector has increased by an average of 9% per year since 1990 representing 31% of TFC in 2004 (compared to 41% for industry). The share of transport was 10% in 2004. The major electricity consumers in 2004 were industry (67%), households (14%), agriculture (5%), commerce and public services (7%), transportation (1%), and construction (1%). China's strategy for increased urbanisation will raise energy consumption because city dwellers use 3.5 times more energy than people in rural areas. The number of vehicles in China is increasing rapidly and creates a demand for more oil. During 1990-2004, the share of coal in TFC has decreased in all sectors. Particularly in the residential sector, it fell from 84 to 14%.

China has considerable *fossil fuel resources*, but they are below the world average when considered on a per capita basis. Coal production grew from 1.3 billion tonnes in 2000 to 1.85 billion in 2004 and 2.19 billion tonnes in 2005. One-third of China's domestic oil production comes from the long-established Daqing oil fields (the world's fourth largest); a new large oil field was discovered in Shengli in 2004. China became a net importer of crude oil in 1996 and by 2004 it imported 40% of its demand of six million barrels per day. High oil prices are stimulating development of offshore oil fields in the South China sea.

Outlook

Between 2000 and 2020, *China's TPES* should roughly double while its GDP would quadruple (DRC, 2004). The growth in supply should be high for all sources (Table 2.3). China's share of the world supply of coal, oil, gas, nuclear and renewables is expected to grow. China's oil imports in 2020 would represent a majority of its total oil supply.

Source: DRC, IEA.

With the high economic growth and large migration from rural to urban areas, *infrastructure* and *building construction* are booming. From 1990 to 2004, per capita living space increased, in urban areas, from 13.7 m² to 25 m². During this period, total investment in fixed assets increased almost tenfold from CNY 452 billion (USD 86 billion) to CNY 7 048 billion (USD 851 billion). In 2004, nearly 3 billion m² were under construction. *Real estate prices are rising*: in 2004 they rose more than 14%. Shanghai has the highest prices in the country, with an average price per m² of more than CNY 10 000.

Table 2.3 **Estimated total primary energy supply in China,**[a] 2000-20

	TPES[b] (Mtoe)			Growth rates (%)	
	2000	2010	2020	2010/00	2020/00
TPES	912	1 443	1 956	58.3	114.5
Coal	66.1%	63.1%	59.0%	51.1	91.4
Oil	24.6%	22.2%	21.5%	42.7	87.3
Gas	2.5%	5.8%	8.6%	267.2	634.4
Nuclear	0.5%	1.6%	2.5%	445.9	1 024.6
Hydro	6.3%	7.2%	8.3%	80.8	180.7
Other renewables[c]	0.0%	0.1%	0.3%	1 050.0	3 700.0

a) Economic growth assumption: annual average growth of 7.55%, 7.35% and 6.9% respectively in the periods 2000-05, 2005-10 and 2010-20; with consequent growth rates of 105.1% and 299.8% respectively for the periods 2000-10 and 2000-20.
b) Based on conversion rate: 1 tce = 0.69955 toe.
c) Excluding biomass and waste.
Source: Development Research Center (DRC) of the State Council, 2004.

3. Social Context

With about 20% of the world's total population, China is the most populous country in the world. China's *population* was 1 307.6 million in 2005. According to the fifth national population census in 2000, among the 56 ethnic groups living in China, the Han group accounts for 92% of the population. Of the 55 ethnic minorities, 53 have their own spoken language and 23 their own written language.

In 2004, the *population density* was 135 inhabitants per km². However, the population is unevenly distributed: population density is higher on the east coast (e.g.

2 747 inh./km^2 in Shanghai, 722 inh./km^2 in Jiangsu); intermediate in the central areas (e.g. 180 inh./km^2 in Shaanxi or Sichuan); and low in the west (e.g. 8 inh./km^2 in the Qinghai).

Further to strict *population control* measures introduced in 1979, population growth has slowed. In 2005, the population growth rate was 0.59%. Since 1990 the natural growth rate has drastically decreased (from 14.4‰ to 5.9‰) due to the large drop in birth rate (from 21‰ to 12.4‰). The fertility rate has dropped from 4.9% (in 1975) to around 1.7% (in 2005[6]), which is below the population replacement level. This has caused an *ageing of the population*: people over 60 now represent almost 10.9% of the total population and will represent 31% in 2050 according to United Nations estimates. As life expectancy is increasing, the younger generation will have to support not only their two parents but also their four grand-parents (i.e. the "4-2-1 family"). In 2005, 51.5% of the population was males and 48.5% females. In 2020, forecasts suggest a very significant gender imbalance.

The "urban" population has more than doubled in less than 25 years and now accounts for 43% of the population. This has generated pressures on infrastructure development, water distribution and environmental protection in urban areas. The urban population is expected to reach around 60% by 2020 (CIIC, 2005b). Meanwhile, more than half of the Chinese population remains rural (57%). Migrations to the urban eastern part of the country have grown parallel to the *increasing income disparity* between urban and rural areas (Table 8.1). National per capita urban income is more than three times as high as per capita rural income. And the urban-rural income gap between the east coast and the western provinces varies by a factor of ten. For instance, in 2004, rural per capita income was CNY 7 066 in Shanghai and CNY 1 722 in Guizhou, while urban per capita income was CNY 16 683 in Shanghai and CNY 7 218 in Ningxia (Chapter 8).

In 2004, 752 million people were *employed*: 47% in primary activities, 22% in secondary activities and 31% in tertiary activities (NBSC, 2005b). Since 1990, the percentage of people engaged in primary activities has dropped by 13 points, in secondary industry has remained stable and in tertiary activities has risen 12 points. The urban *registered unemployment* rate was 4.2% in 2005, but this only includes people officially registered as unemployed (NBSC, 2005b). Including jobless workers laid-off from state-owned enterprises and unemployed rural migrant workers, the unemployment rate is estimated to be around 23% of the total labour force. Working conditions vary widely. In the mining sector, the large number of accidents in public and private mines (e.g. 6 000 deaths in 2004) has prompted the government to respond with mine closures (2 400 in 2005), mining operation suspensions (10 500 in 2005), safety standard enforcement measures and compensation funds (Table 2.4).

Table 2.4 **Selected mining accidents,** [a] 1995-2005

	Date	Location	Type
2005	14.02	Liaoning, Fuxin	Gas explosion
	09.03	Shanxi, Jiaocheng	Gas explosion
	19.03	Shanxi, Shuozhou	Gas explosion
	24.04	Jilin, Jiaohe	Flooding
	28.04	Shanxi, Hancheng	Gas explosion
	12.05	Sichuan, Panzhihua	Gas explosion
	19.05	Hebei, Chengde	Explosion
	20.05	Shanxi, Puxian	Explosion
	08.06	Hunan, Zijiang	Poisonous gas leak
	02.07	Shanxi, Ningwu	Gas explosion
	11.07	Xinjiang, Fukang	Gas explosion
	02.08	Henan, Yuzhou	Gas leak
	07.08	Guandong, Xingning	Flooding
	03.10	Hebei, Xingpai, Huining, Shangwang	Collapse
	21.11	Heilongjiang, Qitaihe	Explosion
	02.12	Henan, Xin'an	Flooding
	09.12	Hebei, Kaiping, Tangshan	Gas explosion
2004	11.02	Guizhou, Liupanshui	Gas explosion
	23.02	Heilongjiang, Jixi	Gas explosion
	01.03	Shanxi, Jiexiu	Explosion
	30.04	Shanxi, Linfen	Gas explosion
	18.05	Shanxi, Jiaokou	Gas explosion
	20.10	Henan, Xinmi	Gas explosion
	11.11	Henan, Lushan, Liangwa	Gas explosion
	28.11	Shanxi, Tongchuan	Gas explosion
	09.12	Shanxi, Yangquan	Gas explosion
2003	11.01	Heilongjiang, Harbin, Fangzheng County	Gas explosion
	24.02	Guizhou, Liupanshui, Muchonggou	Gas explosion
	22.03	Shanxi, Luliang	Gas explosion
	30.03	Liaoning, Fushun	Gas explosion
	13.05	Anhui, Huaibei	Gas explosion
	20.05	Shanxi, Linfen	Explosion
	04.07	Inner Mongolia, Yakeshi	Explosion
	13.07	Henan, Dengfen City	Flooding
	11.08	Shanxi, Datong	Gas explosion
	14.08	Shanxi, Yangquan	Gas explosion
	18.08	Shanxi, Zuoquan	Gas explosion
	14.11	Jiangxi, Fengcheng	Explosion
	22.11	Henan, Ruzhou	Explosion
	07.12	Hebei, Zhangjiakou	Gas explosion

Table 2.4 **Selected mining accidents,**[a] 1995-2005 *(cont.)*

	Date	Location	Type
2002	14.01	Yunnan, Wenshan	Explosion
	26.01	Hebei, Chengde	Explosion
	29.03	Henan, Yuzhou city	Gas explosion
	08.04	Heilongjiang	Gas explosion
	24.04	Sichuan, Panzhihua	Gas explosion
	26.12	Hebei, Wuan	Fire
	04.05	Guizhou, Weining	Gas explosion
	20.06	Heilongjiang, Jixi	Gas explosion
	22.06	Shanxi, Fanzhi County, Yixingzhai	Explosion (gold mine)
	04.07	Jiling, Songshu	Explosion
	08.07	Heilongjiang, Hegang	Gas explosion
	03.09	Hunan, Shuangfeng	Gas explosion
	23.10	Shanxi, Zhongyang	Explosion
	29.10	Guangxi, Nanning	Fire
	08.11	Shanxi, Xituan	Explosion
	06.12	Jilin, Taonan City	Fire
2001	16.07	Guangxi	Flooding (tin mine)
2000	11.01	Jiangsu, Xuzhou	Flooding
	15.04	Shanxi, Yongcai	Gas explosion
	27.09	Guizhou, Shui, Shuicheng	Gas explosion
	11.10	Gansu, Lanzhou	Gas explosion
	24.11	Inner Mongolia	Gas explosion
	03.12	Shanxi, Hejin	Gas explosion
1999	16.01	Guizhou	Gas explosion
	18.03	Gansu	Explosion (gold mine)
	10.05	Shanxi	Gas explosion
1998	24.01	Liaoning, Wangjiaying	Gas explosion
	02.04	Henan, Hebi	Gas explosion
	06.04	Pingdingshan	Gas explosion
	15.05	Sichuan, Dachuan	Gas explosion
	10.08	Shanxi, Zezhou	Gas explosion
	25.10	Guangxi, Heshan	Flooding
	29.11	Yunnan, Xuanwei	Gas explosion
	12.12	Zhejiang, Changguang	Gas explosion
1997	04.03	Henan, Pingdingshan	Gas explosion
	02.05	Shandong, Laiwu	Gas explosion
	19.05	Inner Mongolia, Wuhai City	Gas explosion
	28.06	Fushun, Beilong-Feng	Explosion
	25.10	Henan, Pingdingshan	Gas explosion
	04.11	Guizhou, Liangtian	Gas explosion
	08.11	Chongqing, Nanchuan	Explosion
	13.11	Anhui, Pansan	Gas explosion

Table 2.4 **Selected mining accidents,**[a] 1995-2005 *(cont.)*

	Date	Location	Type
	27.11	Anhui, Xie'er	Gas explosion
	27.11	Shaanxi, Jieping	Gas explosion
1996	21.05	Henan, Pingdingshan	Explosion
	21.05	Gansu	Flooding (lead and zinc mine)
	27.11	Shanxi, Datong	Gas explosion
	02.12	Pingdingshan	Gas explosion
1995	13.03	Yunnan, Jiuwuji	Gas explosion
	16.03	Anhui, Xieyi and Guangde	Gas explosion
	29.04	Xinjiang, Donfeng	Explosion
	29.06	Guiyang	Gas explosion

a) Includes mining disasters involving more than 20 deaths. Refers to coal mines unless otherwise indicated.
Source: SIGMA.

Since 1990, government appropriations for *education* have increased much to reach CNY 385 billion, but nevertheless remain quite low at 2.8% of the revised GDP (OECD, 2006). Student enrolment in junior secondary schools has increased by 67% since 1990. Today, 98% of graduates of primary schools enter junior secondary schools and 63% of graduates of junior secondary schools enter senior secondary schools. Since the 1982 national census, the *illiteracy* rate has dropped from 22.8% to 6.9%. Among the 89 million illiterate people, 71% are female, and 16% are located in the provinces and autonomous regions of Tibet, Qinghai, Guizhou, Gansu, Yunnan and Ningxia, although these regions and provinces represent only 9% of the national population.

The national *government expenditure for social, cultural and educational development* increased from CNY 73.8 billion in 1990 to CNY 749 billion in 2004 and will continue to increase with the growing size of the elderly population and related health expenditure and further increases in educational expenditure. Since 1995, total public health expenditure has tripled, but remains very low at 0.6% of the revised GDP (OECD, 2006).

In its fight against *poverty*, China has achieved very significant progress (Chapter 8). However, in 2003, China experienced, for the first time in many years, an increase in the poor population (i.e. people with per capita annual income less than

CNY 637), which rose by 800 000 compared to 2002. In 2003, while the number of rural poor fell to 3% of the total rural population (compared to almost 9% in 1993), the number of urban poor, particularly women and children, increased. In 2005, people in absolute poverty in rural areas decreased by 11.6% compared to the previous year.

4. Institutional Context

4.1 National institutions

The People's Republic of China (PRC) was established on 1 October 1949. The present *Constitution* came into force in 1982 and has since been amended four times. The six central state institutions are the National People's Congress (NPC), the Presidency, the State Council, the Central Military Commission, the Supreme People's Court and the Supreme People's Procurator. The latter five institutions are responsible to the NPC and its Standing Committee (CIIC, 2005a).

The *National People's Congress* (NPC) is the highest institution of power of the country, with members elected for five years. It is responsible for creating and supervising all administrative, judicial and procuratorial units at every level of government. It examines and approves national economic and social development plans, as well as the yearly state budget.

The *President*, as the head of State, is elected by the NPC and promulgates laws. The President appoints the premier, vice premiers, state councillors, ministers of various ministries and state commissions, the auditor-general, and the secretary-general of the State Council, according to decisions of the NPC (and its Standing Committee).

The *State Council* is the executive governing body and is responsible to the NPC (Figure 2.2). The Premier of China chairs the State Council and is nominated for five years.

The *Communist Party of China* (CPC) was founded in 1921 and is the party in power. It exercises ideological and political leadership. It prepares laws and decisions that are adopted by the NPC. It conducts activities within the framework of the Constitution and laws. The CPC has both central and local organisations. Besides the CPC, there are eight other political parties which also participate in the political life of the nation.

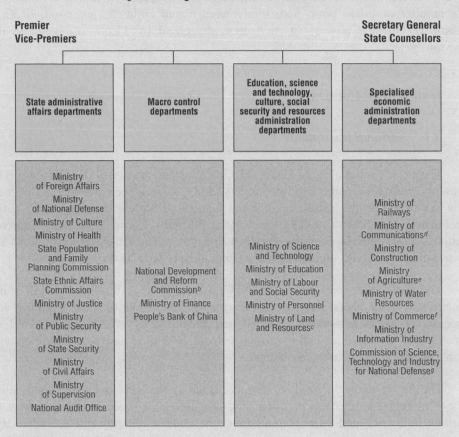

Figure 2.2 **Organisation of the State Council**[a]

a) Excluding administrations and bureaus directly under the State Council, such as the State Environmental Protection
   Administration (SEPA), the State Forestry Bureau, the National Tourism Administration, the State Administration for
   Industry and Commerce, the State Administration of Taxation and the State Statistics Bureau, and other institutions
   such as the Chinese Academy of Sciences, the State Electricity Regulatory Commission and the National Natural
   Science Foundation.
b) Includes State Bureaus on grain reserves, coal industry, machine-building industry, metallurgical industry, light
   industry, textile industry, petrochemical industry, energy and internal trade.
c) Supervises land administration, mining and the State Oceanic Administration.
d) Supervises transport by road and inland waterways.
e) Supervises fisheries.
f) Supervises foreign trade.
g) Includes the State Atomic Energy Agency.
Source: OECD, World Bank Development Research Group.

4.2   Territorial institutions

The Constitution stipulates that the *administrative division* of the country is based on a three-tier system of *provinces, counties and townships*. More precisely, China is first divided into provinces, autonomous regions and centrally-administered municipalities (Table 2.2). There are also special administrative regions (Hong Kong and Macao). Secondly, provinces and autonomous regions are divided into counties, autonomous prefectures, autonomous counties and cities. Thirdly, counties and autonomous counties are divided into towns, townships and ethnic townships. The five autonomous regions as well as autonomous prefectures and autonomous counties are all ethnic autonomous areas (CIIC, 2005a). Between the provincial and the county levels, the prefecture level (also called municipal level) plays a role for some environmental issues.

At the end of 2004, China had *37 334 towns and townships* (i.e. 19 883 towns and 17 451 townships). This is a decrease of 956 compared to the end of 2003, reflecting an *effort to streamline the "primary-level" administration.*

4.3   Environmental framework

The 1982 Constitution establishes: i) in its Article 9, that the *State must ensure the rational use of natural resources and protect rare animals and plants*, and that the appropriation of or damage to natural resources is prohibited; nd ii) in its Article 26, that the *State must protect and improve the living environment and the ecological environment*, must prevent and control pollution and other hazards, and must institutionalise and encourage afforestation and the protection of forests.

The *Environmental Protection Law* of China (1979, amended 1989) sets forth the basis of national policy and defines national government and territorial responsibilities for environmental protection. This law has been formulated for "the purpose of protecting and improving people's environment and the ecological environment, preventing and controlling pollution and other public hazards, safeguarding human health and facilitating the development of socialist modernisation".[7] The law requires the State to adopt economic and technological policies and measures for environmental protection so as to co-ordinate the work of environmental protection with economic construction and social development.[8] The State must also encourage the development of education in the science of environmental protection.[9] In addition, China has a range of more specific environmental laws or environmentally-related laws (Table 2.5).

China's *10th Five-Year Plan (FYP) (2001-05) for National Economic and Social Development* was approved in 2001. It defined objectives, guiding principles and strategies for China's economic and social development, including the planned allocation of 1.2% of the country's GDP to environmental protection (an increase from the 0.93% of the 9th FYP). Under the heading "Co-ordinating Economic Development with Social Development", the 10th FYP stresses that "the State needs to pay close attention to and solve issues of population, resources and the ecological environment, take further steps to implement the strategy of sustainable development, and stimulate co-ordinated economic and social development" (Chapter 7).

Table 2.5 **Selected environment related legislation**

1979	Environmental Protection Law (amended 1989 and 2001)
1982	Marine Environmental Protection Law (amended 1999)
1984	Forest Law (amended 1998)
1984	Law on Prevention and Control of Water Pollution (amended 1996, implemented 2000)
1985	Grassland Law
1986	Fisheries Law
1986	Mineral Resources Law
1986	Law on Land Administration (amended 1998)
1987	Law on Prevention and Control of Air Pollution
1988	Water Law (amended 2002)
1988	Wildlife Protection Law
1989	City Planning Law
1991	Law on Animal and Plant Quarantine
1991	Law on Water and Soil Conservation
1993	Agricultural Law
1994	Regulations on Protected Areas
1995	Law on Prevention and Control of Environmental Pollution by Solid Waste (amended 2004)
1995	Law on Prevention and Control of Air Pollution (amended 2000 and 2002)
1996	Law on Prevention and Control of Pollution from Environmental Noise
1996	Law on Coal Industry
1997	Law on Protecting Against and Mitigating Earthquake Disasters
1997	Law on Energy Conservation
1997	Construction Law
1997	Flood Control Law
1998	Fire Control Law
1998	Law on Promotion of Cleaner Production (amended 2003)
1999	Administrative Reconsideration Law
1999	Meteorology Law
2001	Law on Prevention and Control of Desertification
2001	Law on Administration of Sea Areas
2002	Law on Popularization of Scientific Technology
2002	Law on Safety Production
2002	Law on Environmental Impact Assessment
2003	Law on Radioactive Pollution Prevention and Control
2003	Law on Administrative Permission
2005	Law on Renewable Energy

Source: SEPA.

The *11th Five-Year Plan (FYP) (2006-10) for National Economic and Social Development* was approved in 2005. In this new plan, priority should be given to achieving a better balance between economic, social and environmental development, by building a "harmonious society" with a narrower gap between rich and poor, and by curbing widespread environmental degradation. During the period of the 11th plan, a total of CNY 1 300 billion (on the order of 1.2% of China's GDP) is expected to be spent for environmental protection throughout the country (Chapter 7).

4.4 Environmental institutions

The Environmental and Resources Protection Committee (ERPC) of the NPC is responsible for developing, reviewing and enacting environmental laws (OECD, 2005b).

Under the State Council, the *State Environmental Protection Administration*, (SEPA), the highest administrative body for environmental protection, is responsible for developing environmental policies and programmes. SEPA deals with:

- policy and regulatory matters;
- enforcement and supervision of pollution prevention and control laws and regulations;
- cross-cutting and regional coordination issues;
- standards for environmental quality and discharges;
- environmental management and environmental impact assessment;
- research and development (R&D), certification and environmental industry;
- environmental monitoring and information disclosure;
- global environmental issues and international conventions;
- nuclear safety.

SEPA supervises *Environmental Protection Bureaus* (EPBs) at the provincial, prefectoral and county levels. EPBs are part of the provincial administration (Governor's Office). They implement national and provincial environmental protection laws and are involved in monitoring pollution.

A variety of *sub-national administrative units* also play a role in environmental protection:

- *Mayors' offices* (e.g. of the four centrally-administered municipalities) take decisions on large investment projects involving industrial development and environmental protection;

- *Planning commissions* (at the county level and above) are responsible for reviewing the environmental protection plans of EPBs and for integrating them into local economic and social development plans;

- *Industrial bureaus* play a significant role in day-to-day industrial pollution abatement;

- *Finance bureaus* manage city revenues and expenditures and play important roles in the pollution discharge fee system;

- *Urban construction bureaus* oversee the construction and operation of waste water treatment plants.

Table 2.6 **Environmental-related administrations**

Responsibilities	Entities
Macro-coordination and control	National Development and Reform Commission (NDRC) Ministry of Finance (MOF) Ministry of Foreign Affairs (MOFA)
Pollution Control	State Environmental Protection Administration (SEPA) Ministry of Construction (MOC) Ministry of Railways (MOR) Ministry of Communications (MOCO) Ministry of Water Resources (MOWR) Ministry of Health (MOH)
Ecosystem protection	State Environmental Protection Administration (SEPA) Ministry of Agriculture (MOA) State Forestry Administration (SFA) Ministry of Land and Resources (MOLR)
Natural resource management	State Environmental Protection Administration (SEPA) Ministry of Land and Resources (MOLR) Ministry of Water Resources (MOWR) Ministry of Agriculture (MOA) State Forestry Administration
Others	Ministry of Science and Technology (MOST) Ministry of Education (MOE) State Oceanic Administration (SOA) National Audit Office General Administration of Civil Aviation General Administration of Customs State Administration of Taxation

Source: Changhua Wu; "Governance in China", OECD.

A number of *ministries and agencies* under the State Council are involved in environmental management (Table 2.6). In particular, the Ministry of Communications, Ministry of Railways, Ministry of Water Resources, Ministry of Agriculture and the General Administration of Civil Aviation play a role in environmental protection associated to infrastructure construction (including environmental impact assessment) and management, water and soil conservation, natural resources protection, pollution prevention and control, environmental monitoring and implementing international agreements.

Notes

1. Proved reserves are reserves that are confidently considered to be not only recoverable but also recoverable economically, under current market conditions. In other words they take into account what current mining technology can achieve as well as the economics of recovery (mining, transportation and other relevant recovery costs such as government royalties and coal prices). Proved reserves will therefore fluctuate according to economic conditions and factors, especially their price. It generally follows that proved reserves will decrease when prices are too low for the coal (or the oil) to be recovered economically, and increase when prices are higher.

2. GDP figures are based on the revisions of data on the Chinese economy done by the National Bureau of Statistics of China in January 2006.

3. 1 CNY = 1 RMB = 1 yuan.

4. PPP figures come from the International Monetary Fund, World Economic Outlook Database, April 2006.

5. China's share of global TPES is around 12%, approximately the same as its share of global GDP.

6. According to the United Nations Department of Economic and Social Affairs, Population Division.

7. Environmental Protection Law, Article 1.

8. Environmental Protection Law, Article 4.

9. Environmental Protection Law, Article 5.

Selected Sources

The government documents, OECD documents and other documents used as sources for this chapter included the following. Also see list of websites at the end of this report.

Chinese Government's Official Web Portal (2005), "Wen Explains Proposal on 11th Five-Year Plan", www.gov.cn, Beijing.

CIIC (China Internet Information Center) (2005a), *China Facts and Figures 2005*, CIIC, Beijing.

CIIC (2005b), *Urban Population Set to Boom*, CIIC, Beijing.

DRC (Development Research Center of the State Council) (2004), *China's National Comprehensive Energy Strategy and Policy 2020,* CNESP 2004, Beijing.

Energy Information Administration (2005), *International Energy Outlook*, Washington DC.

IEA (International Energy Agency) (2006a), *Energy Balances of non OECD Countries,* OECD/IEA, Paris.

IEA (2006b), *World Energy Outlook,* OECD/IEA, Paris.

IEA (2005), *Coal Information 2005,* OECD/IEA, Paris.

IEA (2004), *Key World Energy Statistics,* OECD/IEA, Paris.

NBSC (National Bureau of Statistics of China) (2006a), "Announcement on Revised Result about Historical Data on China's Gross Domestic Product", NBSC, Beijing.

NBSC (2006b), "China Statistical Yearbook on Environment", China Statistics Press, Beijing.

NBSC (2006c), "Statistical Communiqué of the People's Republic of China on the 2005 National Economic and Social Development", Beijing.

NBSC (2005a), "National Economy Maintains Stable and Fairly Fast Growth in 2005", Beijing.

NBSC (2005b), *China Statistical Yearbook, 2004*, China Statistics Press, Beijing.

NBSC (2005c), "Statistical Communiqué of the People's Republic of China on the 2004 National Economic and Social Development", Beijing.

NBSC (2005d), *China Compendium of Statistics 1949-2004*, China Statistics Press, Beijing.

NDRC (National Development and Reform Commission) (2005), 国家发展改革委员会 节能中长期转向规划, *China Medium and Long Term Energy Conservation Plan (2004-2020)*, 中国环境科学出版社, 北京, China Environmental Science Press, Beijing.

OECD (2006), *Challenges for China's Public Expenditure Policies – Towards greater equity and effectiveness*, OECD, Paris.

OECD (2005a), *OECD Economic Surveys: China,* OECD, Paris.

OECD (2005b), *China in the Global Economy: Governance in China,* OECD, Paris.

SEPA (State Environmental Protection Administration) (2006a), Annual Statistics: Report on Environment in China, China Environmental Science Press, Beijing.

SEPA (2006b), *China Environment Yearbooks (1995-2005)*, China Environment Yearbook Press, Beijing.

SEPA (2006c), State of the Environment Report, SEPA, Beijing.

Stockholm Environment Institute and United Nations Development Programme (UNDP) (2002), *China Human Development Report 2002: Making Green Development a Choice*, UNDP, Stockholm.

World Bank (2004), Development Research Group and Green Development Institute, *Urbanization, Energy and Air Pollution in China: The Challenges Ahead – Proceedings of a Symposium*, World Bank, Washington DC.

World Bank (2003), Development Research Group and Green Development Institute, *Environmental Institutions in China*, World Bank, Washington DC.

World Energy Council (2005) *China's Energy Supply. Many Paths – One Goal*, English Edition of "Energy for Germany" by WEC's German member Committee, London.

3

AIR MANAGEMENT

Features

- Setting more ambitious emission reduction targets
- Solving some of the world's worst urban air quality problems
- The need for clean coal technology
- Setting more ambitious energy efficiency targets
- The challenge of urban rapid transit in China's large cities

---

Recommendations

The following recommendations are part of the overall conclusions and recommendations of the environmental performance review of China:

• translate the *energy intensity* improvement target into more ambitious *energy efficiency* targets in all sectors; use a mix of instruments to achieve them, including pricing policies, demand management, introduction of cleaner technologies, and energy-efficient buildings, houses and appliances;

• bolster the adoption of *cleaner fuels* (including cleaner coal technology, coal washing and flue gas desulphurisation) and cleaner fuels for vehicles, as well as cleaner cars;

• implement more ambitious *air emission reduction targets* capable of achieving ambient quality objectives already adopted; manage a *wider range of air pollutants*, including VOCs and toxic substances;

• further improve the quality of *monitoring* data needed for effective air quality management and widen their scope (e.g. sources and pollutants);

• develop and implement a *national transportation strategy* that recognises the environmental externalities of transport and takes an integrated approach to private and public transport; streamline the institutional framework for developing sustainable transport systems; use a mix of regulation and economic instruments (e.g. taxes) to give citizens incentives for rational transport decisions;

• strengthen *mass transport in urban areas*, and take measures to encourage the urban use of cleaner transport modes (e.g. bicycles).

Conclusions

During the review period, China achieved improvements in *ambient air quality* (e.g. lowering the concentration of SO_2 in urban areas and designated control zones) and in decoupling emissions of SO_2, NO_2 and CO_2 from economic growth. The overall emission reduction targets for SO_2, soot and dust from stationary sources set out in the 9th FYP (1996-2000) were met and surpassed; those for *stationary source emissions* of soot (–10%) and industrial dust (–20%) stipulated in the 10th FYP (2001-05) are also likely to have been met. The *legislative and regulatory framework* was updated with the tightening of some emission limits, the introduction of total emission control, and the designation of special control zones (covering 39% of the population). The rate of *emission charges* was trebled. A start was made with flue gas desulphurisation at large emission sources. A nationwide *air quality monitoring* network was put in place. *Energy policy and institutions* (including a renewable energy law) were strengthened, and efforts to diversify energy sources had some success. In the domestic sector, the dependence on coal was reduced from 69% to 30% during 1990-2004. Concerning *transport*, environment-related efforts included the adoption of fuel-efficiency standards for light-duty passenger vehicles in 2004, the adoption of the various EURO standards for vehicle emissions at set dates, and the development of bus rapid transit systems in some cities.

Despite these efforts, air quality in some Chinese cities remains among the worst in the world. About 60% of cities above county level are likely to have met the grade II *ambient air quality* standard by 2005. The SO_2 concentration in urban air, after dropping steadily since the early 1990s, began to climb again in 2002. Nationwide SO_2 emissions from stationary sources increased by 13% during 2000-04, and therefore are not likely to have reached the 10% reduction target of the FYP. In the special air pollution control zones, SO_2 emissions fell by 2% instead of the targeted 20%. Likewise, the proportion of cities suffering from highly *acid rain* (i.e. pH under 4.5) rose again to 10% in 2004, after a low of 2% in 2000. Current emission reduction targets are not sufficient to meet ambient air quality standards. To date, insufficient attention has been paid to *VOCs and toxic air pollutants*. Air pollution regulations and permit conditions are not well enforced. China's *energy intensity* per unit of GDP is about 20% higher than the OECD average and, after a decline early in the review period, began to grow again in 2001. Reducing the energy intensity of the Chinese economy is rightly seen as a top priority by national authorities, especially in light of estimates that a doubling of total primary energy supply will be needed to satisfy a quadrupled GDP by 2020 (compared to 2000). But reducing energy intensity by 20% during the 2006-10 period will be a major challenge despite the associated potential multiple benefits (e.g. reducing traditional air pollution, reducing greenhouse gas emissions, increasing energy independence and security, and improving the efficiency of the economy). China did not meet its target of *washing 50% of the coal* it burns and the implementation of flue gas desulphurisation has been low to date. Although car ownership is still low, *vehicle numbers* doubled in the five years up to 2000, and motor vehicle traffic already represents the largest source of *urban air pollution*; the efficiency of urban transportation is showing a downward trend. Urban mass transit has not received sufficient emphasis and the use of bicycles has been allowed to decline. The *quality of vehicle fuels* is low (e.g. sulphur content).

◆ ◆ ◆

1. Policy Objectives

The main texts of *national legislation* concerning air management are the umbrella Environmental Protection Law (1979, revised in 1989 and 2001) and the Law on Prevention and Control of Air Pollution (1995, revised in 2000 and 2002). Other laws relevant to air quality or emissions include the laws on the Coal Industry (1996), Energy Conservation (1997), Promotion of Cleaner Production (1998, revised in 2003), and Renewable Energy[1] (2005). Under Article 6 of the air pollution law, the State Environmental Protection Administration (SEPA) is charged with establishing national ambient air quality and emission standards. The governments of provinces, autonomous regions and municipalities directly under the central government may establish their local standards for items not specified in the national standards; local emission standards may be more stringent than national ones if the circumstances so require (Chapter 7). Also, in areas with severe air pollution problems, Article 15 of the air pollution law allows for the designation of special control zones in which more stringent measures can be taken.

In 1996, three classes (called grades) of national *ambient air quality standards* were promulgated for ten pollutants (Table 3.1). The three grades are structured as follows:

- grade I: air quality under which no adverse impact will be observed on human health and long-term ecological health;
- grade II: air quality under which no harmful impact will be observed on human health and plant and animal health in the short or long term;
- grade III: air quality under which neither acute nor chronic poisoning of humans will occur and plant and animals in urban areas (except susceptible species) can grow normally.

Grade I standards are applied to nature protection, scenic and other protected areas; grade II is applied to residential, commercial, cultural, general industrial and rural areas; grade III is applied to designated industrial areas. Due to different averaging periods, not all Chinese standards can be directly compared with those of OECD countries, but where they can, Chinese grade II standards tend to be somewhat more lenient.[2]

Regulation of emissions from stationary and mobile sources takes place by setting emission limits and through various so-called technical policies, e.g. on pollution abatement of sulphur dioxide emission from coal combustion (2002), the prevention and control of pollution by motor vehicle emissions (2003), and on the prevention and control of pollution by diesel oil vehicle emissions (2003). Emission limits for new and existing thermal power plants became more stringent in 2004 and 2005, respectively, and will be further tightened as from 2010; the emission limits in effect mean that new coal-fired plants need to be fitted with flue gas

Table 3.1 **National ambient air quality standards**[a]

Pollutant	Averaging time	Chinese standards		
		Grade I	Grade II	Grade III
SO_2 ($\mu g/m^3$)	annual	20	60	100
	24-hour	50	150	250
	1-hour	150	500	700
TSP (μ/m^3)	annual	80	200	300
	24-hour	120	300	300
PM_{10} ($\mu g/m^3$)	annual	40	100	150
	24-hour	50	150	250
NO_x ($\mu g/m^3$)	annual	50	50	100
	24-hour	100	100	150
	1-hour	150	150	300
NO_2 ($\mu g/m^3$)	annual	40	40	80
	24-hour	80	80	120
	1-hour	120	120	240
CO (mg/m^3)	24-hour	4	4	6
	1-hour	10	10	20
Ozone ($\mu g/m^3$)	1-hour	120	160	200
Pb ($\mu g/m^3$)	seasonal	1.5		
	annual	1.0		

a) China also has national air quality standards concerning B(a)P and fluoric compounds.
Source: OECD, Environment Directorate.

desulphurisation (FGD) equipment. Initially, emission controls were limited to pollutant concentrations, but this was changed to a system of total emission control, i.e. limits were placed on both the concentration and the total mass of pollutants allowed to be emitted by an installation. There are also fuel quality standards.

More specific, *numerical national objectives* for both emissions and ambient air quality applicable to the review period are set forth in the 9th (1996-2000) and 10th (2001-05) Five-Year Plans for Environmental Protection (FYEPs) (Table 3.2). During the rapid economic growth throughout the review period, the 9th FYEP sought to contain emissions, whereas the 10th proposed to actually reduce emissions by about 10% in the case of SO_2 and soot (i.e. particles from coal burning), and by 20% for industrial dust (particles from other industrial sources). For the present, however, the FYEPs' ambient air quality objectives and emission reduction targets are not yet consistent. Much greater emission reductions will be necessary to achieve the ambient air quality objectives: for example, SO_2 emissions will need to be reduced by 57% and NO_x emissions by 44%, to meet ambient quality standards (DRC, 2004).

Energy and transport policy

The 10th Five-Year Plan for National Economic and Social Development (FYP) also contained *energy policy objectives* and placed energy conservation at the top of the energy policy agenda. Other environment-related energy objectives in the plan included the diversification of the energy mix; development of renewable energies and new sources of energy; improvement of energy efficiency and dissemination of energy-saving technologies; increased use of high-quality coal; construction of a nationally unified power transmission network; and rationalisation of the energy structure to improve energy efficiency and protect the environment. The *Plan for Energy Conservation and Comprehensive Resource Utilisation*, attached to the FYP, set specific goals for energy conservation (Table 3.2).

The National Development and Reform Commission's (NDRC's) first *Medium- and Long-Term Energy Conservation Plan (2004-20)*, approved by the State Council in 2004, is the guiding document for energy conservation in China and is changing the approach of only paying attention to energy supply, into one that gives priority to energy conservation and energy efficiency. This trend is confirmed in the 11th FYP (2006-10), approved in March 2006, which stipulates that energy efficiency should improve by 20% during the plan period (including a 4% improvement for 2006). The plan proposes that the eastern, more developed parts of the country take the lead in achieving a more efficient use of resources, including energy.

The 9th and 10th FYPs, as well as the medium- and long-term energy plan, contain general *transport policy objectives* such as building and improving road and railway networks, strengthening urban road and rail transportation, and developing water transportation. The 10th FYP also states that:

- new cars, light and mini-vehicles, large and mid-sized passenger cars, and mid- to heavy-duty vehicles should meet Euro II emission standards;
- some mid- to top-grade cars, as well as large and mid-sized top-grade passenger cars, should meet Euro III standards;
- gas emissions from four-wheel agricultural vehicles should be gradually reduced and the vehicles equipped with multiple cylinders should meet Euro I standards; and
- gas emissions standards for various vehicles produced around 2010 should reach the international level.

Table 3.2 **Performance regarding selected emission reduction and air quality targets**

Objective		2000 target 9th FYP	Achieved 2000	2005 target 10th FYP	Achieved 2004
Limit SO_2 emissions from stationary sources of which from:	(million t)	24.60	19.95	17.96	22.55
industry			16.13	14.50	..
domestic			3.82	3.46	..
Reduce SO_2 emissions from coal burning[a]	(% compared to 2000)			−10	not met
Increase share of coal washing	(% of total production)			50	not met
Contain SO_2 emissions from stationary sources[b] of which from:	(million t)		13.16	10.53	12.84
industry					11.42
domestic					1.42
Contain industrial soot emissions of which from:	(million t)	17.50	11.65	10.60	10.95
industry			9.53	8.50	8.87
domestic			2.12	2.10	2.08
Contain industrial dust emission	(million t)	17.00	10.92	8.99	9.05
Share of cities[c] meeting at least grade II	(%)		35.60	50	39.5
Share of cities[d] meeting at least grade II for SO_2	(%)			80	..
Share of urban households using gas	(%)			92	..
Reduce the burning of straw (stalks) in "Two-Control" zones	(%)			95	..
PLAN FOR ENERGY CONSERVATION AND RESOURCE UTILISATION – Reduce energy intensity to (tce CNY 10 000 GDP, 1990 prices)				2.2	27%[e]
– Reduce energy intensity in key industries	(%)			20	..
– Total energy conservation during plan period	(million tce)			340	..
– Annual energy conservation rate	(%)			4.5	..
– Conservation and substitution of fuel oil	(million t)			16	..
– Conservation and substitution of finished oil	(million t)			5	..
– Fuel efficiency improvements for different types of vehicle	(%)			10-15	..

a) Notice Regarding Enactment of SO_2 Emission Control Policy for Coal Burning (regulation applying to the coal industry).
b) In "control" zones.
c) Above county level.
d) In "control" zones.
e) 27% above the 2005 target.
Source: SEPA; 9th and 10th FYPs.

Further transport objectives are spelled out in the 2001 Notice on Public Roads and Waterways of the Ministry of Communications. The 11th FYP also gives orientations for building new infrastructure (e.g. new railway lines for coal transport, intercity passenger transport and urban rapid transit systems) and optimising transport systems (e.g. multimodal container transport, the use of information technology).

2. Ambient Air Quality and Emissions

2.1 Emission trends

During the review period, China managed to *significantly decouple emissions of SO_2, NO_x and CO_2 from economic growth*[3] (Figure 3.1). Nevertheless, in the early part of the current decade the emission intensity of these three pollutants remained 2.9, 1.2 and 1.4 times, respectively, higher than the OECD average. Given that the official emissions data include stationary sources only, the real intensities will be greater still.

Official data show that SO_2 and NO_2 *emissions from stationary sources* reached a peak in the mid-1990s. The falling trend for SO_2 emissions was reversed early in the current decade. NO_x emissions from stationary sources also appear to have resumed their rising trend, though not at the same steep rate as in the 1980s and early 1990s. CO_2 emissions grew by 56% in the decade to 2004, when emissions reached about 4 800 million tonnes, accounting for around one-sixth of global emissions and making China the world's largest emitter after the United States.[4]

The *emission targets* of the 9th FYP were not only met but were bettered by a large margin, thus confirming that the "low-hanging fruit" can be harvested at modest cost (Table 3.2). Reaching the emission targets of the 10th FYP already appears to have been more difficult. Based on the latest available data for 2004, the 2005 targets for soot and dust are likely to be met, but those for SO_2 will not, as the previously downward trend reversed in 2003 and in 2004 emissions were 25% above the 2005 target as a result of the rapidly rising coal consumption and the low implementation thus far of coal washing and FGD measures (Box 3.1).

The *structure of air emissions* is dominated by the coal sector. Midway through the review period, coal burning accounted for about 90% of SO_2 emissions, but as of 2005 this figure was closer to 70-80%. As in other countries, vehicle emissions are the main source of urban NO_x and ozone pollution. Little attention has been paid so far to emissions of toxic substances such as mercury, which are causing concern both inside China and beyond its boundaries.[5] No overall quantitative data are available about emissions of SO_2, NO_2, particles and volatile organic compounds (VOCs) from mobile sources.

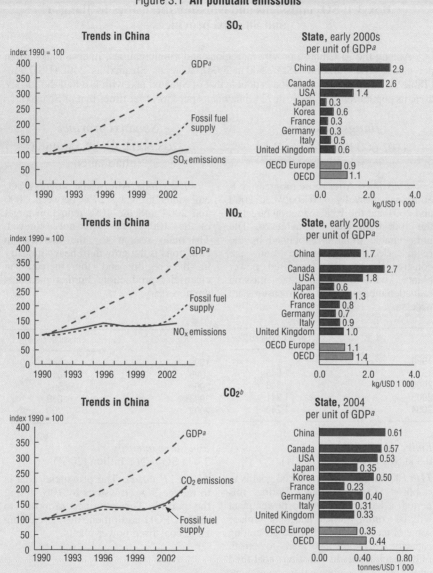

Figure 3.1 **Air pollutant emissions**

a) GDP at 2000 prices and purchasing power parities. Based on the revised historical data on China's GDP done by NBSC, 20 January 2006.
b) Emissions from energy use only; excludes international marine and aviation bunkers; sectoral approach.
Source: World Bank; OECD, Environment Directorate; OECD-IEA.

Box 3.1 SO_2 emissions and emission intensities in Jiangsu and Shaanxi provinces

Among the 31 Chinese provinces, regions and municipalities, Jiangsu ranks 5th in per capita GDP (around USD 2 500 in 2004), whereas Shaanxi is 24th (USD 940) (Table 2.2). Jiangsu's population is double that of Shaanxi and with just half the surface area, its population density is, at 724 inhabitants per km^2, four times that of Shaanxi.

Jiangsu province
(102 600 km^2, 74.3 million inhabitants)

Emissions. After some progress, SO_2 emissions grew by 47% between 2000 and 2004 and the 10% reduction target of the 10th FYP was not achieved. The unfavourable trend is explained by the rising electricity demand and the construction of new coal-fired power plants without adequate emission reduction measures. Further measures are necessary.

Shaanxi province
(205 600 km^2, 37.1 million inhabitants)

Emissions. After some progress, SO_2 emissions grew by 27% between 2000 and 2004 and the 10% reduction target of the 10th FYP was not achieved. The main reason for the increase in emissions is the growth of heavy industry in the region and the increase in electricity production. Further measures are necessary.

	Emissions (in kt)
1990	..
1995	1 045
2000	843
2003	1 241
2004	1 240

	Emissions (in kt)
1990	587
1995	652
2000	554
2003	638
2004	706

Emissions intensity:
0.8 kt SO_2/CNY billion (2004).

11th FYP target: The provincial goal is to stabilise SO_2 emissions. During this period, a large new nuclear power plant will be commissioned and will replace some of the current coal-based generating capacity. There are also plans to retrofit FGD equipment in existing coal-fired power plants. The potential for natural gas in the province is limited.

Emissions intensity:
2.4 kt SO_2/CNY billion (2004).

11th FYP target: The provincial goal is to reduce SO_2 emissions by 10% during the period. The province proposes to retrofit FGD equipment in five existing large coal-fired power plants.

Source: SEPA; provincial Environmental Protection Bureaus.

2.2 Air quality trends

Air quality varies greatly across a large country like China. The large cities suffer *severe air pollution*, sometimes some of the most serious in the world. An examination of *air quality in individual cities* shows that in 2005, 315 of 552 monitored cities (60.3%) met grade II air quality standards for total suspended particles (TSP), SO_2 and NO_2, whereas 152 cities (29.1%) achieved grade III; in 55 cities (10.6%), air quality was below grade III. Between 2003 and 2004, the proportion of cities meeting individual grade II standards for TSP and SO_2 remained stable, but the percentage of cities with lower than grade III air quality fell markedly for TSP and somewhat less so for SO_2 (Table 3.3). The ambient air concentration of NO_2 is still relatively low in most urban areas, but in the large cities, NO_2 pollution is already severe. Little information is available about rural areas, where ozone and the burning of straw affect air quality.

Concerning *trends*, overall, the mean annual concentration of SO_2 in ambient air in urban areas in both the north and south of China decreased steadily on average from the early 1990s until 2003, when the previous trend was reversed[6] (Figure 3.2). On average, the grade II standard was achieved from the late 1990s onwards, but in 2003 most northern cities no longer met this criterion. Although the air quality in cities located in the special control zones was, by definition, worse than elsewhere, a significant improvement was recorded forSO_2 over the period 1998-2004 (Table 3.4).

Health effects of air pollution

China's poor urban air quality continues to *affect public health*, as well as causing *large economic and environmental losses*; various studies (using human resource and contingent evaluation methods) estimate that the cost of air pollution falls in the range of 3 to 7% of GDP. A 1997 World Bank report (i.e. carried out at the start of the review period and before the switch to natural gas in some areas) estimated that smoke and fine particles from coal combustion caused 50 000 premature deaths and 400 000 new cases of chronic bronchitis in 11 large cities[7] each year; the cost of air pollution in these 11 cities was calculated to be greater than one-fifth of urban income. While the situation will have improved since then in larger cities, this is not necessarily the case in medium and smaller cities, which mostly are not yet served by natural gas. Another study carried out some years later in Beijing showed an increase in hospital outpatient admissions with declining air quality. These studies show clearly that investment in air pollution control will bring important health and economic benefits (e.g. increased labour productivity, reduced health expenditure).

Figure 3.2 **Trends in urban air quality**[a]

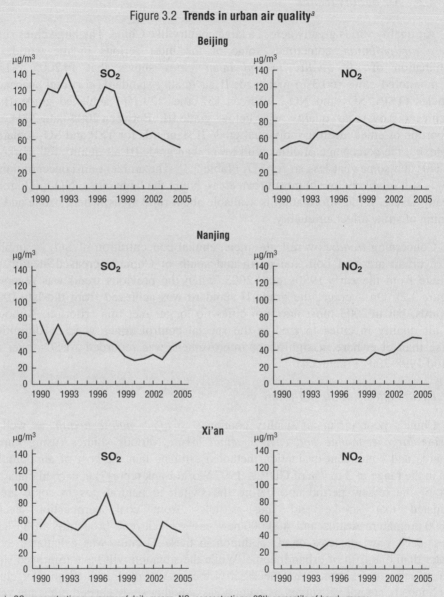

a) SO₂ concentrations: average of daily means. NO₂ concentrations: 98th percentile of hourly means.
Source: OECD, Environment Directorate.

Table 3.3 **Percentage of cities meeting annual air quality standards**[a]

(%)

	TSP		SO_2	
	2003	2004	2003	2004
Grade II (standard for TSP ≤ 200 µg/m³, and for SO_2 60 µg/m³)	45.6	46.8	74.4	74.3
Grade III (200 < TSP < 300 µg/m³; 60 < SO_2 < 100 µg/m³)	33.2	38.9	13.5	16.6
Non-compliant with grade III (> 300, > 100 µg/m³ respectively)	21.2	14.3	12.1	9.1
National average (µg/m³)	..	..	49.0	49.0

a) Sample of 342 cities.
Source: SEPA : Annual Report on Environmental Statistics of China, 2004.

Table 3.4 **Percentage of cities in air pollution control areas meeting air quality standards, 1998-2004**

(%)

	1998	1999	2000	2001	2002	2003	2004
SO_2 Control Areas (62 cities)							
Grade II	32.8	32.8	47.7	49.2	40.6	39.1	40.3
Grade III	29.7	37.3	24.6	28.6	28.1	25.0	32.3
Non-compliant with grade III	37.5	29.9	27.7	22.2	31.3	35.9	27.4
Acid Rain Control Areas (111 cities)							
Grade II	70.6	76.1	81.2	80.0	79.5	75.0	71.2
Grade III	13.7	9.4	6.3	7.8	13.7	14.7	21.6
Non-compliant with grade III	15.7	14.5	12.5	12.2	6.8	10.3	7.2

Source: China Central Environmental Monitoring Station (2005).

Acid rain

Acid rain (with a pH less than 5.6) *falls on about 30% of China's territory* and mainly occurs to the south of the Yangtze river and east of the Qinghai-Tibet plateau, as well as in the Sichuan basin. In mid-, southwest, east and south China, acid rain has reached or passed critical loads. In 2004, acid rain was recorded in 215 out of 526 cities (Table 3.5). In most cases, the acidity did not fall below a pH of 4.5, but in Jishou in Hunan province the annual average pH was as low as 3.05. The number of cities experiencing acid rain fell early in the review period, but then began to increase again; for instance, the proportion of cities suffering highly acid rain (pH of 4.5 or lower) gradually decreased from 1995 to reach 2.0% in 2000, and then climbed again to 10.3% in 2004 (Table 3.5). Acid rain is the focus of international regional co-operation (Chapter 9).

Recent monitoring shows that the concentration of sulphuric and nitric ions in rain falling in both northern and southern China remains high (e.g. when compared with acid rain in Japan). Such acidity cannot fail to *damage forests, crops, water bodies and buildings*. Some damage has been reported in the recent past, notably in the south-western provinces of Guizhou and Sichuan, while in northern China, the presence of high concentrations of alkaline ions (calcium and magnesium) from sand and dust particles will tend to neutralise the rain's acidity. Given the extent of coal use for the foreseeable future, a more stringent control of acid emissions is necessary to avoid endangering agricultural production and nature.

Table 3.5 **Number of cities experiencing acid rain**, 2004

| | With acid rain (pH < 5.6) | | | | | No acid rain total |
	Total	< 4.0	4.0-4.5	4.5-5.0	5.0-5.6	
Number of cities	215	5	49	93	68	311
Ratio (%)	40.9	1.0	9.3	17.7	12.9	59.1

Source: China Central Environmental Monitoring Station (2005).

Dust and sand storms

Strong westerly winds blowing in arid and semi-arid areas such as the Taklamakan and Gobi deserts and the Loess plateau in inland China pick up and carry mineral and soil particles to eastern China and sometimes beyond, as far as the

Republic of Korea and Japan. Storms causing severe damage to transportation systems and public health were reported in 33 out of the 52 years from 1949 to 2000, affecting some 140 regions.

For example, in a severe storm in northwest China in May 1993, 85 people died and 264 were injured, 370 000 hectares of agricultural crops were damaged, and 120 000 head of livestock either died or went missing. Damage to infrastructure and the manufacturing industry was also reported. In total, the direct *economic loss* was estimated at approximately CNY 560 million.

3. Air Quality Management

Monitoring and enforcement

A *significant effort was made during the review period*, and in particular since the passing of the 2000 air pollution law, to improve air quality monitoring and forecasting. Overall, there are now more than 2 000 monitoring stations (of which more than half belong to the national network operated by SEPA's China National Environmental Monitoring Centre) measuring SO_2, NO_2 and PM_{10}; about 600 urban stations are equipped with automatic monitors. More than 350 cities conduct routine urban air quality monitoring of TSP, SO_2, and NO_2. A start has been made in Beijing and other big cities with monitoring $PM_{2.5}$, but this activity is still in the research phase and no results are available yet. China also has a national acid deposition monitoring network comprising 113 automatic monitoring stations and a further 300 sampling sites throughout the country.

Since 2000, 42 cities have issued daily *air quality reports* and a further 47 cities now also publish a daily ambient *air quality forecast*. Ambient air quality is reported in terms of an air pollution index (API), based on the concentration of SO_2, NO_2 and PM_{10} (Box 3.2). Reports and forecasts are sent every day at 3 p.m. to television networks (CCTV), the Xinhua news agency and the newspaper China Daily.

While China has put in place a legislative and regulatory framework to combat air pollution, *enforcement of the rules by the local Environmental Protection Bureaus (EPBs)* is still far from adequate (Chapter 7). Emissions from large power plants are continuously monitored, but those from small sources without monitoring equipment are calculated using fuel data. The EPBs have the mandate but not the financial means to monitor coal quality (ash and sulphur content) themselves. They therefore depend on reports from small factories, which have a strong incentive to under-report in order to reduce their emission charges.

Box 3.2 **The air pollution index and its uses**

Because air pollution affects human *health*, authorities need simple and effective communication tools to warn citizens of any health risks associated with an imminent pollution episode. The same communication tools can also serve to raise public awareness of air quality issues. To this end, SEPA introduced a general air pollution index (API).

In principle, the API is calculated on the basis of *ambient air quality* data for SO_2, NO_2, PM_{10}, CO, and ozone, but as the latter two pollutants are not routinely monitored, they are, for the time being, not included in the calculation. The value of the API is taken as the highest of the three monitored pollutants and is expressed as a number between 0 and 500. Given the importance of ground level ozone and very fine particles ($PM_{2.5}$) to human health, these two parameters should be included in air quality monitoring and the calculation of the API as soon as possible, at least in the cities with the worst air quality. The API is then linked to a health-based grading system.

Air Pollution Index (API)

Grade	API	SO_2 ($\mu g/m^3$)	NO_2 ($\mu g/m^3$)	PM_{10} ($\mu g/m^3$)
I	0-50	0-50	0-80	0-50
II	51-100	51-150	81-120	51-150
III	101-200	151-250	121-280	151-350
IV	201-300	251-1 600	281-565	351-425
V	301-400	1 601-2 100	566-750	421-500
V	401-500	2 101-2 620	751-940	501-600

The API can also be used to rank cities according to air quality and then publish the results as part of a *"name and shame"* approach. In 2004, SEPA publicly disclosed the names of the ten (out of 113) Chinese cities with the worst air quality in 2003. The three worst cities (Linfen, Yangquan, and Daton) were all in Shanxi province. Nine of the ten cities were in northern China, and six were mining cities in the north-western region. Urban air quality is also affected by natural factors such as climate and topology. The only southern city in the ten-worst list, Zhuzhou, is the home of many heavy industries and construction. Beijing was not on the list, but it suffered grade V air quality during eight days in 2004 and six days in 2005. On 5 November 2005, Beijing experienced its worst air pollution episode for the year: the city was enveloped in a haze and the API reached 472. Weather conditions, vehicle emissions and construction dust all contributed to the episode.

Source: SEPA.

Special control zones

In 1998, the government initiated a specific policy to control air pollution in areas suffering from a high SO_2 concentration or acid rain (abbreviated in China to "*Two-Control Zones Policy*"). In all, 173 cities in China's 27 provinces were designated as special control zones, together covering 11.4% of the country's total land area and 39% of the population, and representing 67% of GDP and almost 60% of total SO_2 emissions. National and regional plans containing stringent anti-pollution measures are being implemented by local governments in the control areas, aiming to reduce local SO_2 emissions by up to 90% compared to the 2000 level. The measures include:

- total SO_2 emission control from large-scale sources;
- ban of coal use at small-scale boilers and restaurants and promoting fuel switch to cleaner energy such as liquefied natural gas (LNG) and liquefied petroleum gas (LPG);
- establishment of "no coal areas";
- ban of the use of high-sulphur (> 3%) coal through the introduction of coal quality control;
- installation of FGD at newly established power plants;
- compulsory continuous emission monitoring at newly established sources in urban areas; and
- increasing the SO_2 emission fee.

Application of air pollution control technologies and cleaner production

Washing coal before burning is a cheap and easy technology that will remove 40% of the pyrite and up to 20% of the sulphur content. It requires attention so that wash water is adequately treated before discharge (Chapter 4). China fell well short of meeting the 10th FYP *target of washing 50% of the coal* it burns, for the actual figure was closer to 30% (NDRC, 2005). Given that coal will remain a dominant part of China's energy future for some time, minimising the health and environmental impact of coal production[8] and use must be a major concern for air quality management. In 1995, some 280 million tonnes of coal were washed, equivalent to approximately 22% of total production, and the rate had increased to 33% by 2002 (Energy Foundation, 2005), which remains very low by international standards. Meeting the 50% target (about 550 million tonnes) would have required an additional capacity of 217 million tonnes per year, involving a considerable investment.[9] Improving the performance of the 30 000 or so small coal mines (together responsible for one-third of production) will be a particular challenge, as very few wash the coal they produce.

Replacing boilers in industry also contributes to air pollution control. There is a trend towards phasing out smaller steam-producing boilers in favour of larger ones that are better able to meet air emission standards. Official figures indicate that 85.6% of the 1.06 million tonnes of steam produced in 1995 was generated in boilers meeting standards; in 2003, the corresponding figures were 95.0% and 1.36 million tonnes. Compliance with air emission standards has also improved for industrial kilns and ovens: the compliance rate increased from 54.3 to 66.6% between 1995 and 2003 (SEPA, China Environment Yearbook, 2004).

Flue gas desulphurisation is another technique to reduce SO_2 emissions. By 2005, power stations with a combined generating capacity of 45 GW (out of a total installed thermal capacity of 350 GW) were equipped with FGD. Retrofitting will be carried out at power stations with a further capacity of 146 GW by 2010. Both the capital[10] and operational[11] costs of this technology are high; in 2000, a Japanese study estimated the overall capital costs of equipping all existing Chinese coal-fired thermal power plants (with a combined capacity of almost 200 GW) with FGD at CNY 119 billion,[12] and the annual operational cost at CNY 68 billion. Yet FGD is an indispensable technology to China if it is to achieve its air quality goals; increasing electricity prices to internalise environmental externalities would raise the required finance.

Cleaner production processes are promoted by several government programmes, including clean energy technologies. At the central level, the High Tech Research and Development Programme (the "863 Programme") has been part of successive FYPs since 1986. The 10th FYP included the National Key Technologies R&D Programme administered by the Ministry of Science and Technology. The provinces also encourage cleaner technologies; for instance, Jiangsu province provides incentives for the adoption of clean coal technology and its regulations require boilers to have a certain minimum size. Foreign direct investment is also a driver for the introduction of cleaner technologies.

Emission charges

Under the Environmental Protection Law, *China introduced a pollution levy system (PLS) as early as 1979*. Initially the charges on emissions to air were calculated on the excess concentration (above emission limits) of soot, dust and SO_2; actual charges were then levied only on the largest of the three pollutants. Also, the charges had no incentive value as enterprises found it cheaper to pay up rather than invest in pollution abatement and control. Revenue from the charges could be used to pay for the running cost of the regulating agencies (i.e. the Environmental Protection Bureaus at the provincial and municipal levels), so the latter had a perverse incentive not to pursue their pollution abatement functions too vigorously.

In 2003, *both the structure and the level of the charges were amended*: charges were levied on the total amount of emissions and on all substances included in the system; unit rates were increased. For example, the rate for SO_2 emissions was increased in 2004 from CNY 0.21/kg to CNY 0.42/kg. In 2005, the rate was further increased to CNY 0.63/kg, and a new charge of CNY 0.6/kg on NO_x emissions was introduced. Revenue from the charges is now transferred to the Ministry of Finance, which then submits proposals to the National People's Congress on how to spend it; the revenue remains earmarked for the environment.

Expenditure

The 9th and 10th FYPs anticipated CNY 208 billion and CNY 280 billion, respectively, for *air pollution control investment*, representing 46% and 40% of the total pollution control investment expenditure under the two plans and around 0.4-0.5% of GDP. No overall national figures are available on *air pollution operating expenditure*. However, operational costs are gradually increasing, as pollution control installed capital is accumulating.

4.   Integrating Air Management Objectives into Energy Policy

4.1   Developments in the energy sector

Institutional arrangements

The Chinese authorities have in recent years put *energy policy at the centre of China's economic development model* and have strengthened political leadership on energy matters. In 2003, NDRC was given responsibility for both energy supply (handled by the NDRC Energy Bureau) and energy consumption and efficiency issues (handled by the NDRC Department of Environment and Resources Conservation). The Premier, the two Vice-Premiers and the NDRC Minister have since 2005 been part of the ministerial National Energy Leading Group which meets annually, and under the Leading Group, the NDRC heads an office in charge of overall energy strategy. Also, as of March 2006, a new energy law was in preparation. However, many observers believe that further strengthening of energy institutions is needed to effectively and efficiently implement the energy programmes already adopted.

The review period also saw a gradual *reform of the energy and power sectors* with the purpose of achieving greater efficiency. The governance and commercial functions of the Ministry of Electric Power were separated and the ministry itself was abolished in 1998. The State Power Corporation was established in 1997 with a mission of further reform towards adopting a more market-based approach and creating competition among generators. In 2003, grid operation and power generation functions were separated. Five large and seven smaller state-owned power companies were established following the abolition of the State Power Corporation. A nationwide electricity grid connecting current regional grids is being developed. The State Electric Regulatory Commission, established in 2003, is the sector's regulatory body charged with setting prices, overseeing market operations, transmission and distribution, and project approval.

Trends and potential in energy supply

During the review period, *China remained heavily dependent on coal* as its dominant energy source (Figure 3.3, Box 2.1), although its efforts to diversify energy sources succeeded, with a decrease in coal consumption between 1996 and 2001. However, as a result of the rapid economic growth in recent years, absolute coal consumption has been on the rise again, even if the share of coal in total primary energy supply (TPES) evolved from 61.0% in 1990 to 61.7% in 2004. In 2004, the country consumed 1.97 billion tonnes of coal, an increase of about 90% over 1990 (IEA, 2006b).

While coal will remain a large part of China's energy future for a long time (Table 2.3), the country has a high potential to use *forms of energy that produce less air pollution*. Concerning *electricity*, nuclear energy accounted for 2.3% of electricity generation in 2004 (against 78% for coal, and 16% for hydro), but China is planning to boost the share of nuclear to 6% by constructing 35 more reactors by 2020. Although China already is the world largest producer of *hydropower*, at 105 GW, it has thus far exploited only a small part of its total hydro potential, which at 675 GW[13] is the largest hydroelectric potential in the world. Economic, environmental and social factors[14] will no doubt prevent this theoretical potential from being fully exploited. Concerning *natural gas*, its share is still small at 2.6% of TPES (2004), but a new 3 900 km pipeline is now bringing gas from western China fields to the eastern provinces, mainly for use in the residential sector, and two further pipelines are planned to bring gas from Russia. Infrastructure for imported liquefied natural gas is also being developed, but at a limited scale. Finally, *among renewable sources*, *methane* from biomass is being promoted in the 11th FYP for rural areas (notably in villages not connected to the electricity grid), as is *wind power*, planned to grow from 1 GW in 2005 to 5 GW in 2010. Thermal solar energy use is developing (e.g. for water heaters). Under the 2005 Law on Renewable Energy, the government has set targets to generate 10% of the total energy supply from renewable sources (including hydro) by 2010 and 15% by 2020.

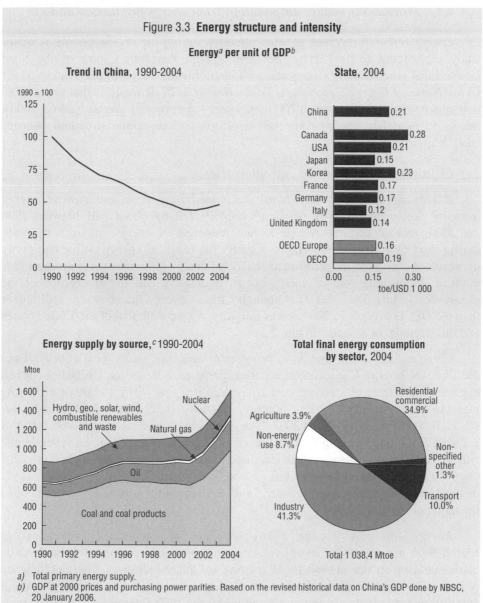

Figure 3.3 **Energy structure and intensity**

Energy[a] per unit of GDP[b]

Trend in China, 1990-2004

State, 2004

China	0.21
Canada	0.28
USA	0.21
Japan	0.15
Korea	0.23
France	0.17
Germany	0.17
Italy	0.12
United Kingdom	0.14
OECD Europe	0.16
OECD	0.19

toe/USD 1 000

Energy supply by source,[c] 1990-2004

Hydro, geo., solar, wind, combustible renewables and waste

Nuclear

Natural gas

Oil

Coal and coal products

Total final energy consumption by sector, 2004

Residential/commercial 34.9%

Agriculture 3.9%

Non-energy use 8.7%

Non-specified other 1.3%

Transport 10.0%

Industry 41.3%

Total 1 038.4 Mtoe

a) Total primary energy supply.
b) GDP at 2000 prices and purchasing power parities. Based on the revised historical data on China's GDP done by NBSC, 20 January 2006.
c) Breakdown excludes electricity trade.
Source: World Bank; OECD-IEA.

4.2   Measures to reduce air pollution from energy production and use

Energy research received growing attention during the review period. A major study carried out in 2004-05 by the Development Research Centre of the State Council and other leading energy research institutions resulted in publication of the *China National Energy Strategy and Policy Report 2020*. It suggests that while *GDP will quadruple* over the period 2000-20, *energy supply will double* (DRC, 2004), thanks to important energy conservation efforts and economic structural changes (Box 3.3, Table 2.3).

Improving energy intensity and efficiencies

Energy demand was significantly decoupled from economic growth: energy intensity (i.e. energy supply per unit of GDP) fell by about half between 1990 and 2004, representing an average annual reduction of energy intensity by 5.8% during the 1990s. This improvement is partly the result of energy policy and partly the result of economic restructuring (for 70% of energy efficiency gains during 1980-2000, a diminishing role of energy-intensive industry and the emergence of the electronics sector) (DRC, 2004). Although China's energy intensity was still higher than the OECD average in 2004, it was roughly on a par with that of the United States and the Republic of Korea[15] (Figure 3.3).

China initially *made significant progress towards the goals of the 10th FYP* and the Plan for Energy Conservation and Comprehensive Resource Utilisation. When expressed in tonnes of standard coal equivalent (tce) per CNY 10 000 of GDP (the measure used in the FYPs), energy intensity fell from 5.32 to 2.68 tce/CNY 10 000 between 1990 and 2002;[16] the latter figure is close to the 2005 target of 2.2 tce/CNY 10 000. Due to the trend reversal of the past four years, however, by the end of 2005 energy intensity had increased again to a level 27% higher than the target[17] (Table 3.2). This result, when viewed in the light of the 20% reduction target of the 11th FYP, means that the latter, in effect, represents a broader restatement of the target of the 10th FYP.

Energy efficiency became a key indicator of economic development in March 2006, and figures for energy consumption per unit of output will henceforth be made public on an annual basis for all regions and major industries. Energy efficiency is now also considered a key element of the country's energy security.[18] About 160 different standards were promulgated under the 1997 Energy Conservation Law. Current proposals for increasing the efficiency of energy transformation and end-use[19] should therefore be pursued with great resolve (Box 3.3). Reviewing energy prices (reducing subsidies, internalising environmental externalities) is key to further improvements in energy efficiency in all sectors (Chapter 7).

Box 3.3 **Reducing energy intensity**

Energy shortages and air pollution, and the fear that these will constrain economic growth, are behind the government's recent recognition of the *urgency of reducing energy intensity* in China. By setting a 20% reduction target for 2010, the 11th FYP aims to reverse the trend of recent years (2002-04), in which investment in energy conservation dropped as a proportion of total energy investment and energy intensity increased (Figure 3.3). In the longer term, China plans to *quadruple GDP while only doubling energy use by 2020* (base year 2000) (Table 2.3). In fact, reducing the energy intensity of the overall economy (through progress in energy efficiency in different sectors) would lead to multiple benefits including: i) improved energy security and independence, ii) economic benefits associated with improved energy efficiency in relevant sectors, iii) reduced emissions of traditional air pollutants and related health benefits, and iv) reduced emissions of greenhouse gases.

So far, China's main response to power shortages (which occurred in 24 out of 31 provinces/autonomous regions/large municipalities in 2004) has been to increase the energy *supply* by building new power plants (66 GW in 2005, 77 GW expected to come on line in 2006). Spending on energy conservation amounted to CNY 23 billion during 2001-05; as a rule of thumb, it is often assumed that the cost of energy conservation is about one-third to one-half that of producing energy. However, the government is keenly aware of the fact that in the long term, supply management alone cannot solve the country's energy problems.

The Chinese authorities have many arrows in their quiver to help meet the country's energy intensity targets. The 1997 *Law on Energy Conservation* set forth the broad principles of energy conservation and a variety of policy instruments is being implemented. For example, fiscal instruments such as reducing coal and petroleum subsidies, and various incentives have been adopted (e.g. interest payment rebates, differential interest rates, revoking import taxes, and reducing income tax) to encourage energy conservation projects. Clearly, further improvements of the energy, *price signals* are key to promote reductions of the energy intensity of China.

The *industrial sector has made* good progress in reducing *energy consumption per unit of output, achieving* a 64% reduction across industry between 1980 and 2000.[a] The objective of the 10th FYP was to improve the energy efficiency of key energy-intensive industries by 20%. The cement industry achieved an improvement of about 10% by reducing energy consumption per tonne of cement produced from 0.201 to 0.181 tce. Energy consumption per unit of output in the steel and copper industries fell by more than 30%. Much of the improvement was achieved by closing down small, inefficient facilities. Nevertheless, energy consumption in eight major sectors in 2000 was on average 40% higher than in OECD countries: for example, coal consumption for thermal power supply was 22.5% higher, and energy consumption per tonne of steel, cement, and pulp and paper was, respectively, 21.4%, 45.3%, and 120% higher. Even as new technology and equipment is installed as the economy grows, much old and backwards technology remains in operation. The potential for energy savings therefore remains large.

Box 3.3 **Reducing energy intensity** *(cont.)*

Concerning *end uses*, many of China's efforts to *improve efficiency* are carried out under the China End Use Energy Efficiency Project (in co-operation with GEF and UNDP). The project promotes energy conservation policy, the associated regulatory system, and voluntary government-industry approaches. Other efforts include setting energy standards and product labelling; for example, the Green Lights Project (also backed by GEF and UNDP) has had some success in developing energy standards for lighting products and electrical household appliances. To improve the hitherto poor implementation of energy conservation in buildings, the Ministry of Construction in 2006 set more stringent energy standards and is running a pilot implementation programme, in which Beijing and Shanghai will soon promulgate more stringent performance standards to cut energy consumption in buildings by two-thirds; the standards will apply nationwide by 2010.

a) Versus 20% for the OECD area and 19% worldwide.
Source: Sinton, J. *et al.*; Sugiyama and Oshita.

Promoting alternative sources of energy

In *urban areas*, progress was made with respect to the systematic phasing out of coal in favour of gas, including the use of gas in district heating in the residential sector; this achievement was the single largest contributor towards the improvement in air quality experienced in recent years. The gradual tightening of emission limits for new and existing thermal power stations in 2005 and 2010, if rigorously implemented, will bring further improvements in air quality.

In *rural areas*, traditional energy sources include firewood, crop residues and manure. The use of each of these for burning impairs the environment in some way,[20] and the authorities have promoted more efficient or *alternative ways of cooking and heating*. The national improved stove programme (NISP) was an early programme under which around 130 million high-efficiency cooking stoves were installed by the early 1990s. Other programmes promote solar energy panels (52 million square metres installed as of 2003);[21] solar plants[22] and micro-hydro power. Biogas from human and animal waste is promoted as an alternative source for cooking under the 11th FYP, as part of a much larger effort to improve the quality of life in rural areas.

5. Integrating Air Management Objectives into Transport Policy

Transportation is the fastest growing economic sector in China. Air transport is booming, while the average annual growth in *vehicle numbers* has amounted to 13% since the early 1990s (Zhou and Szyliowicz, 2005), reaching 27 million[23] passenger cars and commercial vehicles, 79 million motorcycles and 25 million agricultural vehicles in 2004. China is already the world's second largest car market (after the United States.) and new car sales are expected to reach 4 million in 2006. Given that private car ownership still is very low compared to that in OECD countries (Figure 3.4), further rapid growth can be expected. Yet, as the standard of living of its citizens increases, China will have to find its own model for managing mobility, given the environmental, energy and space constraints imposed by its high population density. As urbanisation increases and new urban areas are rapidly built, there is both an opportunity and a necessity to leapfrog current transport patterns and find new solutions with lower energy requirements and environmental impact.

5.1 Developments in the transport sector

Plans and institutions

In 2000, China began to deal in earnest with the *environmental impact of transport*. Leaded petrol was banned, catalytic converters and electronic fuel injection engines were made compulsory, and national EURO I exhaust emission standards were adopted. Transport issues (road, rail, public transport, aviation and water) are also addressed in the medium- and long-term energy conservation plan (2004-20), which sets goals to accelerate development of public transportation, such as by rail. In large cities, the plan makes public road transportation the first priority, rail the second, and private vehicles a supplementary mode. The plan also encourages the use of bicycles. In small and medium-sized cities, the plan focuses on development of public and private transport. It also emphasises the need to speed up the development of electrified railway systems and high-efficiency locomotives, while reducing the use of steam-engine powered trains and thereby the consumption of coal. The plan further aims to promote gas-powered vehicles in city buses and taxis in the 11th FYP period (NDRC, 2005).

As in many countries, a *range of government agencies* have responsibility for different aspects of transport policy in China (e.g. for infrastructures, vehicles, traffic, fuels, finance). Road and water transportation are under the purview of the Ministry of Communications, railroads under the Ministry of Railways, civil aviation under the General Administration of Civil Aviation, and urban transportation under the Ministry of Construction, the NDRC Construction Management Commission, and the Ministry of Public Security. In addition, the Ministry of Finance, the Bureau of Land

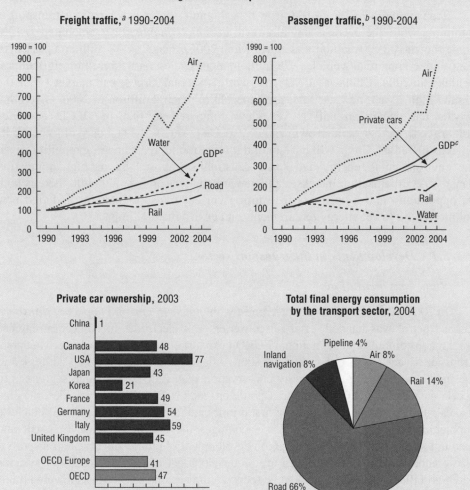

Figure 3.4 **Transport sector**

Freight traffic,[a] 1990-2004

Passenger traffic,[b] 1990-2004

Private car ownership, 2003

Total final energy consumption
by the transport sector, 2004

a) Index of relative change since 1990 based on values expressed in tonne-kilometres.
b) Index of relative change since 1990 based on values expressed in passenger-kilometres.
c) GDP expressed in 2000 prices and purchasing power parities. Based on the revised historical data on China's GDP
   done by NBSC, 20 January 2006.
Source: NBSC; OECD-IEA; OECD, Environment Directorate.

Resources, the State Forestry Administration and SEPA also have a role. This division of responsibility and the attendant inter-agency competition are not conducive to integrated transportation management or an efficient transport system. China would benefit from establishing a *national transportation strategy*, for instance with a 2020 perspective, that recognises environmental and energy costs from transport, including streamlining its institutional framework to develop sustainable transport systems. This strategy could build on the medium- and long-term energy conservation plan.

Current state of transport

The increases in *freight and passenger traffic*, except for air traffic, were (relatively) decoupled from GDP growth during 1990-2003 (Figure 3.4). In passenger transport, air traffic showed the most significant increase, followed by private car traffic, which increased almost as much as GDP over the same period.

The use of public transportation did not develop nearly as fast as private vehicle use. The *decreasing share of public transportation* is exemplified by the case of Beijing,[24] where use of the public transit system fell from 35 to 26.5% of intra-city trips (3.6% of the trips on the subway and 22.9% on trolley buses) during 1986-2000; during the same period, use of private cars rose from 6 to 23.2%. These figures should be seen in the context of the dramatic fall in the use of bicycles for commuting to work.

The transport sector is, with 10% of total final energy consumption (TFC) in 2004, China's *third largest energy consumer* (Figure 3.4); the share of transport in TFC is expected to rise to 16 or 17% by 2020. Road transport, which in 2003 used one-third of the country's total oil consumption of 310 million tonnes, is expected to be the primary cause of growth in the consumption of petroleum products. To limit the country's growing dependence on oil, the medium- and long-term energy conservation plan stipulates that petroleum should be used preferentially in the transportation and petrochemical industry for which there currently is no alternative to petroleum.

China's transport *infrastructure* is lagging behind the demands of the country's economic development, but a large effort is being made, notably in road construction, to make up the shortfall. The total length of road grew from 1.2 to 1.9 million km (including 34 000 km of motorway) in 1995-2004 and is expected to reach 2.3 million km (including 65 000 km of motorway) by 2010. This growth, however, was not mirrored in rail transport, as only 100 km of rail transit were completed during the past 40 years; the total length of the railway network, at 74 408 km, ranks, third worldwide after the United States and Russia. China also has 123 300 km of inland waterways and 133 airports for civil aviation.

The *environmental impact* of transport is evident in urban air quality. During the review period vehicles became the largest source of urban air pollution. For example, in Beijing during the warm season, automobiles are responsible for 92% of carbon monoxide (CO) emissions, 94% of hydrocarbons (HC), and 68% of NO_x (Gallagher, 2006). Similarly, in Guangzhou and Shanghai, on average more than 80% of emissions of CO and 40% of NO_x come from vehicles.

5.2   Measures to mitigate air pollution from transport

Vehicles

China's adoption in the current decade of the EURO *exhaust emission standards* has been an important achievement. EURO I was introduced in 2000 and from 2004 all light-duty vehicles (mainly passenger cars) in China[25] were required to meet EURO II standards. On 1 January 2006, EURO III came into effect nationwide for light- and heavy-duty petrol vehicles as well as for heavy-duty diesel engines. EURO IV emission standards will come into effect nationwide for petrol and diesel engines in 2007 and 2010, respectively. Beijing will apply EURO IV standards to both petrol and diesel light-duty vehicles starting in 2007. Beijing has also enforced a mandatory vehicle retirement policy under which some 38 000 vehicles, including 14 000 microbus taxis, have been forced to "retire" since the end of 1998. New light-duty vehicles have been required to have on-board diagnostic (OBD) systems since January 2006, and vehicle models that received type approval before this date are allowed a grace period of one year before they must comply (Energy Foundation, 2005).

The Chinese government has recently, under its *small car initiative*, started encouraging car buyers to purchase smaller, low-emission and more efficient cars and has called on local governments to abolish all restrictions on the licensing of such cars.[26] A new vehicle tax scheme announced in March 2006 will cut taxes on small-engine vehicles and raise taxes on larger ones (e.g. the tax on vehicles with an engine size greater than two litres has been increased from 8 to 20% of the purchase price). Other measures under consideration include lower parking fees for small vehicles and the use of small vehicles for taxi services. In light of these and other initiatives, a growing number of manufacturers are increasing their investment in the development and production of environmentally friendly and low-emission engines and automobiles.

Fuels

The *fuel efficiency of motor vehicles* in China is low compared to the vehicle fleet in the OECD area.[27] In 2004, the State Council adopted a plan to improve the fuel efficiency of cars, minibuses and light buses. In 2005, new motor vehicle fuel-efficiency standards that are 8% more stringent than the ones adopted five years previously came into effect.[28] The China Automobile Technology and Research Centre (CATARC) is developing technologies (including hybrid-electric vehicles) that will support the implementation of the standards; CATARC is also carrying out work on ways to enforce them. Although China phased out lead in 2000 and set limits on various hazardous substances in petrol, the sulphur content remains high and is limiting the ability of catalytic converters to lower emissions of CO, HC, and NO_x. Several programmes promote the use of alternative transport fuels (Box 3.4). Reviewing *road fuel prices* is key to energy efficiency, reduced emissions of air pollutants and control of CO_2 emissions in the transport sector (Chapter 7).

Box 3.4 **Promoting alternative transport fuels**

The promotion of alternative transport fuels is an important part of China's energy and transport policies.

The *National Technology Plan* under the 9th FYP listed the commercialisation of clean vehicles as a key technology for China. Beijing, Shanghai and eight other cities were designated as demonstration areas for the plan; in 2002, eight other cities joined the programme. As of 2002, 153 440 vehicles were running on *compressed natural gas* (CNG) or *liquefied petroleum gas* (LPG) and 486 filling stations had been built. In Shanghai, 74% of taxis and in Sichuan, 85% of buses had been converted to CNG/LPG. Yet the programme is not as successful as had been anticipated, probably because the price difference between CNG/LPG and petrol is not large enough to persuade car owners to switch. Other barriers include the small number of filling stations and the cost for retrofitting existing petrol vehicles as only 16% of CNG/LPG vehicles are newly manufactured as such. As a result, the number of vehicles using alternative fuels has not risen since 2002.

China is also promoting *ethanol and methanol-fuelled vehicles*. In 2001, the NDRC announced the ethanol fuel development plan. In 2002, an ethanol-petrol blend (10% ethanol) was sold in five cities, and the programme was expanded to nine provinces in 2004. Remaining challenges are developing more efficient technologies and bringing down the price of ethanol, which is still much more expensive than petrol.

China's policy during the review period has also been to attempt to leapfrog conventional technologies in the promotion of alternative fuels. Under the 10th FYP, an investment of CNY 2.4 billion was allocated to the special key programme for electricity-powered automobile research and development, including CNY 880 million to develop *fuel cell and hydrogen vehicle technologies*. The research and development of technologies have been successful, but this success has not been followed up with policy and regulations to implement and market the technology. It may still take two decades for this to happen.

Source: CCICED; Zhao and Melaina.

Traffic

Heavy congestion now is common in China's large cities, resulting in increases in travel time, fuel consumption and air pollution. In order to reverse the trend towards private vehicle use, the Ministry of Construction in 2004 issued a directive stipulating that local governments should endeavour to reach a situation where i) no less than 30% of all trips in the large cities use public transport and ii) the speed of public transportation vehicles is above 20 km per hour (Zhou and Szyliowicz, 2005). Several cities are developing or planning *bus rapid transit systems* (Box 3.5).

Box 3.5 Bus rapid transit systems

Although China has no specific national policy to promote *bus rapid transit systems* (BRT), individual cities are planning or implementing such systems. BRT systems can be defined as comprehensive transport systems improving bus transport networks through a range of infrastructural measures (e.g. reserved bus lanes, transfer stations) and managerial measures (e.g. upgrading of bus fleets, changes in schedules and routes, subscription passes). BRT represents a less costly option than underground rail or light rail solutions.

Cities that are already developing BRT include *Kunming, Beijing, Hangzhou and Jinan*:

• Kunming had three bus-dedicated lanes before 2003 and is a pioneer city in the development of BRT in China.

• Beijing plans to develop a BRT network of 30 km by 2010. The first section of 16 km from Tiananmen Square to the southeast of the city was opened in December 2005 and carries 80 000 passengers per day. Construction will begin in 2006 on a second and third corridor. In addition, Beijing has a total length of 121 km of exclusive bus lanes (Business Beijing, 2006).

• Hangzhou is building its first BRT corridor and is planning to add eleven more corridors over the next 15 years with a total length of 165 km.

• Jinan has launched a BRT project and its first BRT corridor of two km, will be completed in autumn 2006.

Other cities are in the process of planning BRT systems: Xi'an has integrated the BRT principles in its urban master plan and plans to develop eight bus-dedicated lanes. Chengdu has introduced a transit-oriented development (TOD) concept that gives priority to public transportation. Chongqing is contemplating the use of BRT and hybrid bus technologies that can be developed by a Chongqing bus manufacturer. Shanghai intends to develop alternative public transportation systems and has completed several studies on BRT.

Source: China Transportation Programme, Energy Foundation.

Notes

1. Under this law, the government can set national and provincial targets for renewable energy sources, which the law defines as hydro, wind, solar, geothermal, marine and other non-fossil-fuel resources (i.e. including nuclear).

2. For example, the Chinese 24h grade II standard for SO_2 is 150 $\mu g/m^3$, compared to the EU limit value of 125 $\mu g/m^3$; similarly, the 24 h grade II limit for NO_x is 100 $\mu g/m^3$, whereas the Japanese standard is 76-115 $\mu g/m^3$.

3. In all three cases, however, the *decoupling was relative* in the sense that emissions were higher at the end of the review period than at the start.

4. The IEA expects that under current policies, China's CO_2 emissions will surpass those of the United States by 2020.

5. For example, mercury and other pollutants from China's more than 2 000 coal-fired power plants are subject to long-range transport.

6. The rise is in part a statistical effect caused by the inclusion in the sample of some additional cities with a high SO_2 concentration; in four northern cities the mean annual concentration exceeded 200 $\mu g/m^3$, *i.e.* it was more than twice the grade III standard.

7. Beijing, Chengdu, Chongqing, Guangzhou, Harbin, Jinan, Shanghai, Shenyang, Tianjin, Wuhan, and Xi'an.

8. The increasing demand has put strain on China's coal mines, and the lack of safety during coal production has grave social consequences. Mining accidents (Table 2.4) are common in China's 30 000 small coal mines, and many accidents occur in illegal mines that are not officially registered. In 2004, 6 000 people died in mining accidents.

9. The 2010 target has been set at 880 million tonnes per year, so the effort to install new capacity must continue.

10. Less than 11% of the world's thermal power plants are equipped with FGD, compared to 20% in the OECD area.

11. Installed FGD equipment (often financed with foreign money) is not always operated in order to save running cost.

12. Total projected investment (2001-05) on air pollution control in the 10th FYP amounted to CNY 280 billion.

13. For comparison, the Three Gorges dam on the Yangtze river, due to be fully commissioned in 2008, has an installed capacity of 18.6 GW. The dam will produce up to 84 TWh annually (in 2002, total hydrogeneration was 288 TWh).

14. For example, NGOs are lobbying against a proposed large hydro scheme on the Nu river on the grounds of adverse environmental and biodiversity impacts, as well as resettlement and social issues.

15. With GDP at 2000 prices and purchasing power parities.

16. 1990 prices.

17. For example, although 50% of new buildings were to comply with the insulation standards of the building design code, it is estimated that less than 5% of new buildings actually do so.

18. Vice-Premier Zeng Peiyan spoke to the Standing Committee of the National People's Congress on 27 December 2005 on this issue.

19. Electricity transmission and distribution losses are already low at less than 7% (versus about 6% in many OECD counties).

20. For example, depletion of forest resources, partly caused by cutting trees for firewood, was blamed for the floods that struck China in 1998. A logging ban was imposed in many areas of the country, including along the Yangtze river.

21. The municipal government of Shanghai launched an initiative in 2005 to install photovoltaic (PV) systems on 100 000 of the city's six million rooftops at an initial cost of about CNY 150 000 (EUR15 000) per roof, which will be subsidised by the government. Jiangsu province also initiated a 1 000-rooftop PV programme in the trial city of Wuxi.

22. As part of China's western development strategy, the country has invested CNY 2 billion in solar energy power plants in rural towns in western China, with a total electricity generation capacity of 18 MW.

23. Of which, 2.7 million are in Beijing, half of them private vehicles.

24. The Beijing City Comprehensive Master Plan (1991-2010) had envisaged an integrated approach between urban spatial development and transport development with satellite cities and smaller suburbs surrounding the city. The goals of the master plan were not achieved, however, and the city centre has become the dominant employment centre, with the associated necessity for people to commute to and from the city centre. The average speed on Beijing's second ring road is now less than 10 km/h. In other large cities, the average speed is about 12 km/h.

25. Beijing introduced Euro II standards as early as January 2003 and Shanghai followed suit in April 2003.

26. Also, 84 Chinese cities have banned small cars, regarded as less attractive, more polluting and noisy as well as having poor power and safety records. In 2001, for example, Guangzhou stopped issuing license plates for cars with engines of less than one litre.

27. About 25% lower than in Europe, 20% lower than in Japan and 10% lower than in the United States.

28. They are also 20% more stringent than US standards.

Selected Sources

The government documents, OECD documents and other documents used as sources for this chapter included the following. Also see list of websites at the end of this report.

Asian Development Bank (ADB) (2005), *Electricity Sectors in CAREC Countries, A Diagnostic Review of Regulatory Approaches and Challenges*, ADB, Manilla.

China National Environmental Monitoring Centre (2005), *Trends Analysis of National Environmental Quality Change*, Beijing.

DRC (2004), *China's National Comprehensive Energy Strategy and Policy 2020* (CNESP 2004), Beijing.

Energy Foundation (2005), *China National Energy Strategy and Policy Study, Sub-project 11: Policy Study on Development and Utilization of CCT*, Beijing.

Gallagher, K.S. (2006), Limits to Leapfrogging in Energy Technologies? Evidence from the Chinese Automobile Industry, *Energy Policy 34*, Cambridge, MA.

Horii, N. (2005), "Evaluation of Air Pollution Control in China: Examining Its Cost Effectiveness and Cost of Implementation", in Terao, T. and Otsuka, K. (eds.), *Environmental Policy in a Changing Asia: Industrialization, Democratization, and Globalization* (in Japanese), Institute of Developing Economies, Tokyo.

IEA (International Energy Agency) (2006a), *Energy balances of OECD and non-OECD countries 2003-2004,* OECD/IEA, Paris.

IEA (2006b), *World Energy Outlook*, OECD/IEA, Paris.

IEA (2005), CO_2 *emissions from fuel combustion,* OECD/IEA, Paris.

Japanese Ministry of the Environment (2005), *Report of the Committee on Dust and Sandstorms* (in Japanese), Ministry of the Environment, Tokyo.

NDRC (2005), *China Medium and Long Term Energy Conservation Plan*, China Environmental Science Press, Beijing.

OECD (2005), *Economic Outlook No. 77*, OECD, Paris.

SEPA (State Environmental Protection Administration) (2006a), Annual Statistics: Report on Environment in China, China Environmental Science Pres, Beijing.

SEPA (20006b), State of the Environment Report, SEPA, Beijing.

SEPA (2005), *China Environment Yearbook,* China Environment Yearbook Press, Beijing.

Sinton, J. *et al.* (2005), *Evaluation of China's Energy Strategy Options*, Lawrence Berkeley National Laboratories, Berkeley.

Sugiyama, T. and S. Oshita (eds.) (2006), Virtuous Asia: *Policy Development Cooperation on Energy Efficiency for the Better World*, to be published by the International Institute for Sustainable Development, Winnipeg.

UK Department of Trade and Industry (2002), *Review of the Coal Preparation Sector in China*, London.

Walsh, P.W. (2004), "Motor Vehicle Pollution and Fuel Consumption in China", *Urbanization, Energy, and Air Pollution*, The National Academy Press, Washington DC.

World Bank (2005), *World Development Indicators*, World Bank, Washington DC.

World Bank (1997), *Can the Environment Wait? Priorities for East Asia*, ISBN 0-8213-4158-8, World Bank, Washington DC.

World Energy Council (2005), "China's Energy Supply: Many Paths – One Goal", in *Energy for Germany* by WEC's German member Committee, Berlin.

Xu, Jianbo, Yin Li and Wenhua Wang (徐剑波，李胤，王文华) (2005), 1990-2003 年上海经济发展和环境污染状况初步定量的分析 (A Preliminary Quantitative Analysis on the Relationship of Shanghai's Economic Development and Situation of Environment Pollution during 1990-2003), 上海交通大学环境科学与工程学院 (School of Environmental Science and Engineering, Shanghai Jiaotong University), 上海环境科学, 2005年，第24卷，第2期 (Shanghai Environmental Sciences, 2005, Vol. 24, Issue 2), Shanghai.

Yang, Kunhong and Junhui Chen (杨坤红，陈军辉) (2004), 中国NO×控制制度研究（A Study on Control System for Nitrogen Oxides in China), 成都市环境监测中心站, 成都 (Chengdu Municipal Environmental Monitoring Center, Chengdu), 上海环境科学, 2004年，第23卷，第4期 (Shanghai Environmental Sciences, 2004, Vol. 23, Issue 4), Shanghai.

Zhang, Renjian *et al.* (张仁健 *et al.*), (2001), 中国二氧化碳排放源现状分析（Analysis on Present Emission Status of Carbon Dioxide in China), 中国科学院大气物理所 (Institute of Atmospheric Physics, Chinese Academy of Sciences), 气候与环境研究, 第6卷第3期, 2001年9月 (Climatic and Environmental Research, Vol. 6, Issue #3, September 2001).

Zhou, Wei and Joseph, S. Szyliowicz (eds.) (2005), *Towards a Sustainable Future. Energy, Environment and Transportation in China*, China Council for International Co-operation on Environment and Development, China Transportation Press, Beijing.

4

WATER MANAGEMENT

Features

- Water quality challenges
- Water resources under stress
- The role of economic instruments
- Private sector participation in water services
- The water transfer mega-project
- Balancing urban, industrial and agricultural water supply

Recommendations

The following recommendations are part of the overall conclusions and recommendations of the environmental performance review of China:

*   increase investments and management efforts in *urban water supply and sanitation* (including in new urban development projects) to meet China's long-term objectives (concerning health and ambient water quality); increase user charges and cost recovery (of operating and investment costs); improve the *operational performance* of treatment plants; clearly distinguish the responsibilities of water utilities and local authorities;

*   continue efforts to improve water pollution control and efficiency in water use *by industry*; increase the rate of pollution charges and abstraction charges; ensure that treatment plants are efficiently managed; link abstraction and discharge permits to total load planning, while maintaining minimum flows and river quality objectives;

*   continue efforts to improve water pollution prevention and water efficiency in *agriculture*, and to establish water user associations responsible for recovering the cost of providing irrigation water; improve monitoring and collection of groundwater abstraction charges; take measures to halt overexploitation of groundwater aquifers; prevent agricultural run-off into aquifers, rivers and lakes (e.g. buffer zones along rivers and lakes, treatment of intensive livestock effluents, efficient application of agro-chemicals); phase out fertiliser subsidies;

*   strengthen and further develop an *integrated river basin management approach* to improve water resources and water quality management, and to provide environment-related services more efficiently (e.g. flood and drought prevention, soil and water conservation, biodiversity protection, support for recreation and tourism); give greater weight to the protection of *aquatic ecosystems* (e.g. renaturation of rivers and lakes banks, protection of wetlands); foster stakeholder participation (e.g. representatives of economic sectors, environmental NGOs, experts, administration);

*   further encourage *sustainable water use* through: i) *institutional integration* of water quality concerns and of water investments (e.g. at national and other relevant levels of government); ii) *market*-based integration with further progress in the transition towards full cost pricing of water services, while giving attention to the special needs of the poor and of the West; and iii) clarifying and securing the *rights* to extract, allocate and use water, in the context of water legislation and land tenure reform;

*   pursue efforts to provide the *rural population* with safe water supply and sanitation to meet domestic objectives and international commitments (e.g. Millennium Declaration and WSSD); continue to install meters and collect user charges, taking account of social factor.

Conclusions

China has a *comprehensive legal framework* for water resource and pollution management, with clear mechanisms to control abstractions and to set water quality objectives. The 2002 Water Law opens the way for integrated river basin management, stakeholder participation and the use of market mechanisms in water management, in other words for a *major reform* of the water sector. Water supply and waste water treatment utilities have already undergone considerable reform: in many areas, companies providing water services have been established. Basic institutions for river basin management are in place. A range of *economic instruments* (user charges for water services, pollution charges for industry, abstraction charges) are used, although often with relatively low rates. Over the period of the 9th and 10th FYPs (1996-2005), *total loads discharged to watercourses* were reduced in some areas, representing a decoupling of pollution discharge from economic growth. Concerning *floods*, very large investment has been made in *infrastructure to protect against flood damage*, flood risks have been reduced in many areas, and communities are more informed about the risks they face. Physical planning laws are being strengthened to prevent further development on flood plains, and there has been some return of reclaimed areas to flood storage functions. In some government departments, the criteria used to assess performance are incorporating water resource use and pollution reduction targets. These are an addition to the economic growth and population control targets commonly set.

However, China's water situation is of high concern. First, many *water courses, lakes and coastal waters are severely polluted* as a result of agricultural, industrial and domestic discharges. The pollution has severely degraded aquatic ecosystems, is a major threat to human health, and may limit economic growth. The use of untreated water affects development especially in the poorer, more disadvantaged regions. Large investment in water services will continue to be necessary: i) in urban areas to address the investment backlog and meet the needs of the large influx of rural migrants; ii) in rural areas, taking into account affordability issues; and iii) in the least developed areas in the form of development assistance and transfers. Secondly, China has *very low water resources per capita* (one quarter of the world average), and they are unevenly distributed (e.g. one tenth in northern and western areas). Among the 600 larger cities, 400 suffer from water shortages. Thirdly, with surface water polluted and scarce, demand for *groundwater* far exceeds the rate of replenishment in many areas in both rural and urban areas. It will be impossible to maintain the high (and inefficient) levels of urban and agricultural water consumption. The country is undertaking a *major project* to transfer more than 40 billion m^3 a year from the southern Yangtze basin to the North China plain by 2020. However, this will still not meet the needs for economic growth and ecological recovery, without determined demand management and sustainable use by urban, industrial and agricultural users. Finally, around 70% of water withdrawal in China is for *agriculture*, with 40% of farmland being irrigated. Agriculture and the rural communities (that lack sewer systems) are also major sources of pollution. To make water management more sustainable, the demand for water by agriculture must be reduced and diffuse pollution must be identified and prevented.

♦ ♦ ♦

1. Policy Objectives

Acknowledging the serious threat that water pollution and scarcity pose to the health of the Chinese people and to prospects for economic development, Wang Shucheng, Minister of Water Resources, in a keynote speech at the 3rd World Water Forum in March 2003, expressed China's *overarching principles for water management* as "water saving before water transfer, pollution control before water delivery, and environmental protection before water consumption" (Wang, 2003). In 2005, the State Council set the key objectives of water policy as: strengthening river basin management, protecting drinking water sources, addressing trans-provincial water pollution, enhancing water saving in agriculture, and achieving 70% urban waste water treatment by 2010 (State Council, 2005). The legal framework for water management was expanded over the review period (Box 4.1).

Water management holds a high position in China's overall environmental priorities, referred to as "33211" for three rivers (Huai, Hai and Songhua), three lakes (Tai, Chao and Dianchi), two "control zones" (for air pollution by SO_2 and acid rain), one city (Beijing) and one sea (the Bohai). Related quantitative targets were set as part of the *10th Five-Year Plan for Environmental Protection (FYEP)* covering the period 2001-05 (Table 4.1). The 10th FYP also set quantitative targets to: improve surface water quality and meet required standards in over 60% of Chinese cities; treat 45% of household sewage in urban areas at a waste water treatment plant; treat 60% of sewage from large-scale poultry farms; meet required standards for drinking water from public water supplies; meet required standards for irrigation water; and improve food safety. Other key water management objectives of the 10th FYP are: to restore water quality in the upper reaches of the Yangtze river, in the middle reaches of the Yellow river and in the whole Songhua river basin; and to make all efforts to prevent

Box 4.1 **Legal framework for water management**

Implemented primarily by the Ministry of Water Resources, the *2002 Water Resource Law*[*] reflects current thinking on integrated water resource management and demand management. It enshrines the principles that everyone should have access to safe water and that water conservation and environmental protection are a priority, and it focuses on four topics: water allocation, rights and permits; river basin management; water use efficiency and conservation; and protection of water resources from pollution. For the first time, the law defines river basin management institutions and functions, and strengthens the administrative rights of river basin management organisations. It provides for co-ordination and information-sharing on water quantity and quality. The law encourages the establishment of water user associations with defined rights of access to water, ownership of water infrastructure, and conflict resolution mechanisms. The law requires integration of water resource and economic development planning. Water is identified as an economic commodity, with provision for abstraction charges and cost recovery of user charges. The law sets out a system for defining river reach functional use zoning. The *1997 Flood Control Law* sets the framework for managing floods, preventing and mitigating damages, preserving the safety of people's lives and property, and ensuring that flood defence is integrated into the national planning process.

Implemented primarily by SEPA, the 1996 Amendment to the *1984 Law on Prevention and Control of Water Pollution* (WPPC) provides for the spread of urban waste water treatment. It establishes the framework for industrial and municipal discharge pollution controls and charges. Associated regulations introduce discharge standards for industry (in 1996), as well as water quality objectives for surface waters (in 2002). As is the case for the Water Law, the WPPC also sets out a system for zoning the environmental functions of river reaches. However, in this and other areas, WPPC conflicts with the Water Law, resulting in duplication in MWR and SEPA line administrations. WPPC is currently under revision.

The *1982 Marine Environmental Protection Law* (amended 1999) aims to protect and improve the marine environment, conserve marine resources and prevent pollution damages, including from land-based sources. It requires that environmental protection and river management departments (at all government levels) work to prevent rivers from being polluted, so as to ensure the good quality of waters emptying into the sea. It also requires that the competent State Administrative Department in charge of marine affairs works to prevent the marine environment from being damaged by over-fishing or pollution.

These laws are implemented together with secondary legislation and regulations and are supported by State Council decrees. They are enacted in each province by translation into the provincial legislation, a process that can result in some *regional variations on interpretation and implementation.*

---

* Updates the original 1988 Water Law.

water pollution in the Three Gorges dam reservoir area (on the Yangtze) and along the proposed south-north transfer route (from the Yangtze to the Yellow river).

Key water management *objectives of the 11th FYP* (covering the period 2006-10) are to: reduce the total discharge of major pollutants by 10%; reduce water consumption per unit of industrial value-added by 30%; maintain irrigation water at current levels; retain 120 million hectares of farmland area; and increase forest cover to 20%.

At the international level, as signatory of the Millenium Declaration (New York, 2000), China is committed to the UN Millennium Development Goal of *halving the number of people without access to safe drinking water* by 2015. In Johannesburg, the World Summit for Sustainable Development in 2002 adopted the same objective as well as the objective of *halving the proportion of people without access to basic sanitation* by 2015. The right to water was formally recognised as a *human right* by the UN Committee on Economic, Social and Cultural Rights, in 2002.

Table 4.1 **Performance in meeting the 10th FYEP pollution load targets**

(million tonnes)

	COD[a]			Ammonia nitrogen		
	2000	Target 2005	Actual 2004	2000	Target 2005	Actual 2004
Total	14.45	13.00	13.39	1.83	1.65	1.33
industry	7.05	6.47	5.10	0.78	0.71	0.42
households	7.40	6.53	8.30	1.06	0.94	0.91
Huai river	1.06	0.64	0.99	0.15	0.11	0.12
Hai river	1.58	1.06	1.24	0.26	0.20	0.13
Liao river	0.58	0.33	0.47	0.07	0.05	0.05
Tai lake	0.49	0.38	0.44	13.0[b]	9.91[b]	4.20[b]
Chao lake	0.06	0.06	0.07	1.23[b]	1.14[b]	0.80[b]
Dianchi lake	0.04	0.03	0.03	1.09[b]	0.88[b]	0.20[b]
South-north transfer	0.97	0.55	0.49	0.14	0.07	0.05
Three Gorges reservoir	1.35	1.03	1.01	0.11	0.08	0.08
Bohai sea	1.14	1.03	2.09	0.16	0.13	0.01
Beijing	0.19	0.13	0.13	0.04	0.03	0.02

a) Chemical oxygen demand.
b) Total nitrogen in '0 000 tonnes.
Source: 10th FYEP; SEPA.

2. Water Quality Management

2.1   State

Drinking water quality

Nearly *half of China's major cities*[1] *are not fully compliant* with drinking water quality standards, indicating significant health risks (SEPA, 2006). The main issue is faecal contamination, although industrial and agrochemical pollutants may also find their way into the water supply. Though the situation is improving,[2] it is normal practice for all Chinese to boil water before drinking it. Though this removes bacterial contamination and reduces some volatile hydrocarbons or chloro-hydrocarbon residues, it does not remove the mineral impurities. About 70 million people are drinking groundwater of poor quality,[3] leading to diseases such as chronic arsenic poisoning and fluoride poisoning (MOWR, 2005).

Surface water quality

China's water quality objectives for each major reach of a river are defined on the basis of five categories of chemical quality related to the suitability of the water (protecting a natural source zone, drinking and fish hatchery, drinking after treatment, recreation and course fishery, industrial use and non-contact recreation, irrigation and landscape). China's *water quality standards for rivers* cover 23 parameters, including chemical oxygen demand (COD), biochemical oxygen demand (BOD), ammonia, phosphorus, total nitrogen, metals, petroleum and phenol. The number of Environmental Protection Bureaus (EPBs) monitoring stations decreased from 750 in 2002 to around 400 from 2003 onwards. The Ministry of Water Resources supervises 3 200 monitoring stations on 1 300 rivers, and 229 more at provincial level. River water quality is accordingly reported using five grades. *Water quality standards for lakes* cover the same parameters as the standards for rivers, plus total nitrogen and total phosphorous.

About *a third of China's river length is highly polluted* (grade V or above) (Table 4.2). This excludes pollution by nutrients and micro-pollutants (e.g. pesticides), which China does not routinely monitor in its rivers. The State Environmental Protection Administration (SEPA) evaluation of the ten main river basins shows that water quality in northern and eastern China is much worse than in the southern and western areas. The seven key river basins showed a deteriorating trend in water quality over 1996-2002, but most have since shown strong improvement (Figure 4.1). However, 60% of the Hai river length is still highly polluted, as well as more than 40% of the rivers Liao and Huai. River quality is generally poor within cities due to the lack of urban waste water treatment, although the situation is improving (Table 4.3).

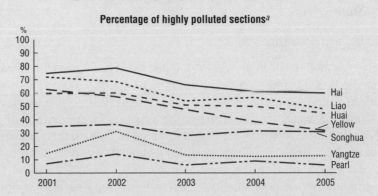

Figure 4.1 **Water quality in main rivers,** 2001-05

Percentage of highly polluted sections[a]

a) As measured at monitoring stations reaching grade V or worse in SEPA classification (including organic and inorganic pollution, excluding pollution by nutrients and micropollutants, *e.g.* pesticides).
Source: SEPA, State of the environment reports.

Table 4.2 **River water quality,**[a] 2005

	Quality class (% of monitoring stations/% of river length)					
	I	II	III	IV	V	> V
7 main river basins (411 monitoring stations)	4	20	17	25	7	27
Total river length (140 497 km)	5	29	27	12	6	21

a) Against surface water quality standards GB3838-2002.
Source: SEPA.

About *three quarters of China's major lakes are highly polluted* (grade V or above) (Table 4.4). In most cases the pollution is caused by excess nutrients (nitrogen and phosphorous), and there are a few cases of excess oxygen demand (as measured by COD). The water quality of the three key national lakes (Dianchi next to Kunming in Yunnan; Chao in Central Anhui; and Tai in southern Jiangsu) has not significantly improved over the last few years despite considerable efforts to treat urban and industrial sewage (Table 4.3).

Table 4.3 **Status of municipal waste water treatment projects, Three-Rivers and Three-Lakes basins**, 2003

	Total	Completed	Under construction	Others
Huai river basin	161	22	33	106
Hai river basin	186	13	40	133
Liao river basin	65	7	15	43
Tai lake basin	147	19	68	60
Chao lake basin	17	2	5	10
Dianchi lake basin	3	0	0	3

Source: SEPA.

Table 4.4 **Water quality of major lakes and reservoirs,**[a] 2005

	Quality class (number of lakes/reservoirs)			
	I-III	IV	V	> V
Lakes (18)	3	2	4	9
Reservoirs (10)	5	1	1	3

a) Against surface water quality standards GB3838-2002.
Source: SEPA.

Table 4.5 **Coastal water quality,**[a] 2005

		Quality class (% of monitoring stations)		
		I-III	IV	> IV
Bohai sea	Liaoning, Hebei, Tianjin, Shandong provinces	81	6	13
Yellow sea	Shandong, Jiangsu provinces	89	4	7
East China	Zhejiang; Fujian provinces	47	12	41
South China	Guangxi, Hainan, Guangdong provinces	94	0	6
Total (293 monitoring stations)		76	6	18

a) Coastal water quality is reported using four classes based on COD levels ranging from 2 to more than 5 mg/l.
Source: SEPA.

Coastal water quality

About a quarter of China's *coastal waters are highly polluted* (grade IV or above) (Table 4.5, Figure 9.1). The situation is particularly acute[4] in the Bohai and East China seas where spring red tides caused by excessive nutrients occur every year (particularly off the mouth of the Yangtze river and the Zhejiang coast). The Yangtze river, which receives 45% of China's industrial effluents and 38% of its municipal sewage, is the largest source of land-based pollution into the East China sea (Li, 2004; Qu, 2005). Coastal waters near Shanghai are at 92% above grade IV, while those around Hainan are in grades I to III. Coastal water pollution is exacerbated by population migrations from western China to the coastal regions (Chapter 9). In 2004 it was estimated that pollution-related losses to the fishing industry[5] amounted to CNY 2.8 billion (SEPA, 2004a). Since 2002, phosphorous detergents have been prohibited as part of the Bohai "blue sea action programme" (launched in 1999). *Bathing water quality* is generally (80%) good at China's 28 bathing beaches (in 16 coastal cities).

2.2 Pressures and responses

Household waste water: urban areas

Waste water discharges from China's cities are a major cause of river pollution. About *a third of the urban population is connected to a waste water treatment plant*, a performance difficult to compare with OECD countries where such an indicator covers the entire population (Figure 4.2). By the end of 2005, 52% of the waste water volume in Chinese cities was treated before being discharged, thus meeting the 10th FYEP target of 45% by 2005. In addition, the 10th FYEP's key pollution control objective of reducing COD discharges from household sewage by 12% over five years was not achieved (Table 4.1, Figure 4.3). These results reflect the fact that China's investment in waste water treatment did not keep up with the high levels of urbanisation and urban population growth over the period. In 2002, approximately 500 municipal treatment works were in operation, providing treatment in 310 of the 660 main cities. Some 17 000 towns have no treatment.

An additional factor is that China's waste water treatment plants *are sometimes operated on an intermittent basis*, particularly in less developed areas. For example, of the ten municipal waste water treatment plants built in Shaanxi, only two are actually in regular operation. When faced with economic difficulties, municipalities (and factories) are tempted to switch off plants to save money, as operating a plant requires a large amount of electricity and chemicals. This practice not only creates

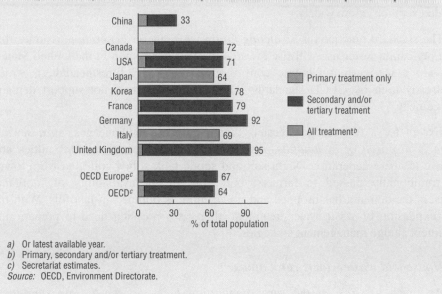

Figure 4.2 **Population connected to public waste water treatment plant,** early 2000s[a]

China 33
Canada 72
USA 71
Japan 64
Korea 78
France 79
Germany 92
Italy 69
United Kingdom 95

OECD Europe[c] 67
OECD[c] 64

Primary treatment only

Secondary and/or tertiary treatment

All treatment[b]

0   30   60   90
% of total population

a) Or latest available year.
b) Primary, secondary and/or tertiary treatment.
c) Secretariat estimates.
Source: OECD, Environment Directorate.

risks to the environment and to downstream users of the water, but can also cause damage to the plant itself.[6] Another issue is that the construction of waste water treatment plants is not always co-ordinated with sewerage development, and plants may not receive full flows until long after first opening. In an attempt to address these issues, government directives have declared that for a treatment plant to be considered operational, and therefore qualified to contribute to the FYP target, it must be certified (by SEPA's provincial bureaus) as having operated for 60% of the time in its first year of operation and 75% of the time by the end of its third year of operation.

Most of China's recent waste water treatment plants provide *secondary level treatment*, incorporating biological nutrient removal. Between 2001 and 2005, however, discharges of ammonia from domestic sewage rose from 0.84 to 0.97 million tonnes, indicating that treatment plants were not sufficiently effective in removing ammonia, which is toxic to fish and exerts high oxygen demand as it decays. It would be more cost-efficient for China to limit the advanced (and costly) nutrient removal techniques to areas sensitive to eutrophication (lakes, other freshwater bodies, estuaries and coastal waters) and to surface freshwaters intended for the abstraction of drinking water, and to spread the more simple (less costly) techniques aimed at reducing ammonia levels (e.g. extended aeration) more widely.

The standard practice for *predicting future waste water generation* assumes that urban per capita water use will rise towards the levels common in the United States; these are twice current Chinese consumption rates (212 litres/head/day), which are already high by OECD standards. This assumption does not support demand management.

Except for a few sludge digesters which generate methane gas, *most sewage sludge is disposed of in landfills*, where it can cause operational difficulties and increase leachate generation. Night soil and septic tank solids collected from towns was traditionally passed to farmers, but with greater availability of inorganic fertilisers the demand has dropped and much more is dumped in landfills. With the rapid expansion of waste water treatment, there is a pressing need to prepare and implement sludge management strategies.

Household waste water: rural areas

As much as *70% of the rural population lacks access to safe sanitation*, compared with 82% in 1995 (UNICEF/WHO, 2004). Part of the other 30% use individual or communal pit latrines. The soil from the latrines is normally applied to land as fertiliser. Where there is no latrine, human excrement, food waste and waste from other animals is often disposed of in the pigsty, where it will be consumed by the pigs. The use of untreated human waste on crops and its recycling through swine or fish represent significant disease vectors.

With the government's increased emphasis on improving rural livelihood, objectives have been set to greatly improve rural sanitation. *Improved sanitation* aims to break direct recycling by allowing waste to compost in storage before disposal. This can be done using improved latrines or septic tanks. Household biogas latrines,[7] a widespread basic technology in China, not only provide energy for cooking and for gas lamps, but also generate high-quality fertiliser and safe food for aquaculture. The amounts of pathogens in effluents are greatly decreased, and the use of biogas reduces pressure on firewood supplies and health-damaging smoke in homes. In 2005, China had 18 million household biogas installations as well as 3 356 medium and large size biogas installations; the 11th FYP aims to raise this to 20 million by 2010, serving 11% of the rural population. Considerable health advances could be achieved at relatively low cost through *targeted education on hygiene practices* and the careful consideration of convenience of hygiene practices in the design of water supply and sanitation improvements.

Industrial discharges

Waste water discharges from industry are also a major cause of river pollution. Over the last ten years, China has targeted the key polluting industries, requiring on-site treatment or closure of the enterprise, if treatment is not economical. *Considerable progress has been made in reducing toxic chemical discharges* from industry to rivers, particularly for mercury, lead and arsenic, with a reduction of 60-70% since 1998 (SEPA, 2006, NBSC, 2006) (Figure 4.3). Similarly, reductions have been achieved in COD-related discharges (Figure 4.3) as well as in industrial discharges of cyanide, petroleum, phenol and sulphide (SEPA, 2006). More than 60 000 industrial waste water treatment plants have been built in China, and it is reported that 90.7% of discharged industrial waste water complie with relevant standards (SEPA, 2004). A reduction in industrial discharges also resulted from the closure of many (inefficient and polluting) industries in small towns and villages, particularly paper mills, which had been established as part of rural labour diversification policies. As part of the strategy for creating more environmentally pleasant cities, many highly polluting industries located in urban areas have been relocated to peri-urban areas or rural areas. This has provided an opportunity to re-establish the enterprises with cleaner processes or effluent treatment.

The chemical spill from the petro-chemical refinery in Jilin in November 2005, and the subsequent river pollution and loss of water supply to the city of Harbin, dramatically highlighted the issue of *emergency protection and planning* (Box 8.3). Since that event, SEPA and its provincial Environmental Protection Bureaus have reviewed 20 000 major industrial installations along the main rivers, identifying those that represent a significant risk and drawing up emergency management plans as well as planning remedial works to protect the rivers. The event also raised the priority of water supply security in the key investment projects of the 11th FYP. CNY 10 billion is to be spent in the Songhua river basin, and other river basins and provinces have also announced major clean-up and source protection schemes.

China has both concentration-based standards for individual discharges and total load-based standards for all discharges in a region, but the two standards are not clearly co-ordinated. In addition, the *discharge standards* are not integrated in any specified way with the ambient water quality standards and objectives. Both the Environmental Protection Bureaus and the Ministry of Water Resources have responsibility for defining the *quality objectives* for each water body and for monitoring compliance with those objectives. The two bodies generally each carry out their activities separately, based on slightly different guidelines.

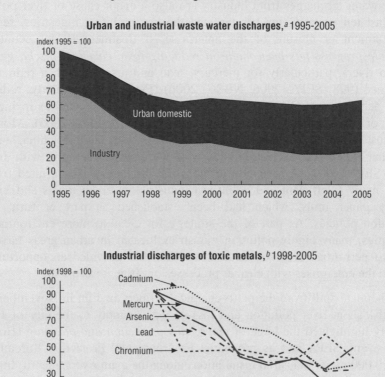

Figure 4.3 **Discharges into surface waters**

Urban and industrial waste water discharges,[a] 1995-2005

Industrial discharges of toxic metals,[b] 1998-2005

a) Based on COD measures in tonnes/year. Index in 1995 = 100.
b) Based on measures in tonnes/year. Index in 1998 = 100.
Source: SEPA.

Agricultural pollution

China's *intensity of use of both chemical fertilisers and pesticides* is three to four times higher than the OECD average but lower than the average in the Republic of Korea and Japan (Figure 4.4), despite controls on the use of pesticides (Box 9.4).

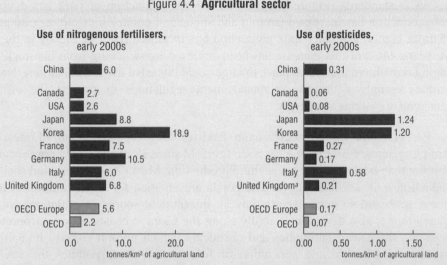

Figure 4.4 **Agricultural sector**

Use of nitrogenous fertilisers,
early 2000s

China	6.0
Canada	2.7
USA	2.6
Japan	8.8
Korea	18.9
France	7.5
Germany	10.5
Italy	6.0
United Kingdom	6.8
OECD Europe	5.6
OECD	2.2

tonnes/km² of agricultural land

Use of pesticides,
early 2000s

China	0.31
Canada	0.06
USA	0.08
Japan	1.24
Korea	1.20
France	0.27
Germany	0.17
Italy	0.58
United Kingdom*a*	0.21
OECD Europe	0.17
OECD	0.07

tonnes/km² of agricultural land

a) Great Britain only.
Source: FAO (2004), FAOSTAT data; SEPA (2004); OECD Environment Directorate.

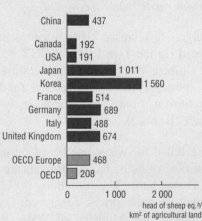

Livestock density, 2005

China	437
Canada	192
USA	191
Japan	1 011
Korea	1 560
France	514
Germany	689
Italy	488
United Kingdom	674
OECD Europe	468
OECD	208

head of sheep eq.*b/*
km² of agricultural land

b) Based on equivalent coefficients in terms of manure: 1 horse = 4.8 sheeps; 1 pig = 1 sheep; 1 goat = 1 sheep; 1 hen = 0.1 sheep; 1 cow = 6 sheeps.
Source: FAO (2006), FAOSTAT data.

Livestock farming has increased dramatically and livestock density is now high by OECD standards (Figure 4.4). Since 1990 the production of pork has doubled, beef production has increased fivefold and sheep and poultry fourfold, egg output is 3.5 times higher, and dairy milk production has increased fivefold, mostly in the last five years. Most of this increase has been achieved by switching from livestock free ranging or being in individual pens, to large-scale intensive livestock facilities. Not all facilities comply with effluent management regulations or operate the effluent management systems effectively.

Part of animal waste is recycled into feedstuff for fish ponds, as part of integrated farming systems. *Fish farming* has risen fourfold since 1990 and China now accounts for more than 60% of world aquaculture production. Most (70%) of fish and shellfish production is in artificial ponds. The ponds are emptied for water renewal at most once a year, and so normally are only an intermittent source of pollution. *Marine aquaculture* is also developing rapidly along the Chinese coasts. Key environmental issues are the use of antibiotics and chemicals, which present risks to human and environmental health, and generation of fish waste, which pollutes the sea and degrades biologically productive coastal habitats.

2.3   Use of economic instruments

User charges for urban waste water treatment

Meeting the FYP targets to increase urban waste water treatment will *require large investments*, estimated (in the 11th FYP) at CNY 83 billion (USD 10 billion) over the period 2007-12. Chinese cities have been investing in domestic waste water treatment, initially with assistance from international funding institutions but now mostly from local government funds, and increasingly by utilising local and international private finance (Box 4.2).

Most urban users are billed for water use based on metered consumption (i.e. *volumetric pricing*). Some 60% of individual apartments are metered in urban areas, a high proportion by OECD standards. The remaining 40% usually share evenly the building's water bill. Domestic water prices consist of: i) a fixed charge (typically CNY 0.13 per m³), payable to the Ministry of Water Resources' Water Affairs Bureau; ii) a user charge for public water supply, payable to the water supply company; and iii) a user charge for waste water services, payable to the waste water treatment company. All charges are collected through a combined bill issued by the water supply company. However, the revenues often pass to the local finance bureau before being reallocated to the water company.

Box 4.2 **Private sector participation in water supply and waste water services**

Private sector participation in public *water supply* has developed rapidly since 2000 and now accounts for 15% of urban water supply (serving 83 million people). International companies own 44% of the supply (in terms of population served), followed by local Chinese companies (37%) and expatriate Chinese companies (19%). Though the local Chinese firms may not have access to as large technical and financial resources as the international investors, they are willing to take a longer-term view to their investments, to operate with much lower margins and to build closer relationships with local officials; they are thus rapidly gaining market share. Few contracts (with terms of 15 to 30 years) have involved full ownership of companies or assets, but the situation is changing with the government policy to develop companies' managerial autonomy[a] (under the control of a board of directors). The companies may be a subsidiary of the municipal Water Affairs Bureau[b] or of an urban development and investment corporation.

Waste water treatment has been practiced for a shorter period than public water supply, and the institutional arrangements are not as advanced. In 2000 only 14% of treatment facilities had been constructed using local funds; the rest were financed by bilateral or international donors. The policy is now that independent waste water enterprises should be established. In some cases, waste water is bundled with the water supply company. Most works constructed in recent years use local finance: the proportion of international donor finance has greatly reduced and the proportion of *private finance, both international and domestic*, is taking off. The government strongly encourages this trend and hopes to leverage USD 30 billion in investments by 2010. Contracts for waste water treatment projects are now being tendered almost daily and waste water equipment manufacturers are setting up production lines in China to serve this market.

However, waste water companies, generally do not incorporate *responsibility for sewerage* (though there is a move towards this now). This can lead to contradictions between waste water treatment capacity and the sewerage capacity to deliver flows to the works. The waste water company is also not in control of industrial discharges to sewers, as these remain the responsibility of the Environmental Protection Bureaus. Thus if an industrial discharge is causing problems for treatment processes, a common occurrence, then the waste water treatment plant operator cannot take direct action but must request an investigation by the Environmental Protection Bureaus.

---

a) Traditionally water companies billed and collected user charges but the proceeds were bundled with local taxation. Where the water company has been established as a financially independent corporate entity, it is responsible for directly financing its operation by collecting and retaining revenues from tariffs.

b) Public water supply is usually led by the Construction Bureau (under the Ministry of Construction) in cities and the Water Resource Bureau (under the Ministry of Water Resources) in rural areas. In larger municipalities like Beijing and Shanghai, and increasingly in smaller ones, public water supply is assigned to Water Affairs Bureaus, with both Construction and Water Resources as parent ministries. By 2004, more than 1 200 Water Affairs Bureaus had been established.

Before 1999, water prices were very low (typically less than CNY $0.2/m^3$), and water supply and waste water treatment (where available) were supported by government subsidy. The 2002 water law introduced a new *policy of cost recovery*, at least with respect to operation and maintenance. Prices have been raised substantially in China's more developed areas, and the process is spreading outwards (Table 4.6). According to the National Development and Reform Commission (NDRC), by the end of 2005 the average water price in China's 36 major large and medium-sized cities was CNY 2.09 (USD 0.26) per cubic metre, including CNY 0.54 (USD 0.068) for treating waste water, an increase of 10% from 2004. Special provisions apply to the registered poor, such as overall price reductions or supply of the first cubic metre per month for free. Any price increase (to achieve cost recovery) must be approved by the local Price Bureau (supervised by NDRC). The merits of the price rise (e.g. better service) are debated in consideration of *affordability*, through a public hearing. Indeed, user charges for public water supply only are typically in the range of 1 to 5% of household income in China, but up to 20% for those on minimum livelihood standard welfare.

Table 4.6 **Water prices for households,** 2004

(CNY/m^3)

City	Province	Fixed and water supply charges	Waste water charge	Total charge
Beijing	Beijing[a]	2.3	0.6	2.9[b]
Jinnan	Shandong	..	..	2.6
Tianjin	Tianjin[a]	..	..	2.6
Changchun	Jilin	..	..	2.5
Harbin	Heilongjiang	1.9	0.5	2.4
Chongqing	Chongqing[a]	2.0	0.4	2.4
Langfang	Hebei	1.7	0.4	2.1
Shijiazhuang	Hebei	1.5	0.5	2.0
Xi'an	Shaanxi	1.6	0.4	2.0
Nanjing	Jiangsu	1.0	1.0	2.0
Guangzhou	Guangdong	0.9	0.7	1.6
Shenyang	Liaoning	1.4	0.2	1.6
Wuhan	Hubei	1.1	0.4	1.5
Wulumuqi	Xinjiang	1.2	0.3	1.5
Yinchuan	Gansu	..	..	1.2
Nanning	Guangxi	..	..	1.1

a) Centrally administered municipality.
b) Beijing water prices are due to rise to CNY 3.9.
Source: MOC.

However, user charges for *waste water services* still do not fully cover all operating and investment costs. More than 100 Chinese cities still have not levied charges on waste water treatment. And where charges are levied, the rate is relatively low, as low as CNY 0.1 (USD 0.0125) per cubic metre in some cities. To increase cost recovery and stimulate investment, it is now common practice to extend the charge to all users, not just those connected to a treatment plant. The charge has also been introduced in some cities with no treatment, to fund sewerage and the future establishment of treatment. NDRC, the Ministry of Construction (MOC) and SEPA have decided that central government funding will not be available for projects with a waste water tariff set at less than CNY 0.8 /m^3.

Pollution charges for industry

The *pollution levy system* (PLS) applies only to industrial sources and covers water discharges as well as air emissions, solid waste, noise and radioactive substances. Sources such as municipalities, hospitals and schools are exempt. From the outset, the PLS, initiated in 1978, was viewed as a means of implementing the polluter-pays principle and providing a (major) source of funding for provincial and local Environmental Protection Bureaus (NCEE, 2004). A key feature of the PLS is that 80% of the funds collected are returned to the enterprises for pollution control investments. Initially imposed on discharges exceeding the effluent standard, since 1993 the pollution charge has been extended to all discharges. The level of the charge was initially based on pollutant concentrations at the point of release, rather than mass or volume. In 1993 volume became a determinant.

While the PLS seems to have been *reasonably effective in reducing pollution*, other factors such as responsibility contracts signed by enterprise managers and local government officials as part of the five-year planning process may have been more important in determining the pollution intensity of industrial activity. The charge rate increase has been much lower than incremental pollution control costs, reducing its influence on polluting behaviour. As a result, the proportion of total charge revenue to the value of industrial output has decreased. The *coverage of the charge* is another issue. Many township and village enterprises are not levied because local Environmental Protection Bureaus do not have the resources to pursue all sources within their jurisdiction or find that the potential revenues from levying smaller sources do not justify the effort. Since such enterprises generally use less advanced technologies, one would expect them to be paying relatively more in pollution levies, not less than average. This suggests the desirability of increasing efforts to impose the pollution levy on a larger proportion of the township and village enterprises. Also, recycling a smaller share (e.g. one-half) of charge revenues for pollution control at the paying facilities should further increase the coverage of pollution control efforts. The issuance of discharge permits on the basis of both national concentration standards and a total load allowance (calculated taking in consideration the assimilative capacity of the river) opens the way for trading of pollution allowances.

Phasing out farm input subsidies?

Production and distribution of *pesticides and fertilisers are subsidised by the government* as an incentive to achieve grain production targets. Price subsidies for fertilisers, chemicals and other farm inputs are estimated at CNY 10 billion (OECD, 2005). They have decreased from more than CNY 30 billion in 1998, when reference prices for fertilisers replaced administered prices, allowing some adjustment for fluctuations in production costs and market demand. However, the 11th FYP proposes to increase subsidies on fertilisers (and road diesel fuel) to promote higher productivity in agriculture.

3.   Water Resource Management

3.1   State of water resource availability

Total available water resource in China is quite large at around 2 800 km^3 per year (MOWR, 2004), but this represents only 1 850 m^3 per capita, less than a quarter of the world average. Water use per capita is relatively low by OECD standards, but not the *overall intensity of use* (Figure 4.5). At 212 litres per capita per day in urban areas, domestic consumption is relatively high by OECD standards, while domestic consumption in rural areas is 68 litres per capita per day. In addition, efficiency of water use by industry is very low.

Severe water stress in northern river basins

The *uneven geographical and seasonal distribution* of water resources translates into severe water stress in the heavily populated northern river basins (Songhua, Liao, Hai, Yellow and Huai). With 44% of China's population and 65% of its cultivated land, this northern region has only 13% of the water resource, averaging out at just 575 m^3 per capita (Table 4.7). Most rainfall occurs in summer, so these areas depend heavily on reservoirs and groundwater for irrigation, especially during the spring planting season (before the summer rain), and for steady supplies to cities and industry. The Yellow river basin, for example, has a total of 3 100 dams providing reservoir storage roughly equal to the total annual river flow of 58 billion m^3. In the northern region, 43% of water supplies are taken from groundwater, and the intensity of water use goes beyond 40%, which most experts regard as the threshold for freshwater ecosystems to remain healthy, (Alcamo *et al.*, 2000). The water resource

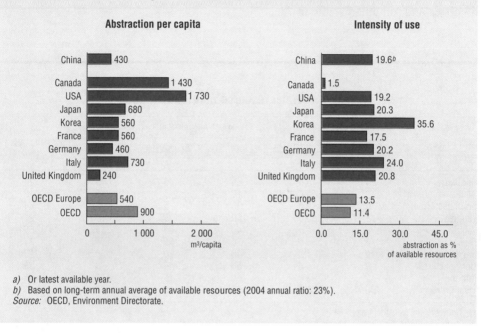

Figure 4.5 **Freshwater use**, mid-2000s*a*

Abstraction per capita

China — 430
Canada — 1 430
USA — 1 730
Japan — 680
Korea — 560
France — 560
Germany — 460
Italy — 730
United Kingdom — 240

OECD Europe — 540
OECD — 900

0   1 000   2 000
m³/capita

Intensity of use

China — 19.6*b*
Canada — 1.5
USA — 19.2
Japan — 20.3
Korea — 35.6
France — 17.5
Germany — 20.2
Italy — 24.0
United Kingdom — 20.8

OECD Europe — 13.5
OECD — 11.4

0.0   15.0   30.0   45.0
abstraction as %
of available resources

a) Or latest available year.
b) Based on long-term annual average of available resources (2004 annual ratio: 23%).
Source: OECD, Environment Directorate.

crisis has become a major constraint to economic development (estimated at between USD 14 and 27 billion per year in lost economic activity) and severely affects the people's health and quality of life.

In the northern region, *groundwater* is often being abstracted much faster than it can be replenished (MOWR, 2004). In the northern plain, groundwater tables have fallen by 10 to 50 metres (at an average of 0.5 metre per year) over the past ten years, sometimes leading to loss of forest cover. As a result of the falling water tables, increasingly larger sections of rivers are drying up during the dry season. Deeper wells are being sunk to access groundwater, and as a result, coastal aquifers are suffering from saline intrusion from marine aquifers.[8] Where salty groundwater is used for irrigation, blending with scarce surface water is required to reduce the salinity. Groundwater overexploitation has also resulted in land subsidence, affecting 61 000 km². This has given rise to drainage difficulties and flooding in seashore cities such as Tianjin and Shanghai. In some areas of Shanghai, land subsidence ranges between 1.5 and 2.5 metres and has resulted in fissures in the land and threatened the safety of houses. In Xi'an, nine rifts traverse the city and

2 600 houses are in danger. Over China as a whole, the groundwater table decreased in 61 of 192 major cities between 2003 and 2004, while it increased in 53 cities and remained stable in 78.

Table 4.7 **Water resource distribution,** 2004

River basin	Share of population (%)	Share of available water resources (%)	Water withdrawal in the river basin			Intensity of water use (%)
			Total (km³)	Groundwater (%)	Non-agricultural uses[a] (%)	
Southern	13	15.0	85	5	40	25
Southwestern	2	25.0	10	5	15	2
Southeastern	6	5.3	25	5	45	24
Yangtze	33	36.0	180	5	45	21
Yellow	9	2.5	40	40	35	59
Huai	16	3.0	60	30	25	74
Hai	10	1.2	40	70	30	123
Northeastern[b]	9	6.6	50	50	30	35
Interior	2	5.4	60	15	5	46
Total China	100	100	550	20	30	23

a) Domestic and industrial uses.
b) Regroups the Shonghua and Liao river basins.
Source: MOWR.

The south-north water transfer mega-project

To address the emerging water resource crisis in Northern China, the country is undertaking a *mega-project to transfer* more than 40 billion m³ a year from the Yangtze basin to the Northern China plain by 2020 (Box 4.3). The eastern and central routes will require a total expenditure of about USD 18 billion (including USD 3 billion for resettlement and 4 billion for pollution control to make the water usable). The western route will require an expenditure of about USD 37 billion. However, even this vast project will still not meet all of China's needs for economic growth and ecological recovery.

Flooding

China is frequently hit by severe flood disasters (Chapter 2). The country's climate, with rain concentrated in a few months, makes flash flooding very common.

Box 4.3 **The south-north water transfer mega-project**

Urban, industrial and agricultural development in the Northern China plain is increasingly restricted by the shortage of water. China is therefore undertaking a vast project to *transfer water from the abundant resources of the Yangtze basin* in Southern China. The project will be built along three routes, with the expectation that transfer flows will increase steadily between 2010 and 2020. Construction of the eastern and central routes commenced in 2003. Costs for the eastern route are estimated at around USD 7.5 billion and for the central route at USD 7 billion, plus as much as USD 3 billion for resettlement costs. The government is exploring domestic and international lending sources to fund the project, but expects to recover costs through increased water use charges.

The *eastern route* will lift water from the lower reaches of the Yangtze, then convey it 1 200 km north (to Tianjin). It will also supply the highly developed regions of Jiangsu and Shandong. By 2010, the scheme should be able to transfer some 9 billion m^3 of water per year, rising to 13 billion m^3 in 2020. Though new canals will be constructed, 90% of the route will be on existing canals (including the 1 500-year-old Grand Canal), on rivers and on lakes, which currently experience severe urban and industrial pollution with many stretches not reaching even the lowest of China's water quality objectives. Unless this issue is addressed, the water transferred will be almost unusable. For this reason, more than half of the investment for this route will be on cleaning up polluted watercourses and constructing 135 waste water treatment plants. The abstraction from the Yangtze will initially be around 400 m^3/s, rising to 700 m^3/s in later phases. This is a relatively small proportion of the annual minimum flows for this part of the Yangtze, which are more than 10 000 m^3/s. Water will also be taken from the Huai basin to produce a total transfer target of around 1000 m^3/s by 2020. Of this, about 400 m^3/s will cross north of the Yellow river and 180m^3/s will reach as far as Tianjin. Also, 75 major pumping stations (740 MW) will be required to lift water to the highest point of 40 metres above sea level at Dongping lake, near the Yellow river crossing. North of the Yellow river, flow will be by gravity.

The *central route* will take water primarily from a greatly enlarged Danjiangkou reservoir (Hubei province), and transfer it along 1 250 km of a new concrete-lined canal (routed to allow flow by gravity) from the canal head (at 147 metres elevation) to its end in Beijing (at 50 metres). A further 142 km of branch canal will transfer the water to Tianjin. Overall, this will mean crossing 219 rivers (including a complicated 7.2 km siphon tunnel under the main Yellow river channel), 44 railways and more than 500 highways. By 2020, *14.6 billion m^3* of water each year will supply cities, industry and agriculture in Hubei (1.2), Henan (4.8) and Hebei (5.4) provinces, as well as in Beijing (1.6) and Tianjin (1.6).

The *western route* is even more ambitious but is still in the planning stage, with construction not expected to start until 2010 at the earliest. This route will attempt to lift water over the mountain ridges separating the upper tributaries of the Yangtze and on to the headwaters of the Yellow river. The project faces huge engineering challenges

Box 4.3 **The south-north water transfer mega-project** *(cont.)*

to construct 10 metre-diameter tunnels for hundreds of kilometres through the mountains at altitudes of over 4 000 metres. Dams of 250 metres or more are required to provide sufficient head for free flows through the tunnels. A staged high lift pumping station (1 380 MW) will make the final transfer to the Yellow river. The costs will be very high – around USD 37 billion – planned in phases over 20 years or more. This route will eventually provide up to *17 billion m³* of water transfer per year.

The construction of the south-north transfer scheme is associated to i) considerable *financial and social costs* (e.g. the relocation of 250 000 people for the expansion of the Danjiangkou reservoir) as well as environmental costs; and ii) to *potential benefits*: enhanced economic growth, reduced environmental pressures on current natural resources (e.g. allowing reductions of groundwater over-abstraction and the maintenance of "ecological flows" of rivers in areas of high abstraction).

China's rapid economic development and high population density have substantially increased the consequences of flooding. The Ministry of Water Resources has estimated the average losses resulting from floods at *CNY 110 billion a year since 1990*, i.e. a cumulated amount equivalent to 3% of GDP in 2004. The year 1998, when more than 1 000 people died and material damages reached CNY hundreds of billions, was particularly dramatic.

Efforts have been made to *mitigate flood risks*. China has around 400 large reservoirs, 2 700 medium-sized reservoirs and more than 80 000 small reservoirs with flood protection functions (normally combined with water resource and power generation), with total storage capacity of 518 billion m³. The Yellow river, which was historically subject to frequent devastating floods, has not flooded for more than 50 years following construction of 3 100 dams. The Three Gorges dam (to be completed in 2009) is being designed to help protect the heavily populated Yangtze plains from potentially catastrophic flooding. In addition, thousands of kilometres of flood embankments along major rivers have been built or improved to provide "1-in-100-years" flood protection for major urban areas and "1-in-50-years" for peri-urban and rural areas. In recent years, large areas of flood plains that had been drained for agriculture have been re-designated as flood plains. Where agriculture still continues in these areas, the farmers are made fully aware of the risks, and warning and evacuation plans are put in place. Summer *typhoons*, quite common in the south and east of the country, continue to cause extensive damage but far less loss of life than they used to, because of development of flood-resilient buildings, warning systems and emergency plans. Overall, more than 465 million Chinese people live behind flood protection systems.

3.2 River basin management

Continued efforts to improve both the availability and the quality of water resources are needed to meet China's objective to achieve food security while protecting the health of the population and of the country's ecosystems. This could be most efficiently achieved through the *establishment of co-ordinated river basin management systems* that are able to make decisions on investment, abstraction and discharge controls, to involve representatives of government institutions and stakeholders, and to balance the needs of rural users with those of urban and industrial users, while protecting the environment.

Such systems could build on existing institutional arrangements. For the seven major trans-provincial river basins, authorities to co-ordinate basin management have already been appointed by the Ministry of Water Resources, with representative offices at provincial and national levels. Smaller intra-province river basins may have co-ordinating basin authorities at provincial levels. These *basin management authorities* are responsible for resolving conflicts between upstream and downstream users in different administrative jurisdictions. At each administrative level, scientific and engineering institutes provide monitoring, surveying, research, policy development and engineering design. "Undertaking units" are mostly now nominally independent companies.

In addition to large-scale supply side engineering solutions to the water resource crisis, China is considering spreading *small-scale local watershed management* projects to improve the retention and utilisation of available rainwater sources in rural areas. However, the local benefits of such schemes must be balanced against possible reduced flows downstream and the hazard of open ponds as disease vectors.

River basin management should greatly facilitate *drought and flood management.* Positive steps have been taken to require drought management plans for each river basin. Such plans should build on plans for water rights management and abstraction permitting, which are being developed, and incorporate mechanisms to fairly restrict and redistribute water resources under drought conditions to ensure that vulnerable users do not lose out disproportionately to large industrial and urban users. In addition to (currently predominating) engineering solutions to flood management, more attention should be paid to the synergies between water and nature management and the ecological services provided by nature conservation on watersheds. Positive steps have been taken with the development of compensation schemes for farmers' easements on flood plains, and the introduction of flood insurance for crops.

3.3   Economic instruments

User charges for public water supply

The user charge for *water supply to households*, though much higher than in the past, still does not fully cover all operating and investment costs (Table 4.6). In Xi'an, for example, water supply is priced at CNY $1.6/m^3$, against a full cost (including investments), estimated by the local Water Affairs Bureau, of CNY $5/m^3$. Leakage levels in urban areas can be quite high, and leakage management techniques[9] are still underdeveloped. Payment defaults are high where the service is poor and intermittent,[10] forcing people to buy water from water seller's tankers or bottles. As an incentive to reduce water consumption, 17 provinces have begun to introduce an increasing-block schedule (i.e. conservation pricing) in the past five years.

Prices for *industrial supply* are typically higher. In the Langfang industrial development zone, for example, the price is CNY $3.6/m^3$. Although construction sites are major water users, they are not charged until the project is completed, by which time the liable party is often no longer on site. Further raising water charges to industry would increase incentives to invest in much-needed modern water-saving and cleaner-production processes.

Around 70% of China's *rural population* have safe and accessible water supplies,[11] up from 60% in 1990. The other 30% have to carry water long distances or only have access to unprotected sources. However, even those with safe sources may lose their water supply during dry seasons. One target of the 11th FYP is for 100 million more rural people to be provided with safe water by 2010. New rural water supply schemes in China use water meters as a basis for charging users a fee – normally CNY 1 to $2/m^3$, sometimes with a minimum charge of CNY $3/m^3$ per month – payable to the village water committee which operates and maintains the infrastructure.

Abstraction charges and irrigation water pricing

From the 1950s to the 1970s, under collectivised agriculture, major investments were made in surface water-based irrigation systems to boost agricultural production. These *irrigation districts* could cover areas of tens of thousands of hectares. However, following agricultural reform and de-collectivisation in the late 1970s, the smaller, village-level organisations of farmers found it harder to raise the capital and co-ordinate the activities required to take over ownership and then to maintain or extend such systems. As a result, many systems have fallen into disrepair.

In their place, entrepreneurs have established *small companies* in co-ordination with the village governments that raise capital to sink wells, buy pumps and construct low-pressure underground distribution pipe networks. Farmers then buy water from such an enterprise on a volumetric basis. *Private well supplies* are often more efficiently managed, as the water suppliers have direct incentives to maintain their assets. Farmers often prefer these sources as being more reliable than district irrigation schemes and offering greater control and autonomy. However, the rural electricity required to operate such systems is subsidised in order to protect farmer income.

This situation of rural water supply entrepreneurs has led to a system under which farmers could be directly paying a volumetric fee for their abstractions.[12] However, with a large number of small abstractions, monitoring, reporting and collection of *abstraction charges* are patchy. In fact, these abstraction charges (e.g. CNY 0.02 to 0.25/m^3) often end up being levied on the village as a whole and then recharged to the farmers bundled in with other local service charges many months later and often pro-rated by land area, thereby breaking the link between water use and charge. This introduces a free-rider incentive for both the well operator (who is not responsible for the sustainability of the common aquifer but only for his own infrastructure assets) and the farmer, who can benefit by taking more than his share of the commonly administered water supply to boost yields while sharing out the additional costs. Collection of abstraction charges and the allocation of water rights are currently being reformed, pursuant to the 2002 Water Law.

With the abolition of agricultural taxes in 2005 (Chapter 7), China now has greater flexibility to implement more effective irrigation pricing. In most remaining irrigation districts, fees charged to farmers are much less than the cost of providing the water. Most irrigation supplies are not metered and management systems are vulnerable to abuse of commons, with those who take more than their share benefiting without sanction. *Water user associations* are being established more widely, pursuant to the 2002 Water Law. These take ownership of the assets and are responsible for setting and collecting user charges for irrigation water. Prices for irrigation water are likely to be much higher than past arrangements.

Water rights

China has started to develop a mechanism to allocate water rights in each river basin. *Provincial water quotas* are calculated and a minimum environmental flow is protected. Provinces can subsequently allocate water rights to different areas and water users. In addition, a "water consumption index system" will be established to measure and enhance water saving by irrigation systems and industry. Many pilots of water rights management schemes are being undertaken around China, with notable success by the Yellow River Basin Authority and the Tarim Basin Authority. In the past, over-exploitation of the Yellow river was so extreme that by 1997 it did not flow to the sea on 226 days out of the year.[13] Water rights management (and water transfer tunnels and storage reservoirs) have resulted in the Yellow river maintaining its flow and not running dry for six years. Similarly, following successful implementation of water quotas (and water-saving agricultural methods), the 363-km lower section of the Tarim river and the Hei river are now flowing again, after 20 years of being dry due to over-abstraction in the upper reaches. Further pilots are being disseminated through the Water Resources Demand Management Assistance Project, led by the Ministry of Water Resources.

However, provincial water quotas do not always take account of the "*governor's grain bag policy*" to maintain high level (95%) of self-sufficiency in grain production at national, provincial and regional levels. This policy, introduced in 1995 and retained under the 2004 regulation on grain marketing, makes it difficult to produce the crops that are best suited to local land and water resource characteristics. It responds to government objectives of food security, national security, social stability and macroeconomic stability rather than sustainable water resource management.

3.4   Challenges for the future

In the future, China will have *to balance the objectives of food security and self-sufficiency against the need for water resources for industrial and urban development* while leaving enough water to meet ecological needs. According to the Chinese Academy of Science, the projected decrease in agricultural water consumption by 2050, attributed to continuously improving irrigation technology, would be more than offset by a dramatic increase in non-agricultural water uses associated with economic development, most prominently industrial use and, to a lesser extent, urban households demand (Table 4.8). Suggestions for water saving in all sectors are set out in the China Water Conservation Technology Policy Outline (NCEE, 2005).

Industrial and urban water use

Industrial water use efficiency (i.e. the rate of water recycling in industrial processes) is low in China. Although it has risen from 40% recycled in 1998 to around 50% today, it is still far below the 75-80% generally found in the OECD area. The 11th FYP set a target to reduce the intensity of water use in industry by 30%, from the current 400 m^3 per CNY 10 000 GDP (MOWR, 2004).

The huge *population migration from the countryside to the cities*[14] will have far-reaching implications for water resource use (and pollution). By 2030, it is expected that more than 50% of the Chinese population will be urban. This process will greatly change the balance of pressures on water resources and will also affect the methods of agricultural production, as fewer farmers will be expected to produce more food.

These trends could potentially be beneficial, as farmers should have available a greater share of land (i.e. land leased from the local authority for a 30-year period, pursuant to the 1998 Land Law) and capital to invest in improved water-efficient farming methods ("implicit structural adjustment"). However, there is also the risk of increased conflict between (rich) urban and (poor) rural communities over access to water resources.

Table 4.8 **Projections of water demand,**[a] 2010-50

	Agriculture		Industry		Urban		Total	
	Amount (billion m^3)	Proportion (%)	Amount (billion m^3)	Proportion (%)	Amount (billion m^3)	Proportion (%)	Amount (billion m^3)	Proportion (%)
2010	465	78	93	16	27	4.6	585	100
2030	453	65	190	28	46	7.0	689	100
2050	416	50	344	41	73	9.0	833	100

a) Around 75% of water abstracted by towns and industry is returned to water bodies, where (though polluted) it is still a potential resource. Only 32% of agricultural water returns to water bodies; the rest is lost to evapo-transpiration.
Source: Chinese Academy of Science.

Agricultural water use

Over the last three decades, *grain production in China has risen roughly in proportion with population*, from 270 million tonnes in 1975 to 485 million tonnes in 2005.[15] The 11th FYP targets are to produce 520 million tonnes by 2010 and 620 million by 2030, when China's population will peak at 1.6 billion. To meet these grain targets, the productivity of the land will have to increase further. It is unlikely that significant new *agricultural land* can be brought into production, given the pace of urban and industrial development. While land-use regulations (administered by the Ministry of Land and Resources) make it difficult to take land out of agricultural production without releasing other land back to agriculture, total farmland in China has been declining over recent years, by 0.8% in 2004 and 0.3% in 2005. Where land has been released back to agriculture, it has often been of lower quality.

Water consumption by agriculture has remained fairly stable since 1990, while China's gross agricultural output has risen by 185% (OECD, 2005), reflecting enhanced efficiency of water use. Meeting the FYP grain targets will require *further improving the efficiency of water use in China's agriculture.* Considerable scope for this remains (NCEE, 2005). Overall, China irrigates 40% of its farmland and produces 0.5 kg of grain per cubic metre of irrigation water, compared to 2 kg/m^3 in OECD countries. Irrigation channel efficiency (i.e. the share of water entering irrigation channels that reaches crops) is only about 45%, compared with 80-90% in OECD countries. China's principal method of irrigation is flood irrigation; more efficient spray and drip irrigation is used on just 5% of farmland. Pilot introductions have been undertaken of promising techniques such as laser levelling of fields, water-retaining ploughing methods, lining of canals, plastic film covering of fields, and use of greenhouses to grow high-value fruit and vegetable crops in controlled conditions. However, the increased production of livestock products may be regarded negatively, as more water is required to produce food energy in the form of meat and dairy products than is required to produce cereals.

Notes

1. 52 out of 110 cities.

2. Though most treatment is still by simple chemically-assisted sedimentation and rapid gravity sand filtration followed by chlorination, advanced drinking water treatment (such as activated carbon, ozonation or membranes) is developing.

3. While most (80%) of groundwater resources can be used to abstract drinking water, deep aquifers are often naturally contaminated with fluoride, arsenic and mineral salts. Shallow aquifers are becoming increasingly polluted because of recharge from polluted river and irrigation water.

4. Red tides resulting from harmful algal bloom rose from 28 in 2000 to 82 in 2005, and the affected area (mostly in the East China sea) grew from 7 800 to 27 070 km^2.

5. Algal blooms include cyano-bacteria which produce toxins that render the fish stock unfit for consumption.

6. An effective biological treatment takes many weeks of constant operation to establish.

7. Costing CNY 1 200 to 2 000 to construct out of readily available brick and concrete.

8. In the Laizhou bay of Shandong province a 400 km^2 area of seawater has intruded; it proceeds inland at a rate of 400 metres every year.

9. Such as district metering to identify and manage leakage in zones, to manage pressure and to use leak detection equipment.

10. 400 of the 600 major Chinese cities periodically suffer from water shortages.

11. Coming from a protected or treated source and piped to the home or obtainable within 200 metres of the home.

12. Charging for abstractions from the main rivers is managed by the river basin management authorities; from the tributaries and catchment areas it is managed by municipal or county authorities. About 30% of industrial enterprises have their own boreholes; an abstraction charge is still due, although state industries are often exempt.

13. The Yellow river basin has a total of 3 100 dams providing reservoir storage roughly equal to the river's total annual flow of 58 billion m^3.

14. 20 million rural people are permanently relocating to cities each year. A much larger number seek temporary work (30 to 60% of rural families have one or more members working away).

15. Recent years have seen major initiatives to increase production after a period of declining grain acreage from 1999 to 2003 (MOA, 2005) and trends do appear to be reversing.

Selected Sources

The government documents, OECD documents and other documents used as sources for this chapter included the following. Also see list of websites at the end of this report.

Alcamo, J., T. Henrichs and T. Rösch (2000), "World Water in 2025", *Global Modelling and Scenario Analysis for the World Commission on Water for the 21st Century*, Centre for Environmental System Research, University of Kassel, Germany.

Changjiang River Water Resources Commission, Ministry of Water Resources, P.R. China (水利部长江水利委员会) (2005), 2004年长江流域及西南诸河水资源公报 (2004 Changjiang and Southwest Rivers Water Resources Bulletin), Beijing.

Changjiang River Water Resources Commission, Ministry of Water Resources, P.R. China (水利部长江水利委员会) 长江水资源质量公报, 2002年1月- 2005年10月 (Changjiang Water Resources Quality Monthly Bulletin, from January 2002 to October 2005), Beijing.

Chinese Academy of Sciences (2000), "Analysis of Water Resource Demand and Supply in the First Half of the 21st Century", in *China Water Resources*, US Department of Commerce, 2005, Washington DC.

China Council for International Cooperation on Environment and Development (CCICED) (2005), *Income Related Incidence and Welfare Impacts of Water Pricing*, CCICED Task Force on Environmental and Natural Resources Pricing and Taxation, Report No. 2, Beijing.

Huaihe River Water Resources Commission, Ministry of Water Resources, P.R. China (水利部淮河水利委员会) (2006), 2004年淮河水资源质量公报 (Huaihe Water Resources Quality Monthly Bulletin, 2004), Beijing.

Huanghe River Water Resources Commission, Ministry of Water Resources, P.R. China (水利部黄河水利委员会) (2006), 2004年黄河水资源质量公报 (Huanghe Water Resources Quality Monthly Bulletin, 2004), Beijing.

Li, Daoji and Dag, Daler (2004), "Ocean Pollution from Land-Based Sources: East China Sea, China", *Ambio*, No. 33 (1-2).

Lohmar, B. *et al.* (2003), *China's Agricultural Water Policy Reforms: Increasing Investment, Resolving Conflicts, and Revising Incentives*, Agriculture Information Bulletin, No. 782, Economic Research Service, US Department of Agriculture, Washington DC.

Ma, Kai (2006), *The 11th FYP: Targets, Paths and Policy Orientation*, NDRC, Beijing.

Mei, Xurong (1999), *Water Shortage and Food Production in China: Issues, Potential and Solutions,* Centre for Water Resources and Conservation Technologies, Chinese Academy of Agricultural Sciences, Beijing.

MOWR (Ministry of Water Resources) (2005), *Evaluation Results of National Groundwater Resources and Environment Investigations*, MOWR, 21 April 2005, Beijing.

MOWR (2004), *Water Resources Bulletin*, MOWR, Beijing.

NBSC (National Bureau of Statistics of China) (2006), China Statistical Yearbook on Environment, China Statistics Press, Beijing.

NCEE (National Center for Environmental Economics) (2005), *China Water Conservation Technology Policy Outline*, Joint Announcement with the Ministry of Science and Technology, Ministry of Water Resources, Ministry of Construction and Ministry of Agriculture, Beijing.

NCEE (2004), *International Experiences with Economic Incentives for Protecting the Environment*, NCEE, Beijing.

OECD (2005), *OECD Review of Agricultural Policies: China*, OECD, Paris.

Qi, Yuzao *et al.* (2004), "Some Observations on Harmful Algal Bloom Events along the Coast of Guangdong, Southern China in 1998", *Hydrobiologia*, No. 512 (1-3).

Qu, J. *et al.* (2005), *GIWA Regional Assessment 36: East China Sea,* UNEP Global International Waters Assessment, University of Kalmar, Sweden.

SEPA (State Environmental Protection Administration) (2006a), State of the Environment Report, SEPA, Beijing.

SEPA (2006b), Annual Statistics: Report on Environment in China, China Environmental Science Press, Beijing.

SEPA (2004a), State of the Environment Report, SEPA, Beijing.

SEPA (2004b), Environmental Statistics Bulletin, SEPA, Beijing.

SEPA (2003), "Circular on the Progress of the Water Pollution Prevention in the Three-Rivers and Three-Lakes Regions", SEPA, Beijing.

State Council (2005), *Implementing the Scientific Concept Development and Enhancing Environmental Protection*, Decision of December 2005, Beijing.

Teng, S.K. *et al.* (2005), *GIWA Regional Assessment 34: Yellow Sea,* UNEP Global International Waters Assessment, University of Kalmar, Sweden.

UNICEF/WHO (2004), *Meeting the MDG Drinking Water and Sanitation Target: A Mid-Term Assessment of Progress*, Joint Monitoring Programme for Water Supply and Sanitation, UNICEF/WHO, New York, Geneva.

Wang, Jinnan *et al.* (1999), "Taxation and the Environment in China: Practice and Perspectives", in *Environmental Taxes: Recent Developments in China and OECD Countries*, OECD, Paris.

Wang, Shucheng (2003), Keynote speech presented at the 3rd World Water Forum, Kyoto.

World Bank (2004), *China Country Water Resources Assistance Strategy*, World Bank East Asia and Pacific Region, Washington DC.

Yan, Tian *et al.* (2005), *A National Report on Harmful Algal Blooms in China*, Institute of Oceanography, Chinese Academy of Sciences, Qingdao, China.

Zhang, Tianzhu *et al.* (1997), *Water Pollution Charges in China: Practice and Prospects in Applying Market-Based Instruments to Environmental Policies in China and Other Countries*, OECD, Paris.

Zhu, Ruixiang (2006), "China's South-North Water Transfer Project and its Impacts on Economic and Social Development", paper presented at the 4th World Water Forum, Mexico City.

WASTE MANAGEMENT

Features

- Reducing materials intensity by building a "circular economy"
- Reducing uncontrolled dumping of municipal, industrial and hazardous waste
- Improving institutional arrangements for waste management

Recommendations

The following recommendations are part of the overall conclusions and recommendations of the environmental performance review of China:

- foster the move *towards a circular economy* by focusing on waste reduction, reuse of waste material and waste recycling, and related targets; require provincial and local governments to adopt and implement comprehensive *waste management plans* (including accurate verification of volumes of waste – municipal, industrial and hazardous – generated and treated) covering elements of the waste hierarchy;

- accelerate the pace of extending *waste treatment capacity* by building treatment infrastructure and establishing systems for the collection, reuse and recycling of waste (e.g. separate collection of household waste), including in rural areas;

- formulate *enforcement plans* for different sectors (e.g. households, large industry, small and medium-sized enterprises) and types of waste;

- streamline the allocation of *responsibility for the management* of different types of waste; ensure that waste facilities operate efficiently and comply with standards; further develop workable regulations and policy instruments for waste management; improve the collection of *waste data* and develop tools to evaluate the effectiveness of waste management policies at national and provincial levels;

- establish *financing mechanisms* with a mix of public and private financing, and move to charging for waste services more progressively in less developed areas; improve the collection rate of waste charges and set them at a level consistent with the government's aim to achieve a circular economy;

- provide the *informal sector* (freelancers) with equipment, organisational assistance and training to continue collection and recycling under improved hygienic and environmental conditions, as part of waste management plans;

- *raise awareness* of waste management and efficient resource use among the public, small and medium-sized enterprises, and industry.

Conclusions

During the review period, China significantly decoupled the generation of municipal and, to a lesser extent, industrial waste from economic growth. Concerning *industrial solid waste*, the country met and surpassed the targets set out in the 9th and 10th FYPs with respect to recovery, reuse of waste material and safe disposal in landfills. China also stepped up its efforts to put in place an adequate *legal framework for modern waste management* by adopting a cleaner production law in 2003 and updating its 1995 waste law in 2004. A series of more specific regulations and standards for various types of waste, such as medical waste, were adopted. A national programme ("Construction Programme of Hazardous Waste and Medical Waste Disposal Facilities") was put in place in 2003 to significantly increase capacity for treating *hazardous and medical waste*, and good progress was made consequently with the treatment of medical waste. Considerable amounts of materials are recycled through informal activities (e.g. by freelancers). The opening of the market to foreign waste management technology is a positive signal for further improvement. Chinese authorities wish to curb the generation of all types of waste by fostering a high quality, *low material intensity economic growth model*. Indeed, given the rapid growth of its economy and of its imports, China's drive to reduce its material intensity parallels the drive to reduce its energy intensity. The *concepts of the "3Rs" (reduce, reuse, recycle) and of the "circular economy"* are part of the 11th FYP.

Nevertheless, the amounts of municipal waste, industrial waste and hazardous waste far exceed what can safely be treated and disposed of. Some of those waste are stored waiting for treatment (e.g. close to 50% of municipal waste) or *are dumped in an uncontrolled fashion*. Human health and the environment are put at risk through a proliferation of uncontrolled dumps surrounding the cities. The 10th FYP target of increasing the capacity of *municipal landfills* to 150 kt/day was not achieved. Total *waste generation* increased by as much as 80% during the review period. Waste management is still the "poor cousin", compared to air and water management, in its share of national government *funding* of investments. Local bodies have trouble collecting waste charges, which remain too low to cover the operational cost of waste management. Overall, the emphasis remains *too heavily on landfilling* (the destination of 44% of municipal waste), and few local governments implement separate collection and recycling. Incineration and composting account only for 3% and 5%, respectively of municipal waste treatment. *Responsibility for waste management* is fragmented across too many agencies. *Enforcement* is inadequate and does not distinguish sufficiently between large industries and small and medium-sized enterprises.

◆   ◆   ◆

1.   Waste Management Framework

1.1   Policy objectives

China's *main laws* covering waste management are the umbrella Environmental Protection Law (1979, 1989, 2001) and the Law on Prevention and Control of Environmental Pollution by Solid Waste (1995, 2004). Other relevant laws include the

Law on Promotion of Cleaner Production (2003), the Law on Prevention and Treatment of Infectious Disease (revised in 2004), and the Law on Environmental Impact Assessment (2002). Nuclear waste is managed separately under a different law.[1]

These laws are implemented through various *national regulations*.[2] The evolution and widening scope of waste policy concerns in China can be traced through the chronology of issues being regulated (Table 5.1). For example, in 2003, the State Environmental Protection Administration (SEPA) and the Ministry of Health jointly defined a hazardous medical waste classification. SEPA also promulgated Technical Rules on the Centralised Disposal of Medical Waste covering packaging, container types, warning instructions, transport and incineration.

Table 5.1 **Selected waste management regulations**

Management Regulations on Litter and Environmental Sanitation	1992
Management of Municipal Refuse	1993
Proposals on Further Developing Resource Utilisation	1996
Interim Provisions for the Administration of Environmental Protection Regarding the Import of Waste Materials	1996
Proposals for Strengthening the Management of Plastic Packaging Waste along Main Roads, in River Basins and at Tourist Attractions	1998
Prevention and Control of Pollution by Mine Tailings	1999
Regulation of Hazardous Waste Transportation Manifest Management	1999
Circular on Strengthening the Administration of Hazardous Chemicals	1999
Technical Policies on the Disposal of Domestic Waste and the Prevention of Pollution	2000
Measures on Pollution by Livestock and Poultry	2001
Safety Management for Hazardous Chemicals	2002
Proposals for Promoting the Industrialisation of Urban Waste Water and Refuse Treatment	2002
Circular on Implementing a Charging System for Municipal Refuse Treatment and Promoting the Industrialisation of Waste Treatment	2002
Circular on Implementing a Charging System for Hazardous Waste Disposal and Promoting the Industrialisation of Hazardous Waste Disposal	2002
Administration of Medical Waste	2003
Technical Policies on the Prevention of Pollution by Hazardous Waste	2003
Circular on Establishing a Charging System for Hazardous Waste Management and Disposal, to Commercialise the Waste Management Industry	2003
Circular on the Environmental Management of Used Electronic and Electronic Equipment	2003
National Construction Programme of Hazardous and Medical Waste Disposal Facilities	2004
Measures on Permits for the Operation of Hazardous Waste	2004
Management of Urban Construction Waste	2005
Prevention and Control of Pollution from Unused Hazardous Chemicals	2005

Source: SEPA.

China's 10th Five-Year Plan for National Economic and Social Development (FYP) (2001-05) contained a number of *general objectives* with relevance for waste management, e.g.:

- establish comprehensive decision-making mechanisms and promote the co-ordinated development of environmental protection and the economy;
- improve the system of environmental laws and regulations and conduct environmental protection according to the laws;
- consolidate government regulations with the market mechanism;
- establish modern environmental management;
- strengthen environmental education and publicity and promote public environmental awareness;
- carry out the responsibility system of environmental protection.

More specific, *numerical national objectives* for waste management applicable to the review period are set forth in both the 10th FYEP and the earlier 9th FYEP (1996-2000) (i.e. Five-Year Plans for Environmental Protection). In the context of China's strong economic growth throughout the review period, the 9th Plan sought to reduce the generation of industrial solid waste by 3%, whereas the 10th Plan proposed a reduction of about 10%. China has also adopted a National Construction Programme of Hazardous and Medical Waste Disposal Facilities.

1.2 Institutional aspects

Responsibility for different types of waste is allocated to several agencies:

- the State Environmental Protection Administration (SEPA) is the central government agency with responsibility for *industrial solid waste*;
- the national Ministry of Construction and the environmental sanitation departments of local governments are in charge of supervision and administration of the cleaning, collecting, storing, transporting and disposing of *municipal waste* (including construction waste);
- the Ministry of Commerce oversees the recycling of *paper and cans*, but private firms carry out the actual recycling;
- the Ministry of Health and the health departments of local governments are responsible for the collection, transport and disposal of *medical waste*;
- water resource departments deal with the treatment and disposal of (often contaminated) bottom *sediments dredged* from the country's rivers.

Local governments are responsible for the collection of municipal waste, either directly or through municipal enterprises. Disposal of municipal waste is also mostly carried out by local governments, but in a small percentage of cases is contracted out to the private sector, including some foreign firms. *Industry* is responsible for treating the waste it generates. A 1999 regulation established a manifest system for tracking hazardous waste.

1.3   Waste definitions

China's waste legislation distinguishes *five types of waste*:

– municipal[3] solid waste (domestic and commercial waste, construction waste);

– industrial solid waste including hazardous waste (e.g. from production and processing metallurgy, coal, electric power, chemicals, traffic, food, light industry, the mineral oil industry and so forth);

– mining solid waste (including waste from dressing and washing);

– agricultural waste (from crops and animal/poultry husbandry);

– radioactive waste (from the nuclear industry and nuclear power production, the nuclear fuel cycle, medical and industrial application of radioactive isotopes, and associated research).

Industrial waste is further subdivided into: i) general industrial waste, as produced by industrial enterprises in production processes; and ii) various classes of hazardous waste defined in a national hazardous waste catalogue (similar to the definitions of the Basel Convention) or specified to be either explosive, ignitable, oxidizable, toxic, corrosive or liable to cause infectious diseases or lead to other dangers. The catalogue presently comprises 47 types of waste or substances. The catalogue is being revised and will ultimately distinguish more than one hundred substances or waste types. A specific catalogue for medical waste[4] distinguishes five types of medical waste. The Chinese definition of solid waste includes dense sludges and gaseous waste in containers, but excludes waste water discharged into water bodies and waste gases emitted into the atmosphere.

2.   Waste Management Performance

China only developed a legislative framework for waste management *since the last decade* (Table 5.1) and has made less progress and, in fact, spent fewer resources in this area than on air and water management. The 10th FYP allocated CNY 90 billion, or about 10% of China's total environmental investment, to the treatment of solid waste. Within the area of waste management, emphasis has so far been on the lower end of the waste hierarchy, although implementation of the 2003 Cleaner Production law and rules on waste from electric and electronic equipment signal greater attention to waste reduction.

Given China's rapid economic growth, the rapid increase in imports of raw materials, and the rapid rise in the prices of certain commodities (e.g. steel, copper), it is in the country's economic interest to drastically reduce the *material intensity* of the economy from its present high level (Chapter 7). A more vigorous pursuit of modern waste management, including adoption of cleaner production techniques and product policies, should be a key part of such an effort. Adopting a product life-cycle approach would not only *reduce domestic waste generation* but will also be essential for maintaining the *competitiveness* of certain Chinese products, such as electronic equipment. Chinese authorities are clearly aware of the need to parallel the country's efforts on energy intensity with efforts on the economy's material intensity.

The waste management goals and targets of the *10th FYEP* were partially achieved. Using resources more efficiently and related waste management targets are prominent features of the *11th FYP* adopted in March 2006. A better integration of environmental and economic policies, improved investment and financing mechanisms, increased funding for solid waste treatment and disposal, streamlining the institutional structure for waste management (including staff capacity building), and more rigorous enforcement should accordingly be the main focus for the coming five years. The "*3R*" (reduce, reuse, recycle) concept, which is linked to the idea of a "*circular economy*", will improve waste management (Box 5.1).

2.1 Generation, treatment and disposal

Waste generation

Overall waste generation on a per capita basis in China is still low compared to that in the OECD area[5] (Figure 5.1). However, official figures indicate that *total waste generation* (including agricultural waste) grew by as much as 80% during the review period (1995-2004): between 1995 and 2004, municipal and general industrial waste grew by 45% and 86% respectively (Table 5.2). Economic growth was even stronger during 1995-2004 (115%), thus industrial waste generation was somewhat decoupled from GDP growth and municipal waste generation was significantly decoupled from the growth in private final consumption (Figure 5.1). Industrial waste intensity (waste generation per unit of GDP) in economically developed areas (e.g. Tianjin, Shanghai, Jiangsu, Zhejiang, Fujian and Guangdong) is below the national average.

Box 5.1 **The circular economy**

The Chinese leadership has in recent years promoted the idea of the "circular economy", and the concept underlies *the 11th FYP* (2006-10) approved in March 2006. The term circular economy denotes an *alternative economic growth model* no longer based on the intensive use of energy and other primary resources and the generation of waste and pollution. The concept has become widely accepted as a result of the rising prices of many materials (e.g. copper, zinc, nickel, aluminium, steel, soja) on world markets and related import costs, and of concerns about sustainable development. The concept of the circular economy includes:

• closing materials loops;

• reducing the material intensity of the economy;

• improving resource productivity;

• waste reduction, reuse and recycling (the "3R" concept);

• changing production and consumption patterns.

In the government's Work Report 2006, delivered to the Fourth Session of the Tenth National People's Congress in March 2006, the Chinese Premier announced a series of circular economy *pilot projects* to be conducted in key industries, industrial zones, and urban and rural areas. The government also announced that it would promulgate tax policies favouring resource recovery, reuse and recycling of waste materials.

Several building blocks for a circular economy, such as the Law on the *Promotion of Cleaner Production*, are already in place or underway. The government proposes to carry out an appraisal of the law's implementation to see where it can be improved. Efforts to promote *resource recovery and recycling* have started and should be strengthened. The government also wants to use the tool of *materials accounting* at the national and enterprise level to provide better insight into what problems should be tackled.

Source: Ren Yong, 2006; CAS, 2006.

In absolute terms, China is already among the world's largest generators of municipal solid waste (World Bank, 2005). The *composition of household waste* has changed significantly over recent years as a result of changing consumption patterns and the move to clean fuels in cities, which is drastically reducing the waste content in coal cinders (Figure 5.2).

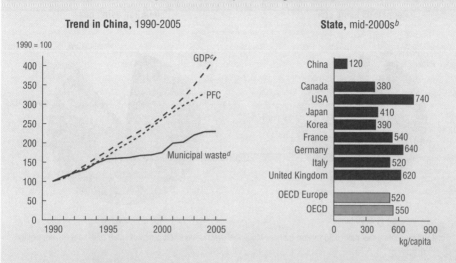

Figure 5.1 **Municipal waste generation**[a]

a) Volume of disposal garbage.
b) Or latest available year.
c) Based on the revised results about historical data of China's GDP done by NBS, 20 January 2006.
d) In interpreting national figures, it should be borne in mind that survey methods and definitions of municipal waste may vary from one country to another. According to the definition used by the OECD, municipal waste is waste collected by or for municipalities and includes household, bulky and commercial waste and similar waste handled at the same facilities. For China municipal waste is volume of garbage disposal in all urban units.
Source: OECD, Environment Directorate, National Bureau of Statistics of China.

Treatment of municipal waste[6]

Modern waste management practices such as separate collection, use of landfill gas and incineration with energy recovery are still the *exception rather than the rule in China*. In 2003, for example, almost one-half of municipal waste was stored waiting for treatment,[7] and aerial surveys indicate that many urban areas are almost surrounded by dumps (Liang, 2004). The same is true in many rural villages, where domestic solid waste contains increasing amounts of plastics and other non-biodegradable matter due to changing consumption patterns.

Controlled disposal of municipal waste mainly involves landfilling. In 2004, 44% was landfilled, 5% was composted and 3% was incinerated (Table 5.3). The calorific value of municipal waste is low due to the presence of a high degree (50-60%) of wet organic matter, mostly kitchen waste (Box 5.2). Hazardous

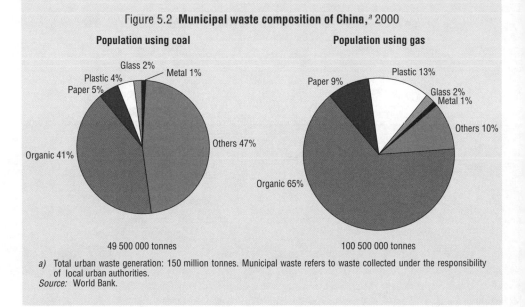

Figure 5.2 **Municipal waste composition of China,**[a] 2000

Population using coal   Population using gas

Glass 2%
Metal 1%
Plastic 4%
Paper 5%
Others 47%
Organic 41%

Plastic 13%
Paper 9%
Glass 2%
Metal 1%
Others 10%
Organic 65%

49 500 000 tonnes   100 500 000 tonnes

a) Total urban waste generation: 150 million tonnes. Municipal waste refers to waste collected under the responsibility
    of local urban authorities.
Source: World Bank.

substances in household waste (e.g. batteries) also are a problem. The shares of both incineration and composting could be increased by collecting these waste separately. Biogas generation could be promoted. Construction waste is mostly used as fill, but some small amounts are reused.

Little quantitative information is available about the extent to which municipal waste (including packaging, batteries, etc.) is *recovered and recycled*, but in general, recycling rates for various materials are estimated to be low compared to those achieved in OECD countries. Paper and cardboard may be an exception, as the authorities believe that much is recycled, and China also imports waste paper (9.4 million tonnes in 2003), partly from countries such as the United States. China imported 9.3 million tonnes of scrap iron and steel in 2003. At present, the collection of recyclable waste with economic value (e.g. metals) is mostly done by private resource recovery enterprises, including freelancers, often migrants from poorer parts of China. Actual recycling rates by the *informal sector* should not be underestimated. It is outside any specific waste policy framework. Legalising informal sector activities in recovery and recycling should be given consideration. Whenever feasible, waste freelancers should be involved in operating modern transfer, stations and recycling centres (Box 5.3).

Table 5.2 **Waste generation,** 1995-2004

(1 000 tonnes/day)

	1995	1996	1997	1998	1999	2000	2001	2002	2003	2004	% increase[d] 2004/1995
Municipal waste[a]	292	295	300	309	312	322	369	374	407	423	45
Industrial waste[b]	1 766	1 800	1 801	2 193	2 149	2 229	2 434	2 589	2 751	3 279	86
non-hazardous waste	..	1 773	1 773	2 167	2 121	2 207	2 407	2 561	2 719	3 252	79
hazardous waste[c]	..	27	27	26	27	22	26	27	32	27	0
Total	2 058	2 096	2 102	2 503	2 461	2 552	2 803	2 963	3 158	3 703	83

a) Volume of municipal garbage.
b) Industrial solid waste produced refers to the total volume of solid, semi-solid and high concentration liquid residues produced by industrial enterprises from the production process in a given time period; includes hazardous waste, slag, coal ash, gangue, tailings, radioactive residues and other waste, but excludes stones stripped or dug out in mining (gangue and acid or alkaline stones not included).
c) Hazardous waste refers to waste included in the national hazardous waste catalogue or specified as having one or more of the following properties in the national hazardous waste identification standards: explosive, ignitable, oxidizable, toxic, corrosive, or liable to cause infectious diseases or lead to other dangers.
d) Versus a 115% growth of GDP (at PPP-2000).
Source: OECD, Environment Directorate.

Table 5.3 **Waste treatment,** 2004

(1 000 tonnes/day)

	Total	(%)	Landfilled	(%)	Incinerated	(%)	Composted	(%)	Recycled or reused	(%)	Stored waiting for treatment	(%)	Non-treated	(%)
Municipal waste	423	100	188	44	12	3	20	4.7	–	0	203	48	–	0
Industrial waste[a]	3 279	100	715	22	–	0	–	0	1 820	56	698	21	46	1
of which: hazardous waste	27	100	7	27	–	0	–	0	11	39	9	34	–	–

a) Quantity landfilled refers to quantity of industrial solid waste that is burnt or disposed of in sites meeting the requirements for environmental protection and not salvaged or recycled (including disposal in the year of some waste of previous years). Disposal includes landfilling (safe landfills should be conducted for hazardous waste), incineration, disposal in containment spaces, deep underground disposal, backfill in mining pits and disposal at sea.
Source: NBSC; OECD, Environment Directorate.

Box 5.2 Of vegetables and chopsticks

Eating culture is generally identified as the main reason for the high ratio of *kitchen scraps in domestic waste* in China (and other Asian countries), which is much higher than in Western countries. Treatment of municipal waste is more difficult when it contains a large proportion of wet organic matter mixed with other materials. The Chinese authorities are therefore considering how to reduce the flow of kitchen scraps into domestic waste.

One experiment found that cleaning vegetables at local markets before sale reduced kitchen waste to some extent. However, it also found that the amount of market waste, which also ended up as general municipal waste, had increased. Now it is proposed to *clean vegetables before transport* to the city and return the waste organic matter to the soil. If successful, such a system would represent a win-win solution by not only reducing municipal waste (and facilitating treatment), but also by improving soil quality and reducing the need for chemical fertilisers.

Another component of municipal waste is *disposable wooden chopsticks*, of which China uses ten billion boxes each year.and exports another six billion, requiring 1.3 million cubic metres of timber from the country's forests. A *5% consumption tax* will be collected on disposable wooden chopsticks to discourage their use, which the Ministry of Finance considers a waste of timber resources. A 5% tax on wooden floor panels will also discourage the consumption of timber resources.

Source: Lin Weigui, 2006.

The government is making efforts to develop a *policy for waste from electric and electronic equipment*. China's stated political goal of moving towards a "circular economy" and reducing its material intensity will require it to more systematically pursue resource recovery and product policies.

Treatment of industrial and other waste

Out of the 1.2 billion tonnes of industrial waste produced in 2004, 56% was recycled or reused, 22% was landfilled, 21% was stored waiting for treatment and the remaining 1.4% (17.6 million tonnes) was dumped in an uncontrolled way (Table 5.3). The recycling of *industrial waste* increased from 40% to 56% over 1995-2004,[8] thereby surpassing the 10th FYP's 50% reduction target for 2005. The target of maximum 28.6 million tonnes per year being dumped in an uncontrolled

Box 5.3 **Waste management in Shaanxi province**

Shaanxi is a large province (37.1 million inhabitants, 205 600 km^2) in central China, located along the Qinling mountain range, which is the divide between the Yangtze basin to the south and the Yellow river basin to the north. The Capital is Xi'an, the city of the first Chinese Emperor, who united China. GDP per capita in Shaanxi province (USD 940) ranks 24th among the 31 Chinese provinces, autonomous regions and municipalities.

Total environmental investment in Shaanxi province amounted to CNY 4.91 billion in 2003, i.e. about 2% of a GDP of CNY 239.9 billion. The dominant share of environmental infrastructure investment comes from the national budget, whereas the provincial contribution is minor: under the 9th and 10th FYPs, it amounted to CNY 120 million and CNY 315 million, respectively; in 2006, the projected provincial contribution amounted to CNY 87 million. No specific waste management investment figures are available.

As in other less affluent parts of China, raising revenue for proper waste management in Shaanxi province is difficult. In Xi'an the *waste fee* is CNY 6 (EUR 0.6) per family (2-3 persons) per month and a significant proportion of households do not pay their charges. The revenue actually received does not allow an environmentally friendly treatment of municipal waste, so the difference is often topped up from local or provincial budgets. The authorities, however, intend to move towards full implementation of the polluter pays principle.

Waste generation and disposal in Shaanxi	2001	2002	2003	2004
Industrial waste (million tonnes):	24.1	28.9	29.5	38.2
– recovered	4.9	7.0	8.6	8.3
– disposed in controlled landfills	0.77	0.79	0.53	0.32
– dumped	18.4	21.1	20.6	29.6
Hazardous industrial waste (thousand tonnes):	68	..	34	35
– disposed in controlled landfills	7	..	6	3
– dumped	61	..	28	32
Municipal waste (million tonnes):	3.39	3.01	..	..
– disposed in controlled landfills	1.64	1.15	..	..
– dumped	1.75	n.a.	..	..

Source: EPB Shaanxi province.

During the 10th FYP period, 34 new landfills and four new *medical waste facilities* (including one in Xi'an) were established. Hospitals in all prefectures now have access to a medical wastes facility; the hospitals pay charges based on the number of beds.

Box 5.3 **Waste management in Shaanxi province** *(cont.)*

Formal systems for the *separate collection of recyclable waste* are still in their infancy, but freelancers collect recyclable household waste they can sell. The separate collection of batteries and electric and electronic equipment will start in 2007 and will include a deposit refund system. For industrial waste, the emphasis is on recycling bulk construction waste and slag from power stations.

The Shaanxi Environmental Protection Bureau is also promoting the development of the *"circular economy"* through demonstration and pilot projects, co-operation with the Shaanxi Development and Reform Committee, and by setting up the Shaanxi Circular Economy Research Committee. A national ISO14000 demonstration area, approved by SEPA, was established In the Xi'an new technological zone.

Source: Shaanxi State of the Environment Reports, 2001-04.

---

way was also achieved. However, there are indications that the targets are not met in all provinces (Box 5.3). In 2000 (latest data available), 6.3 billion tonnes of industrial waste were stored on 58 000 hectares, waiting for treatment.

As for industrial *hazardous waste*, out of the 10 million tonnes produced in 2004, 39% was recycled or reused, 27% was landfilled, 34% was stored waiting for treatment and the remaining 0.1% (11 000 tonnes) was dumped in an uncontrolled way (Table 5.3). Registered companies process less than 4% (1% by landfilling and 3% by incineration) of total annual hazardous waste generation. Proper management of the 1 500 tonnes of medical waste produced every day has been getting much more attention since the Severe Acute Respiratory Syndrome (SARS) outbreak in 2003.

Concerning *agricultural crop waste*, the 2005 FYP target of recovering 80% of straw (stalks) is being implemented by returning it to the soil (e.g. ploughing it into fields, along highways or other places, such as near large airports) or converting it to biogas. By 2003, almost 500 centralised biogas facilities had been built, but it is not clear to what extent the 80% target was reached.

With regard to *international waste policies*, enforcement of the provisions of the Basel Convention on the Control of Transboundary Movements of Hazardous Waste has at times posed problems, notably with regard to the illegal import of electronic waste, and China tightened its control over this type of waste in 2002 (Basel Action Network, 2002). The dumping at sea of often contaminated spoil from the dredging of rivers and harbours should be carried out in accordance with the provisions of the London Dumping Convention (Chapter 9), but presents problems. China does not dump solid waste into the sea. China's waste management practices now have a global impact. For example, secondary materials prices in the United States are now influenced by China's demand for these materials. In addition, substantially increasing the rate of waste incineration in China could significantly add to global ambient levels of dioxin, depending on the effectiveness of emission controls.

2.2   Waste management infrastructure

Although landfilling is China's dominant method of waste disposal, the 10th FYP *target of increasing sanitary landfill capacity to 150 kt/day was not achieved.*[9] As 150 kt/day represents about one-third of current waste generation, it is clear that the lack of landfill capacity is acute, notably in rural areas, where open dumps on the edge of villages pose a public health hazard. Technical standards for the construction and operation of landfills are partially based on those in the OECD area, but less than 10% of municipal waste is disposed of in sanitary landfills that actually conform to such standards. Leaching is a problem at many landfills. It is not clear how much progress has been made in implementing the national landfill gas action plan, but some projects are getting under way, often as part of the Clean Development Mechanism, and the 2005 Law on Renewable Energy is also expected to speed up the utilisation of landfill gas.

Chinese regulations require *closed landfills* to be covered with soil and planted. Nevertheless, many closed landfills, as well as small and large uncontrolled dumpsites, are a threat to public health and the environment and will need to be dealt with, better sooner rather than later. To do this in the most cost-effective manner will require a strategic approach (e.g. an inventory of problem sites, standards for the level of remediation needed, priority setting, funding mechanisms and the expertise to determine suitable techniques). Little information is available about the extent of the problem of China's *contaminated sites*,[10] but the solution will demand a similar strategic approach and will pose another challenge.

Municipal waste *incineration* has so far played only a minor role in China's waste management; by the end of 2004, just 54 municipal waste incinerators[11] with a combined capacity of 33 000 tonnes/day were in operation. Emission limits (including a limit value of 0.1 ng TE/m^3 for dioxins and furans) comparable to European limits in the 1990s have been promulgated, but independent control of emissions is difficult due to the lack of laboratories. Co-incineration of certain waste in coal-fired power plants is practised to some degree.

More capacity and better technologies are also needed to treat the increasing quantities of *sewage sludge* produced by the growing number of waste water treatment facilities. Currently, sewage sludge is mostly deposited in landfills and spread on agricultural soils in the absence of clear quality standards. Small amounts of sludge are incinerated, though this is still too expensive to be widely applied throughout the country. This issue should be addressed on a national or provincial scale.

As of early 2004, just *four hazardous waste facilities were in operation* in China – a landfill in Shenzhen (capacity 50 kt/yr) and one in Shenyang (20 kt/yr), an incinerator (30 kt/yr) and some landfill capacity in Dalian, and a PCB incinerator in Shenyang (7.5 kt/yr) (Adeleke, 2002) – and five more disposal and recovery plants were under construction in Tianjin, Shanghai, Suzhou (Jiangsu province), Hangzhou (Zhejiang province), and Fuzhou (Fujian province).[12] In addition, several large industries operated hazardous waste facilities mainly for their own use. The 2003 National Construction Programme of Hazardous and Medical Waste Disposal Facilities proposed to establish, within three years, 31 centralised facilities for hazardous waste as well as 300 centralised medical waste facilities, involving an overall investment of CNY 15 billion (USD 1.8 billion). It is not clear how the construction of new hazardous waste facilities has advanced,[13] but as of early 2006, 171 incinerators[14] for medical waste had been established.[15]

2.3   Waste management institutions

The *fragmentation of responsibility* for waste management over many agencies at best requires large co-ordination efforts and at worst inhibits the implementation of a coherent and comprehensive waste management strategy for the various types of waste. The experience of OECD countries shows that streamlining the organisation of waste management ensures a more comprehensive approach, reduces the risk of certain types of waste "falling between the cracks", and leads to better results overall. Streamlining should also lead to a more comprehensive effort at improving data acquisition and monitoring of local waste generation and disposal in order to support policy-making.

China's readiness to accept *private (including foreign) investment to build and operate new waste management facilities* (e.g. municipal waste incinerators) or whole waste management systems should enable it to overcome financing difficulties and step up the pace of building an adequate and modern waste management infrastructure. An example is provided by Jinzhou District (Dalian City), where a private company invested CNY 55 million to build a domestic waste facility with a capacity of 300 tonnes/day and producing 50 000 tonnes of organic fertiliser annually. Jinzhou District only needed to provide the roads, water and electricity supply; the company is paid from the sanitary charges levied on households (Ma, 2002).

Streamlining the institutional arrangements for waste management will only yield the desired results if waste management *staff at all levels have been well trained* to carry out their functions (e.g. regulation, operation of facilities, inspection) and adopt a results-oriented approach to their job. A good example is provided by the new rules on medical waste, which require medical institutions to employ personnel with training and education in the relevant areas of solid waste. This example should be extended to other waste areas. Also, every province now has a solid waste centre with a staff of 10-15 people who provide advice to local government.

The success of waste separation and recycling schemes depends on the *participation of a well-informed and motivated public*. Political leaders must therefore ensure that they inform citizens and businesses and invite public participation to obtain "buy-in" of government proposals by all stakeholders. While public awareness of waste issues is still low, there are examples of voluntary private initiatives, such as the Green Star Waste Battery Volunteer Service Team, which has 6 000 volunteers who promote waste battery recycling throughout the country. The government also has a role in raising industry awareness and know-how, for many enterprises possess little knowledge about the hazardous substances contained in their waste and are therefore poorly equipped to adopt the correct treatment methods.

2.4 Expenditure, financing and charges

Overall investment and operational waste management expenditure at all levels were estimated at CNY 30 billion in 2005 (World Bank, 2005), *i.e.* the equivalent of USD 10 per capita or four times that amount in terms of purchasing power parity. Investment in waste treatment facilities programmed in the 9th and 10th FYPs amounted to CNY 50 billion and CNY 90 billion, respectively, representing around 11% and 10% of total environmental investment (Chapter 7). Assuming that operational expenditure reflects the same proportionality, it also is a much smaller share of total pollution abatement and control (PAC) expenditure than is common in the OECD area.[16] No disaggregated figures are available for investment in waste management by private industry or on operational expenditure on waste management in either the public or private sector.

Many provinces and municipalities, notably in the poorer parts of the country, have difficulty *financing* the local share (i.e. 30%) of the required investments, partly because they find it hard to actually collect the waste charges payable by households. In some municipalities, e.g. Beijing, the collection rate is as low as 10%; in others it is much higher, e.g. in Xi'an in Shaanxi province the collection rate is 80%. Increasing fee recovery rates in the longer term will therefore be necessary to at least cover operational costs, but in the meantime investment subsidies from the national government would be appropriate in order to improve the standard of waste management over a reasonable time frame.

The structure of *waste management charges* for households and industry is quite complex, but the overall level is probably too low to cover the actual cost of proper treatment; for example, in Xi'an and Nanjing (Jiangsu province) the charges amount to CNY 6 (EUR 0.63) and CNY 5, respectively, per household per month. Moreover, in many cases the waste charges are subsumed in the water or gas bill, so that householders have little idea about the cost of waste management. Progressive introduction of a volume-based charge for all waste generators (including households) should aim at both implementing the polluter pays principle and reducing the material intensity of the economy.

SEPA is currently investigating the feasibility of various *deposit-refund systems*, but so far there are only some informal systems in Shanghai. Limited use has been made to date of product charges, but a 5% tax on wooden floor panels and disposable wooden chopsticks went into effect in April 2006 (Box 5.2). The potential of this instrument should be further explored, in particular in relation to packaging and hazardous items (e.g. batteries,[17] used oil, tyres). The Chinese system of pilot testing new policy measures in a limited number of places before nationwide adoption is well suited to developing solutions that fit the Chinese situation.

Notes

1. This type of waste is outside the scope of this chapter.

2. National regulations are approved by the State Council; more detailed rules are approved by individual ministries.

3. Municipal waste refers to waste collected under the responsibility of local urban authorities.

4. SEPA and the Ministry of Health were jointly formulating this catalogue as of early 2006.

5. Chinese waste data have to be interpreted with caution. For example, the amounts of municipal waste are estimations only. Also, the collection and recovery of post-consumer recyclable materials are carried out as a normal economic activity and figures on the amounts involved do not enter waste statistics. Also not included in the above figures is the collection of night soil in areas not served by a sewerage system; the amounts of night soil are of the order of 20-30% of municipal waste. Hazardous waste figures cover only hazardous industrial waste and not hazardous municipal waste.

6. Municipal waste refers to waste collected under the responsibility of local urban authorities.

7. When abandoned waste piles are located along rivers, they pose a direct threat to water quality. The problem is also readily visible in the countryside, where littering of white waste, plastic bottles and bags is widespread.

8. For comparison, in the neighbouring Republic of Korea, 84% of industrial waste was recycled or incinerated in 2003.

9. The 11th FYP approved in March 2006 sets a target of 250 kt/day, or 60% of generation.

10. Problems have been encountered where cities have expanded outward; small sites are cleaned up and larger ones are stabilized, but appropriate remediation technologies are lacking. The 1999 Law on Risk Assessment Baseline of Soil Environmental Quality for Industrial Enterprises, while not compulsory, is regarded as the first step towards regulated site clean-up similar to that done under the United States "Superfund" legislation.

11. Most incinerators in China were built under a Build-Operate-Transfer (BOT) arrangement.

12. China Basel Convention Fact Sheet.

13. However, the implementation of this programme appears to be making better progress in the eastern provinces of the country than in the West.

14. China Daily, 25 February 2006.

15. Problems with dioxin emissions from these incinerators were being tackled under a (partly GEF-funded) project costing USD 45 million.

16. For comparison, in the early part of the 2000s, waste management accounted for 36% of total public and business PAC expenditure in the OECD area. Allowance needs to be made for the fact that China is in a different stage of development.

17. China produces about one-third of the world's batteries. Although SEPA made attempts to tackle the problem of safe disposal starting in early 2006, there was no government-sponsored nationwide recycling scheme for used batteries.

Selected Sources

The government documents, OECD documents and other documents used as sources for this chapter included the following. Also see list of websites at the end of this report.

Adeleke, O.F. (2002), *Hazardous Waste Treatment and Disposal: An Examination of the Present Hazardous Waste Management Options Worldwide and in China*, Shanghai University, School of Environmental and Chemical Engineering, Shanghai.

Basel Action Network (2002), BAN Highlights/2002, www.ban.org/hilights/highlights02.html.

Chinese Academy of Sciences Sustainable Development Strategy Study Group (2006), *China Sustainable Development Strategy Report: Build a Resource-Efficient and Environment-Friendly Society*, Government Work Report 2006, Beijing.

Liang, Yunxia (梁云霞) (2004), 固体废物的污染控制及管理 *(Control and Management of Pollution Caused by Solid Wastes)*, 山西省吕梁市行署环境监测站，山西离石 (Shanxi province, Luliang City Administrative Office Environmental Monitoring Station, Lishi, Shanxi), 山西能源与节能，2004年第4期，总第35期 (Shanxi Energy and Conservation, 2004, Issue #4, Total Issue #35).

Lin, Weigui (2006), *Analysis of Present Situation of Solid Wastes and Management and Disposal Measures*, Guangzhou City Fangcun District Environmental Protection Bureau, Guangzhou, Guangdong.

Lin, Weigui (林伟贵) (2005), 固体废物现状分析和管理处置对策 *(Analysis of Present Situation of Solid Wastes and Management and Disposal Measures)*, 广州市芳村区环境保护局，广东，广州 (Guangzhou City Fangcun District Environmental Protection Bureau, Guangzhou, Guangdong), 大众科技，2005年第3期，总第77期 (Popular Science and Technology, 2005, Issue #3, Total Issue #77).

Ma, Xiaolu (2002), *Discussion on the Management and Disposal of Municipal Solid Wastes*, (中华人民共和国固体废物污染环境防治法), China Book Class No. X51, Literature identification No. A, Article No. 1007-245401-0037-02, Qinghai Normal University, Xining, Qinghai.

NBSC (National Bureau of Statistics of China) (2006), China Statistical Yearbook on Environment, China Statistics Press, Beijing.

Ren, Yong (2006), *Policy and Law Framework for Circular Economy in China*, SEPA Policy Research Centre for Environment and Economy, Beijing.

Robinson, A., G.C. Sewell, S. Wu, N. Damodaran and N. Kalas-Adams (2002), *Landfill Data From China: Addressing Information Needs For Methane Recovery*.

SEPA (State Environmental Protection Administration) (2006a), State of the Environment Report, SEPA, Beijing.

SEPA (2006b), China Environment Yearbook, China Environment Yearbook Press, Beijing.

SEPA (2006c), Annual Statistics: Report on Environment in China, China Environmental Science Press, Beijing.

Solenthaler, Balz and Rainer Bunge (2003), *Waste Incineration in China* (中国的垃圾焚烧), Institut für angewandte Umwelttechnik, Hochschule für Technik, Oberseestrasse 10, CH-8640 Rapperswil.

World Bank (2005), "Waste Management in China: Issues and Recommendations", World Bank Working Paper No. 9, May 2005, Urban Development Working Papers, East Asia Infrastructure Department, World Bank, Washington DC.

Zhao, Youcai and Hong Liu (赵由才, 刘洪) (2002), 我国固体废物处理与资源化展望 (A Review on the Solid Wastes Management and Technology in China), 同济大学环境科学与工程学院污染控制与资源化研究国家重点实验室 (Pollution Control and Resources Studies State Key Laboratory, Environmental Science and Engineering College, Tongji University), 苏州城建环保学院学报，第15卷第1期，2002年3月（Suzhou Institute of Urban Construction and Environmental Protection Journal, Vol. 15, Issue #1, March 2002).

Zhao, Weijun (赵维钧) (2003), 中国固体废物管理的目前状况和对策 (*Present Situation and Countermeasures for Solid Waste Management in China*), 国家环境保护总局污染控制司 (SEPA Pollution Control Department), Beijing.

NATURE AND BIODIVERSITY

Features

- A mega-diverse country
- Protecting the giant panda
- Traditional medicine and other uses of endangered species
- Managing the rapidly expanding nature reserves
- Forest policy and biodiversity conservation
- International co-operation in nature protection

Recommendations

The following recommendations are part of the overall conclusions and recommendations of the environmental performance review of China:

- modernise and implement *legislation on nature protection*, in particular adopt a law on the protection and management of Nature Reserves, notably favouring an increase of marine protected areas and of protected areas with higher protection status; consider ratification of the Bonn Convention;
- enhance the *capacity of national, provincial, prefecture and county level agencies* to manage biodiversity protection of existing reserves and integrate nature conservation within economic and social development projects outside protected areas;
- increase the *financial and human resources* for nature and biodiversity protection and further involve local residents in patrolling, monitoring and habitat enhancement, in the context of poverty alleviation; diversify the sources of financing of nature conservation;
- develop the use of *economic instruments related to nature and biodiversity protection*, not only as income supporting measures, but to reward the provision of environmental services;
- integrate long-term plans for rehabilitating and maintaining species and protected areas (including managing alien species) with *land*-use and river basin management plans, and any subordinate provincial, prefecture and country plans;
- integrate the economic and social values of protecting habitats and species (e.g. ecological services, tourism development) within *development decision*-making, in particular as part of EIAs;
- promote *sustainable forest management* through issuance of forest management plans, certification of foresting practices, and labelling of forest products in China; expand co-operation with supplying countries in the *forestry sector*, to ensure that imported wood and wood products are sourced from forests that are managed on a sound, sustainable basis.

Conclusions

China has established a comprehensive *legal framework* for managing nature and biodiversity, which includes wildlife and marine protection as well as terrestrial and marine protected areas. China actively reports on its international commitments and also publishes annual state of the environment reports related to its internal goals and targets. *Protected areas* at the national, provincial, prefecture and county levels have been dramatically increased over the review period, and China has received international recognition for its wetlands, biosphere reserves, and natural and cultural heritage preservation programmes. Outside of protected areas, ecological considerations have led to *afforestation* of large areas. New forestry initiatives have been taken to further develop shelter forests in arid, mountainous and coastal areas, to streamline forest management (e.g. more stringent harvest quotas) and to promote farm forestry on land sensitive to soil erosion (e.g. grain for green policy). Various environmental protection programmes within the country have begun to recognise the value of environmental outreach (alien species, endangered wildlife). China has been proactive in developing bilateral and regional co-operation in the area of nature conservation. There has been a regular increase in the number of world heritage sites and Ramsar wetlands.

However, there is a need for more *institutional co-ordination* and integration of efforts to assess and protect nature and biodiversity inside and outside of protected areas, given the number of agencies and stakeholders involved. There is insufficient monitoring to assess trends and evaluate the *protection status of nature reserves*. The main targets for species and habitat protection are in terms of percentage of land area. There is a need to ensure that the key natural habitat types and ecosystems are adequately protected and that they support species recovery plans. Although China has a relatively high percentage of total area classified as protected, *marine habitats and species* are not sufficiently represented and are subject to land-based sources of pollution and habitat alteration, in addition to exploitation pressures. Management level of reserves needs to be improved and attention should be paid to integrated habitats protection to minimise fragmentation and to *enhance habitats continuity through biodiversity corridors*. There is a need to integrate nature protection concerns into *development plans* especially in impoverished central and western regions with abundant biodiversity. Little has been done to promote biodiversity protection on forestland and to tailor payments to forest owners to the provision of forest ecosystem services. China has not yet ratified the Bonn Convention on migratory species, although it is active in regional co-operation on migratory waterbirds.

◆ ◆ ◆

1. Policy Objectives

Legislation specific to protection of nature and biodiversity includes the 1982 Marine Environmental Protection Law, the 1988 Wildlife Protection Law[1] and the 1994 Regulations on Nature Reserves (Table 2.5). The first provides for preservation of marine habitats. In 1989, the second led to a list of key protected species. The third statute regulates the management of protected areas. In addition, Chapter II of the 1979 Environmental Protection Law contains provisions for protection of the natural environment, including: protection, development and use of aquatic and terrestrial wildlife "in a rational way", taking account of fishing, hunting and forestry regulations;

prevention of soil erosion and damage to ecosystems; and, protection and development of forest and grassland resources. Also key to nature conservation is the 2002 Law on Environmental Impact Assessment, which evolved from the 1986 Administrative Rules of Environmental Protection for Construction Projects.

Adopted as early as 1994, the *National Biodiversity Strategy and Action Plan* outlines broad objectives for the protection of individual species of wildlife and genetic resources, effective management of nature reserve and the "sustainable use of biodiversity". During the same period, other strategies and action plans were developed: China's Agenda 21 (1994), action plans on marine biodiversity (1992), agriculture (1993), urban flora (1994), and forests (1995). Action plans were also developed in the early 1990s (e.g. for the panda, crested ibis and South China tiger).

Quantitative targets for nature conservation were set as part of the *10th FYP* (2001-05). These targets included extending nature reserves to cover 13% of the land area, and marine nature reserves to cover 40 000 km^2, by 2005. Old growth forests were to remain stable, forest cover was to reach 18%, and desertification areas were to decrease by 60% from 2000. Fifteen ecological reserves were to be established at the national level and 40 at the provincial level.

The *"National Programme of Wildlife Conservation and Nature Reserve Development"* set long-term (up to 2050) targets to: expand nature reserves to cover 16% of the land area (155 million hectares) by 2010 and 18% (170 million hectares) by 2050; establish 80 Ramsar sites by 2010; expand forest coverage to 19% by 2010, 24% by 2030 and more than 26% by 2050; and expand forest plantations to 39 million hectares in 2010 and 46 million hectares in 2030.

2.   Species Protection and Recovery

China's size, range of habitats, topography and seven climatic zones support some of the world's richest biodiversity, making China *one of the world's twelve "mega-diverse" countries*. China's diversity of wild plants and animals is greater than that of all of North America or Europe, and is equal to one-eighth of all species on Earth.

2.1   Threatened species

China has 356 threatened (i.e. critically endangered, endangered or vulnerable) animal species[2] (IUCN, 2006), which is 6% of the world total. The *share of threatened species in each taxonomic group is low by both OECD standards* (Figure 6.1) and world standards.[3] Protection of the giant panda is becoming a success (Box 6.1).

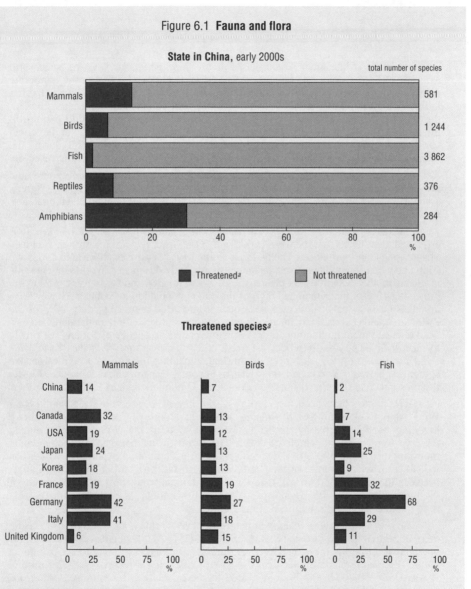

Figure 6.1 **Fauna and flora**

State in China, early 2000s

total number of species

Mammals — 581
Birds — 1 244
Fish — 3 862
Reptiles — 376
Amphibians — 284

■ Threatened*a*   ▢ Not threatened

Threatened species*a*

Mammals
China 14
Canada 32
USA 19
Japan 24
Korea 18
France 19
Germany 42
Italy 41
United Kingdom 6

Birds
China 7
Canada 13
USA 12
Japan 13
Korea 13
France 19
Germany 27
Italy 18
United Kingdom 15

Fish
China 2
Canada 7
USA 14
Japan 25
Korea 9
France 32
Germany 68
Italy 29
United Kingdom 11

a) IUCN categories "critically endangered", "endangered" and "vulnerable" in % of known species.
Source: OECD, Environment Directorate.

Box 6.1 **Protecting the giant panda: a success story**

Protection of the giant panda, listed as an *endangered species*, is a successful example of both species and habitat conservation. Giant pandas live in the temperate forests of the Upper Yangtze, at the elevation of 1 200-3 400 metres. Despite having a digestive system more suited to a carnivorous diet (pandas are classified as bears), they subsist primarily on *bamboo* (10 to 40 kg of bamboo a day per adult panda). In the mid-1970s and mid-1980s, hundreds of pandas died during the large and cyclical bamboo die-offs (138 carcasses were found during the earlier period, 250 during the later). Fortunately many pandas were rescued and transferred to appropriate reserves.

While logging and poaching have declined following China's rural exodus and the 1998 ban on logging in natural forests, panda watching has become a major threat. *Tourism is also exerting increasing pressure* on giant pandas' habitat. Habitat loss attributed to tourist visits has sometimes occurred at high rates within protected areas.

China has been *very proactive in establishing nature reserves for pandas*. It is estimated that China has *between 1 600 and 3 000 wild pandas*, mostly in Sichuan, Shaanxi and Gansu provinces. The first four nature reserves were established in 1963, and China now has 56 panda reserves, including the recently designated World Heritage Site in Sichuan: the "Giant Panda Sanctuaries", home to 300 pandas, covers 9 500 km^2, protecting 45% of the habitat and 60% of the panda population in Sichuan. However, an urgent issue is to better protect pandas outside protected areas (e.g. only 20% of the 2 700 km^2 panda habitat in Sichuan's Pingwu county is protected). In particular, there is a need to create more migration corridors to allow pandas to change elevations seasonally to meet their dietary and reproductive needs. *Close co-operation with the World Wide Fund for Nature* (WWF) has been established, including monitoring activities and an agreement between WWF and the Shaanxi Forestry Department to establish 13 new reserves and create the first habitat corridors in the Qinling mountains (2002).

Giant pandas have been *successfully raised in captivity* (e.g. in the Shaanxi Rare Wild Animals Rescuing and Raising Research Centre). China has 184 captive breeding pandas, including 16 babies born in 2006 in the Wolong Giant Panda Reserve Centre, all of which survived. The dramatic decrease in the baby panda mortality rate (from 100% to 0%) is the result of accumulated experience which has led, for example, to replacing dog milk with high-fat milk and sharing twin babies between the mother and human care-takers (as female pandas often abandon one of the twins). In April 2006, a giant panda raised in captivity was released for the first time into the wilderness, by the Wolong Centre.

The *cost of protecting pandas* in nature reserves is much less than that of raising them in captivity; the difference is on the order of USD 10 000 versus USD 600 000 for one animal during its lifespan. Moreover, tourists' willingness to pay to see pandas in nature reserves is much higher than the cost of operating a reserve (USD 57 million *versus* USD 250 000 in the Wolong Panda Reserve). *Innovative funding mechanisms* have been established to help fund conservation projects in China. For instance, zoo operators abroad must pay USD 2 million to *rent a Chinese panda* for a year. People who want to *name a panda* are charged USD 4 900 for a year or USD 37 000 for the panda's life. Raising pandas in captivity would not be possible without private donations, which currently finance half of the expenditure for the 184 captive pandas.

In absolute terms, however, China is among the countries with a *high number of threatened species, both endemic and non-endemic*, together with Australia, Brazil, Indonesia and Mexico (IUCN, 2004). Between 1996 and 2006, China added 147 threatened animal species, including 9 mammals, 31 fish, 19 reptiles and 90 amphibians, to the IUCN Red List.[4] Multiple Chinese agencies are responsible for the administration and management of protected areas and the management of individual species. For example, the State Forest Administration is responsible for protection of the giant panda and its natural habitat, while the Ministry of Construction is in charge of its ex situ conservation.

Threatened species habitat includes mostly forests and, to a lesser extent, wetlands (Table 6.1). Habitat loss, primarily associated with agriculture and wood extraction, is the most pervasive threat to the majority of threatened species (Table 6.1). Other *key pressures* are use as food, water pollution (primarily from agriculture), and human disturbance via recreation and tourism. This highlights the need to integrate biodiversity considerations into sectoral policies, most notably agricultural, forestry and tourism.

Table 6.1 **Threatened species,**[a] by habitat and threat types

Habitat type	Nb of species	(%)	Threat type	Nb of species[b]
Forest	184	*33*	Habitat loss/degradation	258
Wetland	132	*24*	Agriculture	163
Grassland	41	*7*	wood extraction	126
Artificial terrestrial[c]	40	*7*	Harvesting (hunting/gathering)	151
Shrubland	38	*7*	food: for subsistence and trade	105
Coastline	38	*7*	Pollution	73
Sea	33	*6*	water pollution	44
Artificial aquatic[d]	14	*3*	agriculture	27
Desert	10	*2*	industry	15
Savanna	5	*1*	Human disturbance	52
Caves	2	*0*	recreation and tourism	14
Rocky areas	1	*0*	Intrinsic factors	99
Unknown	14	*3*	restricted range	59
Total	552	*100*		

a) Includes mammals, birds, fish, reptiles, amphibians, invertebrates and plant species. Average over 2000-06.
b) Do not add up, as the same species can be affected by various threat factors.
c) Mainly arable land.
d) Mainly fish ponds.
Source: IUCN Red List of Threatened Species, database.

2.2   Traditional medicine and other uses of endangered species

Traditional Chinese Medicine (TCM) is the most widely practiced traditional medicine in the world. At least 27 animal species are threatened as a consequence of medicinal use (IUCN, 2006). A range of species that are traded from China are associated with TCM (as well as other uses) (Table 6.2). The use of parts of endangered species, such as rhinoceros horn, tiger bone and deer musk in TCM, and consumption of shark fins in soup, has contributed to reducing the numbers of these species to critical levels (Callister, 1995). In many cases the situation is worsening (Table 6.2). The

Table 6.2 **International trade uses of threatened species**

Threatened species	Use	Concern
Musk deer[a]	TCM[b] and perfumes	Population at 3-5% of 1960s levels. Urgent protection needed.
Tiger	bones and derivatives; skins	Export of live tigers increased from 1.4/year in 1995-99 to 13.2/year in 2000-04.
Leopard	bones and derivatives; skins	Exports of derivatives increased from 319 mean units/year in 1995-99 to 507 in 2000-04
Asian Elephant	ivory	Increase in ivory seizures since 1998, linked to China's demand for ivory.
Rhinoceros	horns	Found in 12.5% of 280 businesses investigated in 1996.
Saiga Antelope	horns	Historical large-scale poaching. Re-introduction needed.
Tibetan Antelope	shahtoosh (wool)	Population decreased to 75 000 in 1998: 3 000 poachers arrested, 3 killed.
Bears	gall bladder	Increased enforcement needed.
Pangolin	scales, meat	"Endangered" status in 2000, zero export quota.
Green Sea Turtles	food, shell	Logging and livestock pressures on the only nesting area (Huidong Nature Reserve).
Freshwater Turtles	TCM,[b] Food, shell	Several species possibly extinct due to high demand, including exports to Vietnam.
Seahorses	Whole	24 million harvested/year, almost all from the wild.
Sharks[c]	fins, meat, jaws, oil	China dominates world fin trade; tenfold increase in shark meat imports since 1998.
Snake	TCM,[b] meat, live, skin	Most popular exotic animal for consumption.
Yew[a]	anti-cancer drug	Unsustainable harvest of bark and needles; most species threatened.
Asiatic Ginseng	TCM[b]	No meaningful information on wild growing.

a) 5 species.
b) TCM: Traditional Chinese Medicine.
c) Basking, Whale, Great white species.
Source: UNEP World Conservation Monitoring Centre/CITES trade database; CITES-Elephant Trade Information System; TRAFFIC East Asia (The WWF-IUCN Wildlife Trade Monitoring Network); WWF China; Ramsar Convention Secretariat; Conservation International.

United States is the prime importer (66%) of tiger and leopard products, followed by New Zealand (11%) and Great Britain (6%) (UNEP-WCMC/CITES database). Imports of dried shark fins to China originate mainly from Singapore and Spain.[5]

One positive development is the *TCM labelling system* introduced in 2003 by China's State Forestry Administration and State Administration for Industry and Commerce. Only TCM made from legal wildlife ingredients can be granted the label. TRAFFIC East Asia, China's CITES Scientific Authority, and experts throughout China's traditional medicine community have formed the *Traditional Medicines Advisory Group* to serve as a forum to discuss conservation and sustainability. In 2006, the Advisory Group held a workshop on the clinical effectiveness of various bones, the potential for captive breeding programmes, and the use (versus conservation) of rare animal medicinal resources.

A survey in Hong Kong revealed that the majority (74%) of TCM users would agree to give up certain TCM if it would save wildlife from extinction (Lee, 1998). Similarly, about 65% of "rhino-horn users" and "tiger bone users" would stop if they knew the products were illegal. The study highlights the potential efficacy of *increasing public education and product labelling requirements*, especially if coupled with well-publicised enforcement efforts.

2.3   Invasive alien species

Alien species are *widespread throughout China*,[6] and are found in many ecosystems. Examples include ragweed, water hyacinth and Amazon snails (Xie, 2001). They have either been intentionally introduced for commercial purposes or, as is often the case, have become established incidentally through trade and immigration (Liu, 2005). Damage to the Chinese economy (mainly to agriculture) caused by alien species has been estimated at USD 14.6 billion, or 1.3% of GDP (SEPA, 2005). With increasing trade and transportation comes the increased chance for foreign species to spread and cause long-term ecosystem damage. Freshwater fisheries (aquaculture) and the lawn grass sector have been identified as having a high potential for future risk.

China was one of six countries that participated in the 2000-04 *Global Ballast Water Management Programme* supported by the International Maritime Organisation (IMO), the Global Environment Facility (GEF) and the United Nations Development Programme (UNDP). The port of Dalian (Liaoning) was used as a pilot to inventory biodiversity and control the introduction of invasive species into the marine ecosystem from vessels discharging ballast water. This project contributed to the adoption, in 2004, of the International Convention for the Control and Management of Ships Ballast Water and Sediments (which China has not yet signed).

However, *domestic regulations* to prevent damage to local ecosystems and native species from invasive species[7] are missing. There is no inventory of potential alien species nor a comprehensive tracking system for invasive alien species. However, the Office for the Management of Invasive Alien Species and Centre for the Prevention and Control of Invasive Species were established by the Ministry of Agriculture in 2003. Education and awareness campaigns could also significantly reduce intentional introductions.

3. Habitats and Protected Areas

3.1 Commendable progress

According to preliminary counts published in 1998, China has 212 categories of forests, 36 kinds of bamboo forests, 113 types of shrub lands, 77 types of meadows, 37 marshlands, 55 steppes, 52 deserts, and 17 types of alpine tundra and cushion-like and talus vegetation, making a total of *599 different categories of terrestrial habitats.*[8] China has 16 of 233 globally important eco-regions, three of which (Southeast China subtropical forests, Hainan Island forests, and Central China temperate forests) have been described as "critical or endangered" (Olson, 1998) and should therefore be given priority for future action. Habitat preservation in the East China sea and the Dianchi lake (Yunnan) is also of concern (Chapter 4).

China's *protected areas have dramatically increased* since the early 1980s (Figure 6.2). Inland and marine protected areas[9] cover some 154 million hectares, or about 16% of China's area (Tables 6.3, 6.4). This surpasses the 10th FYP's objective of 13% and is close to the OECD average. Most protected areas are inland, covering close to 151 million hectares. There are 149 marine protected areas, covering 3.7 million hectares. The national protected areas are generally larger than the provincial ones (38% of national protected areas are larger than 50 000 hectares; 66% of provincial areas are under 20 000 hectares). The protected areas at prefecture and county levels are often smaller: 61% of the former and 63% of the latter are under 5 000 hectares. China's national and provincial protected areas account for 90% of the total inland area under protection (national areas accounting for 60% and provincial for 30%).

China's *inland protected area system is well distributed across the territory*, although the less densely populated provinces have much larger areas protected (Table 6.3). In particular, the seven western provinces of Tibet, Qinghai, Xinjiang, Gansu, Inner Mongolia, Sichuan and Yunnan account for 80% of China's total inland area under protection. Significant land protection efforts have also been made in the

Figure 6.2 **Protected areas**[a]

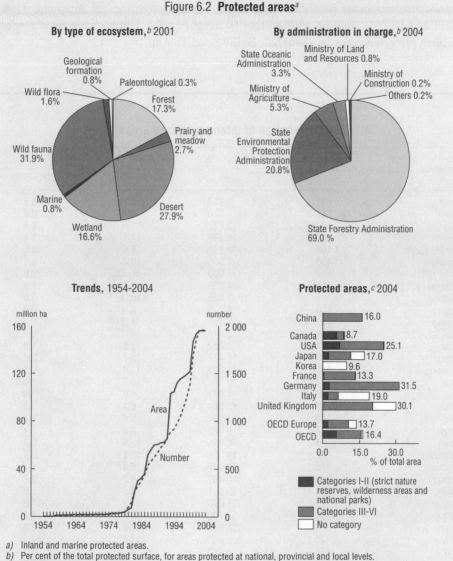

By type of ecosystem,[b] 2001

- Geological formation 0.8%
- Paleontological 0.3%
- Wild flora 1.6%
- Forest 17.3%
- Prairy and meadow 2.7%
- Wild fauna 31.9%
- Marine 0.8%
- Desert 27.9%
- Wetland 16.6%

By administration in charge,[b] 2004

- State Oceanic Administration 3.3%
- Ministry of Land and Resources 0.8%
- Ministry of Construction 0.2%
- Ministry of Agriculture 5.3%
- Others 0.2%
- State Environmental Protection Administration 20.8%
- State Forestry Administration 69.0 %

Trends, 1954-2004

million ha / number

Area
Number

1954 1964 1974 1984 1994 2004

Protected areas,[c] 2004

	% of total area
China	16.0
Canada	8.7
USA	25.1
Japan	17.0
Korea	9.6
France	13.3
Germany	31.5
Italy	19.0
United Kingdom	30.1
OECD Europe	13.7
OECD	16.4

0.0   15.0   30.0
% of total area

- ■ Categories I-II (strict nature reserves, wilderness areas and national parks)
- Categories III-VI
- □ No category

a) Inland and marine protected areas.
b) Per cent of the total protected surface, for areas protected at national, provincial and local levels.
c) IUCN management categories I-VI and protected areas without IUCN category assignment. National classifications may differ.
Source: IUCN; SEPA; Chinese Academy of Sciences.

northern provinces of Jilin and Heilongjiang, as well as on the Hainan island. In contrast, the extent of protected area in the eastern provinces of Zhejiang, Hebei, Anhui and Jiangsu amounts to less than 10 m² per capita. Efforts have been made to establish protected areas in Shanghai, Beijing and Tianjin.

Table 6.3 **Distribution of inland protected areas,** 2004

Provinces[a]	Area (1 000 km²)	Protected areas[b]		
		(km²)	(% of land area)	(ha/'000 inhab.)
Beijing	17	953	5.7	0.6
Tianjin	11	920	8.1	0.9
Hebei	190	2 832	1.5	0.4
Shanxi	150	10 696	7.1	3.2
Inner Mongolia	1 198	102 237	8.5	43.0
Liaoning	150	27 565	18.4	6.5
Jilin	187	78 060	41.7	28.8
Heilongjiang	460	41 730	9.1	10.9
Shanghai	6	1 026	16.3	0.6
Jiangsu	103	6 805	6.6	0.9
Zhejiang	102	1 789	1.8	0.4
Anhui	140	5 544	4.0	0.9
Fujian	121	4 252	3.5	1.2
Jiangxi	167	8 432	5.1	2.0
Shandong	157	21 002	13.4	2.3
Henan	167	9 742	5.8	1.0
Hubei	186	11 077	6.0	1.8
Hunan	212	13 364	6.3	2.0
Guangdong	180	11 900	6.6	1.4
Guangxi	237	6 925	2.9	1.4
Hainan	34	6 735	19.9	8.2
Chongqing	82	10 006	12.1	3.2
Sichuan	485	79 546	16.4	9.1
Guizhou	170	10 497	6.2	2.7
Yunnan	394	35 101	8.9	7.9
Tibet	1 275	435 673	34.2	1 613.6
Shaanxi	206	8 700	4.2	2.3
Gansu	390	138 646	35.6	52.9
Qinghai	720	207 639	28.8	384.5
Ningxia	63	4 193	6.7	7.1
Xinjiang	1 656	202 726	12.2	103.4
Total	9 614	1 506 313	15.7	11.6

a) Province, administered municipality or autonomous region.
b) Of which national (60%), provincial (29%), prefecture (4%) and county (7%).
Source: National Bureau of Statistics of China; Chinese Academy of Sciences.

China's *marine protected areas* are intended to help protect coastlines, river mouths and islands with historic and scientific value, as well as rare and endangered marine animals (dolphin, dugong, turtles) and their habitats (mangroves, coral reefs and coastal wetlands).

The largest natural ecosystems covered by China's protected area system are *deserts, forests and wetlands*, whilst prairies and meadows as well as ocean and coastal ecosystems are under-represented (Figure 6.2). Large areas have been allocated for wildlife reserves. The largest national protected areas are Qiangtang, established in 1993 in Tibet on nearly 30 million hectares, and Sanjiangyuan, established in 2000 in Qinghai province on some 15 million hectares. Both are grassland ecosystems above 5 000 metres, and are home to the Tibetan antelope and the snow leopard. Sanjiangyuan includes the source of the Yellow river.

Table 6.4 Distribution of marine protected areas, 2005

	Coastal province*a*			Protected areas (number)			
Name	Type*b*	Coastal waters	National	Provincial	Prefectural	Total	Total (km²)
Liaoning	P	Bohai sea	5	1	9	15	14 034
Tianjin	M	Bohai sea	1	1	–	2	1 432
Hebei	P	Bohai sea	1	3	1	5	501
Shandong	P	Bohai sea/Yellow sea	4	5	5	14	4 650
Jiangsu	P	Yellow sea	2	1	1	4	5 039
Shanghai	M	Yellow sea	2	2	–	4	938
Zhejiang	P	East China sea	1	2	2	5	1 405
Fujian	P	East China sea	3	7	4	14	1 533
Guangdong	P	South China sea	4	8	48	60	5 986
Guangxi	AR	South China sea	3	1	–	4	756
Hainan	P	South China sea	4	7	11	22	1 310
Total number			30	38	81	149	37 584
(%)			*20*	*26*	*54*	*100*	

a) Ranked from North to South.
b) M: centrally administered municipality; P: province; AR: autonomous region.
Source: State Oceanic Administration.

3.2 Management implications

The massive increase in number and area of nature reserves in the last two decades has *not been matched by a corresponding enhancement of management inputs and capabilities*. This has resulted in insufficient staff and funding across the entire protected area system, despite increased public expenditure on nature conservation since 1998 (Li, 2003). While site design of protected areas has been generally satisfactory, subsequent management efforts (threat prevention, site restoration, wildlife management, involvement of the resident community, visitor management, monitoring, training and research) have been somewhat "diluted", and most protected areas are considered to have low protection status.[10]

According to the *1994 Regulations on Nature Reserves*, protected areas should be managed as category Ia ("strict nature reserves") and should be divided into a core zone, buffer zone and experimental zone. No people or buildings should be allowed in the core zone, and no activities other than tourism in the nature reserves. A *draft law* providing a modern legal framework for nature protection, with implications for the designation, planning and management of protected areas and for the communities living in and around them, is currently being discussed. The State Environmental Protection Administration (SEPA) and the State Forestry Administration (SFA) are both contributing to the drafting of the law, which would replace the 1994 Regulations.

Nearly 70% of China's protected area falls under the supervision of SFA, while 20% is under SEPA (Figure 6.2). Most (around 90%) of the two agencies' protection activities are at the national or provincial level. The Ministry of Agriculture is also quite active at the provincial level, with 13% of the total area protected. The Ministry of Water Resources generally protects small areas and divides its activities equally among provincial, municipal and county levels. Although most protected areas are under the jurisdiction of SFA, *SEPA is responsible for overall co-ordination* (Liu, 2003).

However, various ministries and administrations currently manage nature reserves "within their scope of official duty", with mandates, responsibilities, regulations and policies that may or may not support biodiversity conservation. In 2005, SEPA conducted investigations in national and provincial reserves and found 260 *illegal projects* (mining, hydropower, construction, tourism) in seven reserves, six national and one provincial. Even though such infractions can lead to cancellation of a reserve's status, pursuant to provisions in the 1998 State Council Gazette Notice, the very low fines[11] provided for by the 1994 Regulations on Nature Reserves are little deterrent. The forthcoming law on protected areas may address this issue.

Most of China's protected areas, including the national areas, are managed and funded at the provincial, municipal or county level. The central government provides only *limited financial support* for nature conservation. In 2000, public expenditure (both investment and current) for national protected areas was estimated at only USD 1.13/hectare, and at USD 0.53/hectare for other (local) protected areas (Han, 2000). In a few cases, economic incentives have been granted to residents who agree to move out of the protected area with a view to enhancing biodiversity protection (e.g. to protect giant pandas in the Wolong Nature Reserve). But the protected areas designated at the municipal and county level are inadequately funded (Xie, 2004), which often leads the reserve staff to exploit "protected" resources for economic gain. It is estimated that 80% of the agencies that manage China's nature reserves have engaged in profit-oriented activities, with considerable damage to "protected" areas (Jim, 2003).

China's authorities need to confront several key questions on urgent basis, if the country's environmental goals are to be achieved with broad citizen support. These are: i) *whether the reserves should exclude traditional resource-tapping activities*; ii) whether reserve management bodies should be allowed to earn income from the reserves, i.e. whether they should be the "umpire" or a "player" in the utilisation of natural resources found in the reserves; and iii) whether an integrated management structure should supersede the present compartmentalised arrangement (Jim, 2003).

4.   Integrating Biodiversity Concerns into the Forestry Sector

4.1   State of forest resources

While the world's forest area continues to shrink, *China's forest area is increasing* and now ranks fifth in the world; it covers nearly 200 million hectares, or more than 20% of China's land area (Table 6.5). This more than meets the 10th FYP goal of 18% but is based on a looser definition of forest area.[12] The steady increase since 1990 is partly due to large-scale afforestation efforts. Chinese forests are rich in biodiversity (2 500 native tree species), and the area of primary forest (12 million hectares) has remained stable since 1990, in accordance with the 10th FYP target.

China's Forest Law, as revised in 1998, defines *five classes of forests*: i) *protection forests*, which enhance water storage, stabilise river banks, prevent soil erosion (from wind and water) and desertification, shelter farmland and grassland, and line roadways; ii) *production forests*, whose primary function is to provide wood products; iii) *economic forests*, which provide non-wood products such as foodstuffs and medicinal and industrial raw materials; iv) *fuelwood forests*, which are designated

Table 6.5 **Chinese forestry at a glance**

	Units	1990	2005	Annual change rate (ACR) 1990-2005	Annual change rate (ACR) 2000-05	Comments
FOREST RESOURCES						
World	million ha	891	831	−0.46	−0.50[a]	ACR in %[a]
China	million ha	157	197	1.5	2.2	ACR in %[a]
forest plantations	million ha	18	31	1.3	1.5	ACR in million ha/year
FOREST ECONOMY						
Growing stock	billion m$^{3c,\ b}$	..	13	0.03	−0.52[c]	ACR in m^{3c}/ha/year
Wood removal		159	135[d]	−1.1	−1.4	only 1% of growing stock
	million m$^{3c,\ b}$					(2005)
Non-wood removal						
medicines	000 tonnes	..	11			animal products (live, skin…)
exudates	000 tonnes	..	1 131			not documented
Value total removals	USD billion	..	5			equivalent to USD 25/ha
Employment	million people[e]	2.5	2[f]			
FOREST TYPE						
Primary forest	%	..	6			2 500 native tree species,
Modified natural forest						but 175 are threatened
	%	..	58			(IUCN, 2004).
Semi-natural forest	%	..	20			29% (64%) of growing stock
Forest plantation	%	..				composed of 3 (10) species
production	%	..	15			
protection	%	..	1			
Total	%	..	100			
Primary forest	million ha	11.6	11.6			
FOREST FUNCTION						World average:
Production	%	..	58			34
Protection	%	..	31			9
Biodiversity conservation[g]	%	..	3			11
Social services	%	..	1			4
Multipurpose	%	..	7			34
Unknown	%	..	0			8
Total	%	..	100			100
FOREST HEALTH AND VITALITY[g]						
Insects	million ha	..	6			fires affect relatively small areas
Diseases	million ha	..	1			(50 000 ha in 2000)

a) % change of the remaining forest area each year within the period.
b) Over bark.
c) Reduced productivity of the growing stock, which was already low (67 m³/ha in 2005 compared with world average of 110 m³/ha).
d) 66% industrial roundwood; 34% fuelwood.
e) Excluding wood processing and informal employment.
f) In 2000.
g) Includes, but is not limited to, protected areas.
Source: FAO.

production areas for fuelwood; and v) *forests for special uses*, whose functions are either national defence, research, seed procurement, biodiversity conservation, historical interest or aesthetic value.

In addition to implementing China's forestry policy, the State Forestry Administration manages most protected areas in China including forestland, deserts and wetlands. However, *less than 3%* (5.3 million hectares) *of the country's forest area has been designated for biodiversity conservation* (Table 6.5). The other 17.2 million hectares of forestland that has been included in the protected area system consists of other wooded land.[13] Many of China's native tree species (175 species) are threatened (IUCN, 2004). A small number of common species make up the bulk of the growing stock, resulting in a lack of diversity in forest stands.

4.2   Six key forestry initiatives

Following the major drought and flood disasters in the late 1990s,[14] China's central government decided to *enhance the ecological functions of forests* and to significantly increase the funding of forest policy. In 2000 the State Council launched six key forestry initiatives "to combat desertification and soil erosion while saving endangered species" and "to eradicate major ecological plagues while providing sustainable economic growth". The resulting six programmes cover the period 2001-15 as follows.

Protecting natural forests

China launched major forest programmes in the mid-1990s to protect river catchments and, in particular to rehabilitate the forest ecosystems in the Yangtze and Pearl river deltas, along the Liao river valley, and in the middle reaches of the Yellow river (loess plateau). To benefit from the soil and water conservation functions provided by forests, it was decided to ban logging in natural forests and to limit harvesting in severely degraded watersheds. The *Natural Forest Protection Programme* (NFPP), initiated in 1998, introduced a total ban on logging in the upper Yangtze river basin and mid-to-upper Yellow river basin, and reduced logging in state-owned forests.[15] According to an SFA review of the NFPP in 2003, the logging ban was extended to 27 million hectares of collective forests,[16] accounting for 30% of all collective forests in China, with no compensation for economic losses. The NFPP has also funded the reforestation of 14.6 million hectares of forest and grassland since 2000, and has helped pay the pensions of retired employees of state forestry enterprises,[17] for which it has been allocated an annual budget of USD 1.2 billion (80% from state funds, 20% from local funds). NFPP funds have also been used to help resettle 502 000 loggers away from state-owned forests.

Constructing shelter-forests

Shelterbelt development programmes, some of which were initiated in the late 1970s, have been a major component of Chinese forest policy, in terms of scope and funding. These have included the Three-North programme (building a shelterbelt system stretching some 4 500 km to combat desertification across the north of China), the Yangtze river programme (to protect headwaters), the Coastal programme (to establish windbreaks on 25 million hectares along the coast), and the Farmland programme (to develop agro-forestry in plain areas). These programmes have been united to form the *Key Shelterbelt Development Programme*. By 2010, 3.8 million hectares of forests are expected to be planted to form a "Green Great Wall" in North, Northwest and Northeast China; along the middle reaches of the Yellow river; and in Beijing and its surrounding area.

Turning marginal cropland into forests

Another key measure that uses the soil and water conservation functions of forests, the "grain for green" or Sloping Land Conversion programme, has promoted farm forestry in hilly agricultural areas (gradient of 25% or above). After a pilot phase in Gansu and Sichuan in 1999, the *programme* was broadened to 20 provinces by 2001. The basic idea is to convert cropland on steep slopes prone to erosion to forestland and grassland, "compensating" farmers annually with 100-150 kilograms of grain and CNY 20 in cash for each mu[18] retired. Compensation is granted for two years for land returned to pasture, for five years for land converted to "economic forests" and for eight years for land converted to "protection forests". The programme's costs are borne mainly by the central government; in 2003 they involved annual budgetary transfers of around CNY 15 billion and comprised 20% of the government's total agricultural support payments to Chinese farmers[19] (OECD, 2005). More than 16 million hectares of cropland have been converted under this programme. While Chinese law prohibits converting (designated) forestland to any other type of land use, measures should be taken to avoid the return of such "farm forests" to cultivation when the payments cease. Such measures could include payments for ecosystem services.

Controlling sandstorms around the capital

Afforestation of the Taihang mountains (west of Beijing) and restoration of sand dunes, both major forestry programmes initiated in the early 1990s, have been regrouped under the *Sandstorm Control Programme*. Over two million hectares have been planted as part of this programme (SEPA, 2005). Sandstorm control is an area with active international co-operation (Chapter 9).

Creating new wildlife reserves

As of 2002, *325 reserves for wild fauna* had been established on 41 million hectares and *111 reserves for wild flora* on two million hectares. All protected areas in China are state-owned, involving public funding for investment and current expenditure. In 2001-04, USD 97 million was spent on infrastructure in the overall protected area system (SEPA, 2005). Extra revenues are expected from the booming (domestic and international) tourism industry.

Developing fast-growing plantations as a base for the forest industry

Since 1985 China has had a *policy to preserve domestic forest resources* through the setting of harvest quotas (annual allowable cut) for five-year periods. The quota for 2001-05 (45 million m^3 a year) was more stringent than the quotas for 1991-95 (49 million m^3 a year) and 1996-2000 (53 million m^3). As a result, wood removals have been decreasing and the intensity of timber use (harvest as a per cent of annual growth) is now very low[20] (1%) (Table 6.5). Despite the quota policy, however, the growing stock of Chinese forests is low[21] and decreasing (Table 6.5). This suggests that a substantial amount of illegal logging might be occurring in China (Nilsson, 2004).

The rate of timber and pulp production is clearly below expanding domestic demand. The gap between consumption and domestically produced forest products (expressed in roundwood equivalent) was 106 million m^3 in 2002 and is likely to be in the magnitude of 150-175 million m^3 by 2010 (Nilsson, 2004). To reduce pressure on its forest resources, China has embarked on a *very ambitious forest plantation programme* focusing on development of the domestic forest industry, especially the pulp industry. The target is to establish 13.4 million hectares of new plantations during the period 2001-15. The plantations would be for pulpwood (5.9 million hectares), logs for panels (5.0 million hectares) and large timber (2.5 million hectares). There is concern, however, that China's plantations will not be cost-competitive with plantations in many other producing countries due to frequent cases of shallow and nutrient-poor soils, water shortage and locations far from industrial sites (Barr, 2004). Nearly 30% of China's forest plantations are affected by insects and diseases (Table 6.5).

China's internal policies, such as the ban on harvesting trees from natural forests, coupled with a burgeoning wood manufacturing sector, have led to an *increase in imported wood*, including both legal and illegal supplies of hardwood, softwood and logs, lumber, panels and waste paper, to the extent that China has been identified as the major market for most illegally logged material (Cohen, 2006). To avoid such consequences and to promote sustainable development abroad, China should encourage supplier countries to support sustainable forest practices, including through using a certification system to assure consumers that the materials used in finished wood products came from sustainable forests. Recent steps have been taken in that direction (Chapter 9).

4.3   Payments for ecosystem services

While one of the main (publicised) achievements of the six key forestry initiatives has been extensive afforestation,[22] new emphasis has been given to providing economic incentives as "compensation" for local communities to maintain and improve existing forests. In 2000, new regulations under the Forest Law suggested that 30% of forests be "set aside for public benefit", although there was no legal obligation for provinces to comply. A *forest ecosystem compensation programme* (FECP), set up to support the objective,[23] has so far been implemented on 13 million hectares, mostly (64%) collective forestland, spread over 11 provinces, mostly in the east of China.[24] In many other provinces, forestry bureaus have identified areas of "public benefit", jockeying for inclusion in the programme.

However, participation rate in the FECP remains low, particularly in collective forest areas. With an annual budget of USD 121 million, the programme can only provide an *average level of compensation* of USD 9 per hectare, often less than foregone income from forest management (Miao, 2004). More generally, the cost of inaction is much larger than total government outlays for environmental protection. China's total economic loss resulting from conversion of natural ecosystems has been estimated at 2% of GDP and losses due to deforestation at 12% of GDP (Smil, 1998).

The FECP lacks *clear environmental objectives* for soil and water conservation, biodiversity protection, carbon sequestration, recreation, and provision of non-wood products (Sun, 2002). It would be more cost efficient (and probably more environmentally effective) to pay forest owners (or contractors) for the provision of environmental services, thereby applying the beneficiary-pays principle (e.g. water companies could reduce their cost of purifying water by paying forest owners for protecting drinking water sources).

China could, as appropriate, authorise the seeking of economic returns without compromising the provision of ecological services, in the context of (certified) *sustainable forest management*. Reducing the tax burden on forestry would increase the incentives for sustainable forest management, particularly in collective forests. One positive development was the abolition, in January 2006, of the tax (introduced in 1983) on "special agricultural products", which applied to harvested timber[25] (at a rate of 8%). However, forest owners must still contribute to local social welfare funds as well as paying various provincial charges specific to forest management.[26] Such fees and charges are often determined arbitrarily by local authorities (Tian, 2003), thereby encouraging illegal logging or the selling of (low-value added) fuel wood. Charges and fees should be designed in a transparent way, and rates set to cover the costs of providing services to forest owners (e.g. pest control, fire protection) or as part of social policy.

5.   International Co-operation

5.1   Bilateral level

China's bilateral co-operation on nature and wildlife conservation extends back to the mid-1980s when the Ministry of Forestry entered into an agreement with the *United States Department of Interior*, called the Protocol on Co-operation and Exchanges in the Field of Conservation of Nature. Subsequently extended at five-year intervals, the protocol has provided a framework for Chinese and US. counterparts to collaborate on wildlife conservation, habitat management and environmental education projects in China. Nearly a hundred scientific exchanges have been carried out. China has also been co-operating with *Japan and Australia* under an agreement on the protection of migrating birds and their habitats, and in 1996 designated three major wetlands for habitat and species protection, including one on the Chongming Island in Shanghai. Formal bilateral agreements with *Russia and Mongolia* include co-operation on nature reserves, and an agreement with *Italy* involves ecological investigations in China.

5.2   Regional level

At the regional level, habitat protection and wildlife conservation issues have been discussed within the framework of the *Tripartite Environment Ministers Meeting*[27] (TEMM) and "10 + 3" forums (also involving the Association of Southeast Asian Nations[28]), and constitute components of the regional seas action plans for China's offshore waters. In 2003 a Senior Ministers meeting within the North-East Asia Sub-regional Programme on Environment Co-operation[29] (NEASPEC) approved new collaborative work in the region on nature conservation (e.g. conservation and recovery of large mammals and threatened species; conservation, monitoring and co-operative research on important migratory species). The initial effort is focussing on development of a sub-regional action plan for the conservation of threatened large feline mammals and migratory birds. Chinese experts are also participating in a new regional initiative involving 15 countries, to maintain wetlands in the Himalayan Tibetan plateau.

China has also joined with *Vietnam, Laos, Cambodia, Myanmar and Thailand* to launch a biodiversity protection corridor programme under the larger programme of environmental co-operation in the Greater Mekong sub-region. China's plans to *erect a series of dams on the upper reaches of the Mekong river* for flood control, electricity and barge traffic could complicate future environmental co-operation. China's downstream neighbours are already expressing concerns about altered water flows, suggesting the need for early consultation, information sharing and negotiation.

The Tumen river forms a boundary for China, Russia and North Korea, with implications for Mongolia and the Republic of Korea. Degradation of the watershed led the three boundary countries to jointly develop the *Tumen river basin environmental preservation project*, implemented as a sub-project of the environmental segment of the UNDP Tumen Area development project, which began in 1995. In 2002 the three governments signed a strategic action plan covering 2002-12, which is being implemented with UNDP funding of USD 5.2 million from the GEF. Since 2003, a GEF-sponsored Yellow sea large scale marine ecosystem project has engaged experts from China and the Republic of Korea in exploring strategies to better protect the Yellow sea's exceptional biodiversity.

5.3   Global level

China is an *active participant in key worldwide multilateral agreements*, including the Convention on Biological Diversity (CBD), to which China has been a party since 1993,[30] the Ramsar Convention on Wetlands, and the Convention on International Trade in Endangered Species (CITES). A growing number of protected areas have received *international recognition*, either as biosphere reserves within the framework of UNESCO's Man and the Biosphere Programme or as UNESCO World Heritage Sites (Boxes 6.2, 6.3).

Although China has 272 known species of waterbirds, it has acceded to but not signed the *Bonn Convention on migratory birds*. In addition to its bilateral co-operation on migratory birds with Australia and Japan, China is a member of the Asia-Pacific Migratory Waterbirds Conservation Committee (MWCC) along with Australia, Japan, India, Russia, Indonesia and the United States. In 2002, China actively contributed to gaining approval from the 8th Conference of Parties of the Ramsar Convention for the Asia-Pacific Migratory Waterbird Conservation Strategy: 2001-05. Based on this strategy, the East Asia Australasian Shorebird Network, the North East Asian Crane Site Programme, and the East Asian Anatidae Site Network were established, with Chinese experts playing active roles. China is also a party to the Memorandum of Understanding concerning conservation of the Siberian Crane, together with Azerbaijan, India, Iran, Kazakhstan, Pakistan, Russia, Turkmenistan and Uzbekistan.

Box 6.2 International recognition of China's nature protection efforts

China has *26 internationally designated reserves under the Man and the Biosphere (MAB) Programme.*[a] These are areas which simultaneously serve an ecosystem or species conservation function, an economic or human development function and a logistic function (supporting research, monitoring or education). Each reserve typically consists of a central core area (which requires legal protection), a buffer zone (to protect the core zone) and a transition zone.

China's MAB reserves *cover 6.4 million hectares* and include mountain ecosystems (2.9 ha), grassland (1.8), forest (1.4) and aquatic (0.3) ecosystems. The two grassland reserves are large (covering more than 700 000 ha). The seven mountain reserves are all larger than 100 000 hectares, up to 1.8 million ha for the recently (2004) established Qomolangma reserve (in Tibet next to Mount Everest). The 14 forest reserves cover a range of ecosystems, from boreal to tropical. The three coastal MAB sites include the Yancheng reserve (in Jiangsu province), the largest coastal wetland in China, with high biodiversity,[b] an area of 280 000 hectares and a coastline of over 580 km.

China is one of 137 states that have signed the World Heritage Convention pledging to protect their natural and cultural heritage. Since 1987, no less than 33 Chinese sites (spread across 19 provinces and Macao) have been inscribed on the *World Heritage List by UNESCO.* Pursuant to the World Heritage Convention, China has implemented protection measures to address the threats from tourism, invasive species, non-sustainable use, fire, erosion and natural disasters (e.g. earthquakes). None of the sites are included on UNESCO's "Danger List".[c] Most (24) sites are in the "cultural" category, five are in the "natural" and four are "mixed". Most recently, natural heritage sites were included to protect biodiversity in Yunnan (in 2003) (Box 6.3) and to provide sanctuaries for the giant panda in Sichuan (in 2006) (Box 6.1).

---

a) Three are wholly or partly World Heritage properties and three others are wholly or partly Ramsar sites.
b) Also designated as a Ramsar wetland, it is home to 50% of the world population of the red-crowned crane, the largest over-wintering population of that species in the world.
c) Sites where major corrective action is necessary and for which assistance has been requested under the convention.

Wetlands of international importance (Ramsar)

China has approximately *25 million hectares of wetlands* (marshes, swamps, lagoons, deltas, lakes, riverine and coastal areas), covering about 2.5% of its territory; 80% of the wetlands are freshwater, with lakes and marshes predominating. China's wetland area has shrunk over time. In Hainan, for example, mangroves have been reduced 50% from their original coverage and 80% of the coral reefs have been damaged (Zou, 2003).

Box 6.3 **Nature and biodiversity conservation in Yunnan**

The protected area of *the Three Parallel Rivers* (3PR) covers 1.7 million hectares in the mountainous region of Yunnan and hosts the headwaters of the Yangtze, Mekong and Salween rivers. This protected area has outstanding value for displaying the geological history of the last 50 million years and a wide array of scenic landforms and rock types including gorges, glaciers, and alpine karst formations. The 5 980-metre vertical scale of the landscape, from gorge bottom (760 metres) to mountain top (6 740 metres) provides the backdrop for climate, landscape and biological interactions (and isolation). The area's ice-free status during the Pleistocene glaciation periods provided refuge for many plant and animal species which are now endemic to the area. The area is inhabited by 280 000 people comprising sixteen different ethnic groups (e.g., Tibetan, Naxi, Bai, Lisu, Nu, Pumi, Dulong) with different languages, beliefs and customs, occupying traditional housing and practicing subsistence agriculture..

Although the 3PR protected area covers only 0.2% of China's territory, and 3% of Yunnan's province, it is *one of the most biologically diverse areas in the world* and was inscribed as a World Heritage Site in 2003. It includes between 12 and 28% of China's animal biodiversity (excluding fish). The endemism of the area ranges from 5% for birds (which can fly across deep gorges) to 69% for amphibians; 45% of its mammals, reptiles and fish species are found only here. The area is home to many rare and endangered animal species including the giant panda, the red panda, the Yunnan snub-nosed monkey, the Chinese screw-mole, three species of leopard, the Bengali tiger and the black-necked crane. Similarly, the site contains representatives of about 20% of China's higher plants. About 8.5% of China's rare and endangered plants are also present in the area.

The 3PR protected area encompasses *15 different protected areas, with differing degrees of protective status*, including national and provincial nature reserves, national scenic areas and smaller areas. Conservation of the area was previously under the control of seven provincial and four local administrations. In 1995, a 3PR management office and management committee were set up to help co-ordinate activities. More recently (at the time of World Heritage submission in 2002), an IUCN technical evaluation highlighted the varying levels of protection and administration as a concern that would need to be revisited in the future.

Since China signed the Ramsar Convention in 1992, it has designated 30 wetlands of international importance in 13 provinces,[31] ranging in size from 400 to 740 000 hectares (one site in Inner Mongolia). China's t*otal Ramsar area is 2.9 million hectares*. SFA, the main agency responsible for implementing Ramsar, was recognised by Ramsar in 2004 with the first Wetlands International Global

Recognition for Wetlands Conservation and Wise Use Award. Nine sites were designated in 2004: eight high altitude marshes and lakes in Tibet, Qinghai, and Yunnan (totalling 262 000 ha) and one marine wetland in Liaoning (128 000 ha). A third of China's Ramsar sites protect marine or estuarine habitats. Out of a total of nine sites larger than 100 000 hectares, six are in the coastal provinces of Heilongjiang, Jilin, Liaoning and Jiangsu.

The Global Environment Facility

Since 1991 the GEF has facilitated biodiversity and nature conservation in China by financing ten multilateral co-operation projects, totalling *USD 141 million* (of which 40% grants and 60% co-financing). Most (87%) of the funding was directed at the protection and sustainable use of forest ecosystems and wetlands and at improving the management of nature reserves.

CITES

China has been active in supporting CITES initiatives in East Asia through its CITES China centre (Chapter 9). By becoming a *signatory to CITES* in 1981, China (along with the 168 other convention parties) committed itself to ensuring that international trade in wild animals and plants does not threaten their survival. New regulations on trade in endangered species of wild fauna and flora came into effect on 1 September 2006, which will directly support China's CITES obligations by: i) banning the import and export of endangered wild animals and plants and products made from them for commercial purposes; ii) requiring government approval to import or export endangered wildlife for special purposes (e.g. scientific research, artificial naturalisation, propagation and cultural exchange); and iii) prohibiting exports of valuable unnamed and newly discovered wild animals and plants and products containing them.

China has *20% of the world's animal species*[32] listed in CITES Appendix I (i.e. species that are considered to be threatened with extinction) and 9% of those listed in Appendix II (i.e. species that need trade controls to ensure survival). Trade and use as food is a major threat for 105 threatened species found in China (Table 6.1), as is trade and use as traditional medicine (Table 6.2). There is critical need for enhanced surveillance (Chapter 9).

Notes

1. The law provides for the death penalty for significant damage to wildlife resources.
2. Including mammals, birds, fish, reptiles and amphibians; excluding 34 threatened species in Hong Kong. Reliable data are not available for vascular plants.
3. The world average is respectively 20% of described species for mammals, 12% for birds, 4% for fish, 4% for reptiles and 31% for amphibians (IUCN, 2006).
4. Since 1996, two bird species have been withdrawn from the 2006 Red List: the short-tailed albatross and the crested ibis, for which recovery measures have been successful, involving co-operation with Japan for the latter (Box 9.2).
5. Total imports decreased from 4 500 tonnes in 1996 to 3 500 tonnes in 2002, following a dramatic decline of imports from Japan.
6. They are established in every Chinese province, municipality and autonomous region (SEPA, 2005).
7. Chinese legislation only provides for quarantine of diseases, pests and weeds.
8. There are no similar data on the categories of freshwater and marine habitats.
9. Commonly called "nature reserves". SEPA and the Chinese Academy of Sciences are producing good quality registers and related documents.
10. 96% of inland areas and 97% of marine areas would be in IUCN category V (protected landscape); the rest would be in category VI (protected area with managed resources).
11. CNY 300 to 10 000 for those conducting illegal economic activities and CNY 300 to 3 000 for the responsible agency, for inadequate supervision.
12. The FAO Global Forest Resources Assessment defines a forest area as trees taller than 5 metres and a canopy cover of more than 10%. China's requirement for classification as a forest area is currently 20% canopy coverage; before 1996 it was 30%.
13. The terms forestland and forest area are distinct in China. Forestland encompasses: i) forested land; ii) prospective forest areas; and iii) designated forestland yet to achieve the minimum requirements of a forest. In 1998 China's forestland was about 257 million hectares (Guangping Miao, 2004).
14. In 1997 the Yellow river failed to reach the sea for 267 days, causing great economic losses to industry and agriculture. In 1998 devastating floods in the South (causing over 3 000 deaths and great economic losses) were attributed to damaged natural forests in watersheds of the upper Yangtze river.
15. The NFPP involves 13 provinces, most particularly Sichuan, Yunnan and Shaanxi.
16. Collective forests account for almost 60% of China's total forest area; the remaining 40% are state forests.
17. China's 135 state-run forest enterprises employ one million workers.
18. 1 mu = 1/15 of 1 ha.
19. Other direct payments are based on farm income (50%) and input use (30%), the latter being the most distorting form of support in terms of environmental impact.

20. There are large uncertainties in the data on wood removal for fuelwood consumption.

21. In 2005 the average productivity of Chinese forests was 67 m^3/hectare, compared with a world average of 110 m^3/hectare (FAO, 2005).

22. The increase in forested areas was higher in 2000-05 than in earlier periods (FAO, 2005).

23. While the six key forestry initiatives are primarily led by the SFA, the FECP is directed principally by the Ministry of Finance, with other ministries and agencies participating.

24. China's forest area is primarily (74%) concentrated in the Northeast and Southwest.

25. The tax was partly designed to discourage farmers from shifting their resources away from grain production; it came in addition to other agricultural taxes.

26. The People's Congress may institute charges at provincial level. Township and villages may institute fees for purposes such as rural education, family planning and road construction.

27. Regrouping environment ministers of the Republic of Korea, China and Japan.

28. The ten ASEAN member countries are (in accession order): Indonesia, Malaysia, Philippines, Singapore, Thailand, Brunei Darussalam, Vietnam, Laos, Myanmar and Cambodia.

29. An inter-governmental organisation of six countries of the region (China, Japan, Mongolia, North Korea, the Republic of Korea and Russia).

30. China submitted its Third National Report on Implementation of the CBD in 2005.

31. Plus Shanghai and Hong Kong new territories.

32. Including 105 species of mammals, birds, fish, reptiles, amphibians and invertebrates. The world total is approximate because there are no agreed lists for some of the higher taxa.

Selected Sources

The government documents, OECD documents and other documents used as sources for this chapter included the following. Also see list of websites at the end of this report.

Barr, C. and C. Cossalter (2004), "China's Development of a Plantation-Based Wood Pulp Industry: Government Policies: Financial Incentives and Investment Trends", *International Forestry Review*, Vol. 6, No. 3-4, December 2004.

Cai, Lei, Zhidi Yu, Jie Wang and Dehui Wang (蔡蕾，于之的，王捷，王德辉) (2003), 中国防止外来入侵物种的现状与管理评估 (Control Alien Invasive Species to Conserve Biodiversity), 国家环保总局中国履行《生物多样性公约》工作协调组办公室 (Coordination Office on Fulfilling Convention on Biological Diversity in China, SEPA), 环境保护, 2003.8 (Journal of Environmental Protection, 2003.8), Beijing.

Callister, D.J. and T. Blythewood (1995), "Of Tiger Treatments and Rhino Remedies: Trade in Endangered Species Medicines in Australia and New Zealand", TRAFFIC Oceania, Sydney.

Clarke, S. (2004), "Shark Product Trade in Hong Kong and Mainland China, and Implementation of the Shark CITES Listings", TRAFFIC East Asia, Hong Kong.

Cohen, D. (2006), *Impacts of China on the Global Value Chain for Manufactured Wood Products, in China's Boom: Implications for Investment and Trade in Forest Products and Forestry*, Conference proceedings, 18-20 January 2006, Vancouver.

FAO (2005), *Global Forest Resources Assessment 2005: Progress Towards Sustainable Forest Management*, FAO Forestry Paper 147, FAO, Rome.

Hayes, S and D. Egli (2002), *Directory of Protected Areas in East Asia: People, Organisations and Places*, IUCN, Gland.

Han, Nianyong (2000), 中国自然保护区可持续管理政策研究 (Study of Sustainable Management Policies for China's Nature Reserves), 中国人与生物圈国家委员会，科技文献出版社 (China National Committee on Man and Biosphere, Scientific and Technical Documents Publishing House), Beijing.

IUCN (The World Conservation Union) (2006), *2006 IUCN Red List of Threatened Species*, The IUCN Species Survival Commission, Gland.

IUCN (2004), *2004 IUCN Red List of Threatened Species: A Global Species Assessment*, The IUCN Species Survival Commission, Gland.

Jim, C.Y and Steve Shaowei Xu (2003), Getting out of the Woods: Quandaries of Protected Areas Management in China, *Mountain Research and Development*, Vol. 23(3).

Li, Diqiang *et al.* (2003), *China: Management Effectiveness Assessment of Protected Areas in the Upper Yangtze Ecoregion using WWF's RAPPAM Methodology*, WWF, Gland.

Liu, Jianguo and Jared Diamond (2005), *China's Environment in a Globalizing World*, Nature 30 June 2005, Vol. 435.

Miao, Guangping and R.A. West (2004), Chinese Collective Forestlands: Contributions and Constraints, *International Forestry Review*, Vol. 6 (3-4), December 2004.

Mills, J.A. (1999), *Fashion Statement Spells Death for Tibetan Antelope*, TRAFFIC East Asia, Hong Kong.

Mills, J.A. (1997), "Rhinoceros Horn and Tiger Bone in China: an Investigation of Trade since the 1993 Ban", TRAFFIC East Asia, Hong Kong.

Mills, J.A., Simba Chan and Akiko Ishihara (1995), "The Bear Facts: The East Asian Market for Bear Gall Bladder", TRAFFIC East Asia, Hong Kong.

NBSC (National Bureau of Statistics of China) (2006), China Statistical Yearbook on Environment, China Statistics Press, Beijing.

Nilsson, S., G.Q. Bull, A. White and Jintao Xu (2004), "China's Forest Sector Markets: Policy Issues and Recommendations", *International Forestry Review*, Vol. 6, No. 3-4, December 2004.

OECD (2005), *OECD Review of Agriculture Policies: China*, OECD, Paris.

Olson, D.M. and E. Dinerstein (1998), "The Global 200: A Representation Approach to Conserving the Earth's Most Biologically Valuable Ecoregions", *Conservation Biology*, Vol. 12.

Ouyang, Xihui (欧阳喜辉) (2005), 论绿色食品与农业生态环境保护 (On Green Foods and Ecological Environmental Protection in Agriculture), 北京市绿色食品办公室, 北京 (Beijing Office of Green Foods, Beijing), 食品安全 (Food Safety).

SEPA (State Environmental Protection Administration) (2006a), State of the Environment Report, SEPA, Beijing.

SEPA (2006b), China Environment Yearbook, China Environment Yearbook Press, Beijing.

SEPA (2005a), *China Third National Report on Implementation of the Convention on Biological Diversity*, 15 September 2005, SEPA, Beijing.

SEPA (2005b), Annual Statistics: Report on Environment in China, China Environmental Science Press, Beijing.

Smil, V. and Yushi, Mao (1998), *The Economic Costs of China's Environmental Degradation*, American Academy of Arts and Sciences, Cambridge.

Sun, Changjin and Xiaoqian, Chen (2002), "A Policy Analysis of the China Forest Ecological Benefit Compensation Fund", in *Workshop on Payment Schemes for Environmental Services*, 22-23 April 2002, China Council for International Co-operation and Development, Task Force on Forests and Grasslands, Beijing.

Tian, Weiming and Liqin, Zhang (2003), "Recent Economic and Agricultural Policy Development in China", paper prepared for the Roles of Agriculture International Conference, 20-22 October 2003, Rome.

Wang, Dehui (王德辉) (2003), 中国履行《生物多样性公约》的进展 (Progress of Fulfilling Convention on Biological Diversity in China), 国家环境保护总局自然生态保护司 (Biodiversity Conservation Department, SEPA), 环境保护 (2003.1), (Journal of Environmental Protection, 2003.1), Beijing.

Xie, Yan (2004), "Review on the Management System of China's Nature Reserves", in Xie Yan (ed.), *China's Protected Areas*, Tsinghua University Press, Beijing.

Xie, Yan *et al.* (2004), *China Species Information System*, Institute of Zoology, Chinese Academy of Sciences, Beijing.

Xie, Yan, Zhenyu, Li, Gregg, W.P. and Dianmo, Li (2001), "Invasive Species in China – An Overview", *Biodiversity and Conservation*, Vol. 10, No. 8, pp. 1317-1341.

Zou, Keyuan (2003), "Management of Marine Nature Reserves in China: A Legal Perspective", *Journal of International Wildlife Law and Policy*, Vol. 6.

7

ENVIRONMENTAL-ECONOMIC INTEGRATION

Features

- Development concepts and Five-Year Plans
- Decoupling environmental pressures from economic growth
- Institutional integration: environmental assessments
- Market-based integration: sectoral subsidies and environment-related taxes
- Environmental expenditure
- Environmental planning
- Use of economic instruments
- Enforcement and compliance assurance
- Land use regulations and planning

Recommendations

The following recommendations are part of the overall conclusions and recommendations of the environmental performance review of China:

Integration of environmental concerns in economic decisions

- review *price levels* for energy, water and other natural resources so as to better reflect their scarcity value and internalise externalities; consider mechanisms to compensate or mitigate their impact on poorer sections of the population and regions that would be adversely affected by such price increases;

- consider establishing an inter-ministerial group to examine how *environment-related* taxes might be restructured to help better achieve environmental policy objectives;

- *increase and diversify the sources of environmental finance* by fuller implementation of the polluter pays and user pays principles, and increase the effectiveness and efficiency of allocating public environmental expenditure;

- strengthen the institutional mechanisms for *better integrating environment into economic and sectoral policies*, possibly by establishing a Leading Group on environment or on sustainable development; fully implement the provisions in the EIAs law for assessing the potential environmental impacts of sectoral programmes;

- continue to establish *national targets* to achieve key environmental objectives, taking into account scientific, economic and social analysis.

Implementing environmental policies more effectively and efficiently

- *implement environmental law and regulations nationwide* for products and industrial/energy facilities; strengthen *monitoring, inspection and enforcement capabilities* throughout the country, including through the independence of the enforcement functions of Environmental Protection Bureaus (EPBs);

- consider establishing *SEPA as a ministry*; strengthen SEPA's supervisory capacity of EPBs in local government;

- continue efforts to make *local leaders more accountable* to the higher level government and to local populations for their environmental performance;

- strengthen the *integrated permitting system* and establish it as a more central instrument for pollution prevention and control; strengthen the integration of environmental protection in land-use planning and regulations, as well as in other relevant plans and regulations;

- extend the use of *pollution charges, user charges, emissions trading and other market-based instruments* and their incentive functions, taking social factors into account.

Conclusions

Integration of environmental concerns in economic decisions

China's two digit average economic growth was accompanied by some *decoupling of pollution from economic growth* in the period 1990-2005. This was the case, in particular, for SO_2 and recently NO_x emissions. Energy intensity has improved by about a half since 1990, though the decrease has levelled off. Water withdrawal and municipal waste have also been significantly decoupled from the economic growth. Successive *Five-Year Plans for National Economic and Social Development* (FYPs) have provided an important means for identifying and addressing priority environmental problems: they are underpinned by solid analysis, they establish quantitative targets, and they frame investment programming and budgeting. The Chinese leadership has announced its intention to place environmental protection in a more strategic position. In this perspective, the 11th FYP advocates a *new economic model in which growth is guided by resource conservation* rather than by continued expansion of resource use. Improved energy intensity and the concept of the "circular economy" are recognised as key to help reduce the pollution and resources intensity of the Chinese economy. Various measures have been taken to better *integrate environmental and economic decision-making*: provision has been made in the 2003 EIA law to assess the potential environmental impacts of sectoral programmes. Some energy prices have been deregulated (e.g. some coal prices). The use of environment-related taxes has expanded, but accounts only for about 3% of total tax revenues.

However, the pollution, energy and material intensities of the Chinese economy remain high, as well as its water use intensity, and *pollution remains very serious in many locations*. China generates more pollution and consumes more resources per unit of GDP than OECD averages. There is a high rate of environmentally significant accidents, and resource degradation is constraining economic development. *Health costs and ecological damages of present development are high*. The target of quadrupling GDP between 2000 and 2020 requires *commensurate strengthening of environmental management and finance*, so that economic growth is environmentally sustainable. It is not sure that present policies, although going in the right direction, are *sufficiently ambitious* to meet the strategic environmental objectives identified by Chinese leaders. The *under-pricing of energy, water and other resources* needs to be addressed. More effective arrangements at the level of the State Council are needed to better integrate environment into economic and sectoral decision-making, including a *strengthened role for SEPA*.

Implementing environmental policies more effectively and efficiently

China's comprehensive and modern set of environmental laws, together with its successive Five-Year Plans for National Economic and Social Development (FYPs) and Five-Year Environmental Plans (FYEPs), provide a high-quality framework for pursuing sustainable development and environmental progress. In December 2005, the State Council issued a decision for better implementing environmental policies. In April 2006, the Chinese Premier announced, in the sixth national environmental protection meeting, three new policy directions, including: integrating environmental protection and economic decision-making on an equal footing, further decoupling pollutant emissions from economic growth, applying a mix of instruments to resolve environmental problems. The proposed directions and measures are being implemented and go a long way towards addressing the environmental policy implementation gap. Within their mandates, *departments under the State Council have worked hard to support environmental policy implementation*. A range of regulatory and economic instruments (e.g. pollution charges, user charges, emissions trading) and policy approaches that harness markets and public interest in the environment have been developed. Campaigns and award schemes to support implementation at the local level have been organised; work with non-governmental organisations (NGOs) to develop procedures for public participation in *environmental impact assessment* (EIA) is an important recent example. There is evidence that *local leaders* in some of the richer provinces are responding to demands from the public for better environmental conditions, and are recognising the benefits to the economy and the society. More than 8 000 companies are registered under *ISO 14000*. In 2004, pollution abatement and control (PAC) investment expenditure was 1.2% of GDP.

However, these efforts have not been sufficient to keep pace with the environmental pressures and challenges generated by the very rapid growth of China's developing economy nor to capture the potential economic benefits to be obtained from improved pollution abatement and nature protection. Overall, environmental efforts have lacked effectiveness and efficiency, largely as a result of an *implementation gap*. The weaknesses in the present system are demonstrated by the failure to achieve some of the key objectives of the 10th FYP, and the severity of environmental problems in many parts of China. National environmental legislation and regulations should be *compiled* in an environmental code, to make them more consistent and user-friendly. Environmental policy priorities should be *focused on human health and key natural resources*. Consistent nationwide implementation of environmental regulations for products and industrial/energy facilities should be enhanced and given priority. The biggest obstacles to environmental policy implementation are at the *local level*. The performance objectives of local leaders, the pressures to raise revenues locally to finance un-funded mandates, and the limited accountability to local populations have generally meant that *economic priorities have over-ridden environmental concerns*. There is a need for much stronger *monitoring, inspection and enforcement capabilities* to establish a better mix of incentives and sanctions. In addition, *environmental expenditure* needs to be made more efficiently, and environmental policy instruments need to be made more effective. Implementation of the polluter pays and user pays principles should be strengthened. Special provisions are needed to integrate environment into the development strategies of the less-developed regions and to ensure the affordability of environmental services for the poor. There is an increase in damages associated with disasters of climatic and industrial origin, requiring improved prevention and mitigation measures.

◆　◆　◆

1.　Towards Sustainable Development

1.1　Economic development in China

By any measure, China's *economic development* has been remarkable. The economy has grown at an average rate of 10.1% over the last 15 years, about four times the OECD average in this period. China's GDP in 2005 stood at CNY 18 232 billion (or current USD 2 259 billion or USD 8 536 billion using PPP 2000). China is now the fourth largest economy in the world, after the United States, Japan and Germany. The stated goal of the Chinese government is to increase GDP four-fold between 2000 and 2020.[1] China's growth target for the period 2006-10 is 7.5%.

Rapid economic development has lifted millions of people out of *poverty*. In the 20 years to 2001, 400 million people escaped extreme poverty, and the percentage of the population living in poverty fell from 53 to 8%. Nevertheless, average GDP per capita is still very low compared to the OECD average (USD 5 975 compared to USD 25 300 using PPP 2000), and China ranks beyond 100th in the world on this measure (OECD, 2005a). In nominal terms, income per capita in China is close to 6% of the OECD average, underpinning China's considerable comparative advantage in labour-intensive activities. The income gap between rich and poor is much wider than that in most OECD countries, but it is still lower than a range of countries including Brazil, Mexico, Russia, South Africa and Turkey (Shankar, 2001). Wealth is largely concentrated in the densely populated eastern coastal provinces, where income levels in some areas are comparable to those of some OECD countries. However, in the western region, income levels are more akin to those in lower-middle income or low-income developing countries (Table 8.1, Figure 8.2).

China's remarkable economic performance has been driven by *changes in economic policy* that have given greater rein to market forces. The transformation started in the agricultural sector more than two decades ago and was extended to industry and large parts of the service sector. The role of state-owned enterprises has significantly decreased with the emergence of the private sector which has become the main driver of economic activity. According to OECD estimates, in 2003 the private sector was responsible for about 57% of value-added produced by the non-farm business sector and about three-quarters of all exports. More than one-third of state-owned enterprises currently do not earn a positive rate of return and are targeted for further restructuring (OECD, 2005a).

Economic development has been accompanied by important *structural changes*. Between 1990 and 2004, primary production as a share of GDP fell by half, with the secondary and especially the tertiary sectors expanding. Urbanisation has been increasing at the rate of 4% in recent years, and this has also boosted productivity growth. These trends are likely to continue. Using Chinese definitions, about 43% of the population live in urban areas. Important economy-wide policy differences between urban and rural areas have resulted in significant disparities between the populations in these areas (Chapter 8). Redressing this balance has become a priority for the Chinese government.

China's highly open economy has benefited from a rapid integration into the world economy. China's *accession to the World Trade Organisation* (WTO) in December 2001 reinforced and amplified a policy direction that had been followed for twenty-five years. By 2004, China's trade represented 40% of its GDP, with the European Union, United States and Japan the largest trading partners. In 2005, China's trade surplus with the rest of the world amounted to USD 102 billion.

Since 1993, China has been the largest recipient of *foreign direct investment* (FDI) among developing countries, mostly concentrated in the manufacturing sector. The establishment of a manufacturing base in China by foreign companies has greatly facilitated China's integration into the global economy. Foreign controlled companies now account for more than half of overseas sales (OECD, 2006a).

China's *public finances* are healthy. The rapid growth in tax revenues and tight control over public expenditure has kept the budget deficit below 1% of GDP, and the overall public debt at around 23% (OECD, 2006b). China has one of the highest savings rates in the world, about 40% of GDP. With relatively under-developed stock and bond markets, most of the savings are channelled into banks. In 2005, banks had assets equivalent to more than 200% of GDP, a very high figure by international standards. The strong role of the government in the banking sector has helped stimulate a high rate of investment in fixed assets. However, a high rate of non-performing loans and poor rates of return have led to reforms that will require banking operations to be more prudent and market-based.

Inflation is low, averaging 1.4% in the period 2001-05. Registered *unemployment* was 4.2% in 2005 in urban areas, but is estimated to be as high as 23% in rural areas.

1.2   Decoupling environmental pressures from economic growth

Pollution intensities

Over the review period, *emissions of SO_x, NO_x and CO_2 increased, but at a significantly lower rate than GDP.* While GDP grew by 284% from 1990 to 2004, SO_x emissions grew by 16%, NO_x emissions by 41% (to 2003) and CO_2 emissions from energy use by 110% (Figure 3.1, Table 7.1).

Table 7.1 **Selected economic trends and environmental pressures,**[a] 1990-2004

(% changes)

	1990-2004	1995-2004
SELECTED ECONOMIC TRENDS		
GDP[b]	284	115
Population	14	7
Agricultural production	97	49
Industrial production	452	144
Road freight traffic[c]	154	86
Passenger car traffic volume[d]	410	231
POLLUTION		
CO_2 emissions from energy use[e]	110	59
SO_2 emissions	16	–5
NO_x emissions	41[f]	5[f]
ENERGY		
Total primary energy supply	86	53
Total final consumption of energy	116	31
RESOURCES		
Water withdrawals	..	7[g]
Nitrogenous fertiliser use	36	10
Municipal waste[h]	129	45

a) Per cent changes over the period.
b) At 2000 prices and PPPs, on the basis of the revision done by NBSC (January 2006).
c) Based on values expressed in tonne-kilometres.
d) Based on values expressed in vehicle-kilometres.
e) Excluding marine and aviation bunkers; sectoral approach.
f) To 2003.
g) Growth rate between 1993 and 2004.
h) Volume of municipal garbage.
Source: SEPA; FAO; NBSC; IEA-OECD.

The intensities of these emissions per unit of GDP are higher than OECD averages. SO_x emissions were 2.9 kg/USD 1 000 compared to the OECD average of 1.1; NO_x emissions were 1.7 kg/USD 1 000 compared to the OECD average of 1.4; and CO_2 emissions from energy use were 0.61 tonnes/USD 1 000 compared to the OECD average of 0.44 (Reference IA). Moreover, these data underestimate China's *pollution intensity*, as they only cover stationary sources. Per capita annual emissions of CO_2 were recently 3.6 tonnes compared to the OECD average of 11.1 tonnes, but they grew seven times more than the OECD average during 1990-2004. The inefficient *use of coal* as the primary source of energy, and inadequate pollution control, are major factors in these trends. In addition, *passenger car traffic* volume increased by 410% in the review period, further increasing air emissions, particularly NO_x. *Air passenger traffic* grew about twice more than GDP in the period 1990-2004. As China's per capita income converges with OECD averages, these pressures will intensify.

Energy intensity

China's total primary energy supply grew by 86% from 1990 to 2004, and energy intensity fell by about 50%, reaching 0.21 toe per USD 1 000 of GDP, although the downward trend levelled off and then slightly increased at the end of the period. Energy production and consumption continued to rise, and China's energy intensity is now comparable to the most energy-intensive OECD countries (e.g. Canada, the Republic of Korea, United States). Concern about the social impacts of raising energy tariffs has held back energy price reform and contributed to the high energy intensity of the economy.

Material intensities

Municipal waste increased by 129% in the review period, less than half the rate of GDP growth. Municipal waste generation was estimated to be 120 kg/capita per year, about one-fifth of the OECD average. China's lower rate is related to its different income levels and consumption patterns, but the drivers of municipal waste generation are likely to intensify over time. Industrial waste grew faster than municipal waste but less than GDP.

Water withdrawal grew by 7% from 1995 to 2004, (compared to GDP growth of 115%). Intensity of water use in China is higher than the OECD average. Agriculture accounts for more than 70% of total water consumption. The demand on scarce water resources is exacerbated by the poor efficiency of water use in agriculture, which is partly linked to under-pricing. The agricultural use of fertilisers in sown areas increased by 7% between 1995 and 2002. These trends exacerbate water pollution and scarcity problems (OECD, 2005b).

The overall *material intensity* of China's economy is higher than that in OECD countries. One study by the Chinese Academy of Science indicated that the consumption intensity of five resources (water, primary energy, steel, cement and non-ferrous metals) was 90% higher than the world average (Xu, 2005).

1.3   Development and institutional integration

Development concepts

Building on the thoughts of Deng Xiaoping, who launched the process of economic reform in China in the late 1970s, successive generations of Chinese leaders have elaborated *various concepts of development*. These concepts have shown an increasing convergence with the concept of sustainable development, but with distinctly Chinese characteristics.

The *"scientific concept of development"* was advanced by the current generation of Chinese leaders and adopts a more comprehensive, evidence-based approach than its predecessors. Rather than focusing narrowly on GDP growth, it calls for people-centred development that is comprehensive, co-ordinated and sustainable. It particularly stresses the co-ordinated development of urban and rural areas, of the different regions, of economic and social issues, of man and nature, and of domestic development and opening to the outside world.

The concept of a *harmonious society*, advanced by current leaders, stresses social welfare, more equitable income distribution and the rule of law. The objective of achieving a *well-off society* (xiaokang) has its origins in an ancient Chinese idea and suggests a level of economic development where people are comfortable, and where a balance has been achieved between economic growth, social equity and environmental protection.

FYPs for National Economic and Social Development

China's *Five-Year Plans (FYPs) for National Economic and Social Development* are a major mechanism for co-ordinating the country's public policy priorities. The development and implementation of the plans is overseen by the National Development and Reform Commission (NDRC). The FYPs provide the framework for the development of sectoral plans, including for the environment, and for plans at the sub-national level. The 9th FYP (1996-2000) incorporated China's Agenda 21. The 10th FYP (2001-05) called for continued implementation of the strategy for sustainable development.

The *11th FYP* (2006-10), was adopted by the National People's Congress in March 2006. Contrary to previous FYPs, the 11th is referred to as a "programme", to reflect its greater programming character. It contains fewer targets than past plans, and a few of the targets are environment-related. A distinctive feature is the greater emphasis given to the quality of economic growth, including its social and environmental dimensions.[2] The 11th FYP calls for a new model of economic development driven not by the increasing use of resources, but by more efficient use of resources. One of the main principles guiding the 11th FYP is the development of a "resource-saving and environmentally-friendly society". Under this heading, a major section of the programme addresses the "circular economy", better management of natural resources, energy efficiency and strengthening pollution prevention and control (Box 5.1). The responsibility of territorial governments for implementing environmental laws at the sub-national level is emphasised, as is the need to participate actively in global environmental and development affairs and to implement international environmental conventions.

Institutional arrangements for sustainable development

Following the 1992 Rio Conference on Environment and Development (UNCED), China promulgated the *world's first state-level Agenda 21* white paper on population, environment and development in the 21st century. Subsequently, various Chinese policy statements have given increasing importance to balancing environmental, economic and social policy objectives. In 1996, the Chinese government officially identified sustainable development as a "major national strategy". In 2002, sustainable development was designated as a goal to be met in establishing a "well-off society for all". In 2003, a programme of action for sustainable development in China in the early 21st century was adopted.

Near the end of 2005, the *Central Committee of the National People's Congress* judged that the contradictions between socio-economic development and environment had become more obvious and that environmental degradation and natural resources scarcity had "seriously restricted" economic development. Shortly after, on 3 December 2005, the State Council adopted a Decision on Implementing the *Scientific Concept of Development* and Strengthening Environmental Protection, and explained that the rationale for this action was the need "to place environmental protection in a more strategic position". This Decision:

– presents reasons why greater emphasis should be placed on environmental protection (to realise faster and "better" growth; to foster economic activities related to environmental protection and the related employment opportunities; to improve the ethical basis of Chinese society; to protect human health and the quality of life; and to maintain the resource base for present and future generations);

- elaborates how environmental protection could be integrated with the "scientific concept of development";

- identifies a range of environmental priorities; and

- proposes ways to strengthen China's environmental policy and institutional framework, including new measures to strengthen the implementation of environmental policy.

The State Environmental Protection Administration (SEPA) and the Ministry of Supervision are directed to jointly review implementation of the Decision and to report to the State Council annually.

In April 2006, the Chinese Premier announced at the 6th National Environmental Protection meeting *three new policy directions* (i.e. the "three transformations"): *integrating* environmental protection and economic decision-making, and putting them on an equal footing; promoting pollution prevention and further *decoupling* pollutant emissions from economic growth; and taking a comprehensive approach to environmental protection by applying a *mix of policy instruments*.

In parallel with the strengthening of policy pronouncements on environmental and sustainable development, *institutions responsible for sustainable development* have been consolidated within the Chinese governmental system. NDRC has continued to play a key role as the body responsible for developing and implementing FYPs. In this capacity, NDRC has helped integrate environment into the overall planning system in China, and into specific sectors such as energy. In 1998, as part of a broader restructuring of the Chinese administration, the status of the national environmental protection agency was upgraded and it was renamed the State Environmental Protection Administration. SEPA did not become a ministry, but its top official, the administrator, was given ministerial status. In addition, SEPA took over the function of co-ordinating government environmental policy, which had been exercised by the Environmental Protection Commission, an inter-ministerial body.[3] The 2005 State Council Decision might help to reinforce SEPA's role, but it may not be sufficient. Consideration should also be given to establishing SEPA as a ministry and strengthening the integration of environment with sectoral and economic policies at the level of the State Council. A possible option could be to establish a leading group[4] on the environment or on sustainable development. The new Ministry of Environment might be entrusted with responsibilities concerning pollution control, ecological conservation as well as forest resources and water infrastructure.

SEPA and the NBSC begun work to estimate *economic losses* in China due to: i) changes in the stocks of natural resources (i.e. depletion in land, minerals, forests, water and fisheries resources); and ii) environmental degradation (i.e. pollution and ecological damages). This work is based on the economic and environmental accounting experience gathered domestically and internationally. Concerning *pollution only*, early results show that 2004 economic losses associated with a portion of water, air and waste-related pollution have reached more than CNY 511 billion or more than 3% GDP (SEPA, 2006). Much progress is still needed to estimate the range of economic dimensions associated to natural resources and environmental degradation not included in conventional GDP calculations.

There are several initiatives related to sustainable development at the sub-national level. Since the 1992 Rio Conference, the Ministry of Science and Technology and NDRC have promoted *local Agenda 21s* and sustainable community programmes, and have provided guidance and disseminated results. Resources are provided by the participating authorities. SEPA also has a programme to promote *model environmental cities*.

Environmental assessment of policies and plans

China adopted the 2002 *Law on Environmental Impact Assessment* (EIA) as part of a sustainable development strategy, to prevent adverse impacts on the environment that might result from policies, plans and projects. The law requires relevant departments of the State Council or local government at or above the level of prefecture to assess the potential environmental impacts of policies and plans concerning industry, agriculture, pasturage, forestry, energy, water resources, transport, urban development, tourism and the exploitation of natural resources. The assessments must be submitted to the authority responsible for the policies and plans before they are examined and approved. However, implementation of the law is still pending, awaiting the elaboration of an *appropriate EIA methodology*.

Although trade is not covered by this law, an *environmental assessment of China's accession to WTO* was prepared in 2004, within the framework of the China Council for International Co-operation on Environment and Development (CCICED). This was one of the most comprehensive studies ever done of the environmental implications of trade liberalisation. The study identified the opportunities and threats to China's environment in six particularly sensitive sectors, and proposed recommendations to maximise environmental benefits and minimise environmental threats (Box 7.1). WTO-environment issues have continued as a focus area in CCICED's work.

Environmental assessment of potentially polluting investment projects

The first legal provision for *environmental assessment of pollution prevention and control equipment in investment projects* was in the 1989 Environmental Protection Law. The law requires that equipment for pollution prevention and control

Box 7.1 **Environmental impacts of WTO accession: an assessment by the CCICED**

Agriculture. The impact could be positive if increased trade liberalisation shifted production from products requiring high levels of land, water and chemical inputs to more labour-intensive products. This shift should be supported by measures to reduce subsidies for chemical inputs, increase support for advisory services, disseminate information about foreign environmental requirements for agricultural products, and strengthen domestic standards.

Forestry. Timber imports are projected to increase five-fold from 1995 to 2010, in part to support the production of wood products, notably furniture, for export. While this may have a beneficial impact on Chinese forests, particularly if accompanied by improved forest management, it may also contribute to unsustainable forestry practices in supply countries in Asia and beyond. China should consider reducing escalating tariffs on finished wood products, and should strengthen its international co-operation to combat illegal logging and to promote sustainable forestry throughout the entire product chain.

Aquaculture. WTO accession has accompanied a sharp rise in aquaculture exports whose volume currently is roughly equivalent to China's net imports of agricultural products. Environmental problems have been exacerbated by this trend (e.g. nutrient and chemical pollution, substrate eutrophication and red tides). However, these costs could be outweighed by the economic and environmental benefits if appropriate policies are put in place that aim to: ensure high product standards, strengthen control of land-based marine pollution, manage resources effectively to optimise the quality and quantity of products produced, disseminate information, provide technical support, and participate in international activities related to standards for aquaculture.

Automobiles. WTO accession has had a dramatic impact on automobile sales by reducing tariffs and duties and hence price, thereby boosting demand. Despite the potentially adverse environmental impacts, this creates a window of opportunity to enhance the environmental performance of cars by reducing emission levels and increasing fuel efficiency without increasing their aggregate price. A variety of other measures are needed to minimise the environmental impacts associated with the sharply rising numbers of cars, such as promoting the development of cleaner technologies, developing public transport and applying environmental taxes to automobile use.

Energy. Trade liberalisation has further stimulated economic growth and demand for energy. From a trade perspective this has increased demand for oil and increased dependency on foreign imports. This trend has been somewhat off-set by an increase in world energy prices. Nevertheless, integrating environment into energy policy is a major challenge that will require an appropriate mix of regulatory and economic instruments, as well as the application of cleaner technologies.

Box 7.1 **Environmental impacts of WTO accession: an assessment by the CCICED** *(cont.)*

Textiles. WTO accession, and the intention of WTO members to end quota restrictions on Chinese textiles, would sharply stimulate textile production. While some environmental benefits will derive from technology transfer, the overall environmental impact is expected to be negative unless appropriate policies are put in place. Energy and water consumption in the sector were predicted to double by 2010 and waste water discharge to increase by 60% specifically because of increased trade, assuming quota restrictions ended in 2005. Waste water treatment is the biggest challenge because of the toxic nature of waste products and the large number of dispersed small and medium-sized textile producers.

Source: CCICED (2004).

be taken into account in the design, construction and implementation of potentially polluting investment projects, a procedure referred to as the "three simultaneities", or "3S".[5] Proposed pollution prevention and control technology or equipment must be submitted for approval by local EPBs before the investment project is carried out. In 2004, 79 500 investment projects were the subject of the 3S procedure, out of a total of 127 500 investment projects. In slightly more than 76 000 cases the 3S procedure was approved (SEPA, 2000, 2004). However, there is evidence that sanctions for non-compliance with 3S procedures are often not applied by local authorities. A specific expenditure line is dedicated to the investment projects under this procedure: CNY 46 billion in 2004, up from CNY 26 billion in 2000 (SEPA, 2000, 2004). The projects registered under the 3S procedure are supported by national public financing in pollution prevention and control.

Environmental impact assessment of projects

The 1989 Environmental Protection Law also requires *projects with potentially negative environmental effects to be subject to EIA* before approval by local DRCs. The 1998 Ordinance on Environmental Management for the Construction Projects (OEMCP) prescribed three different levels of assessment depending on the potential environmental impact of proposed projects (i.e. a full Environmental Impact Report for projects likely to have significant adverse environmental impacts, a less detailed Environmental Impact Form for projects likely to have a limited number of significant adverse environmental impacts, and a basic Environmental Impact Registration Form for projects that are not expected to cause significant adverse environmental impacts[6]). The OEMCP also required resubmission of the EIA documents in cases of significant changes in the nature, size, location and process technologies of proposed projects.[7] Nevertheless there was *widespread dissatisfaction with the EIA system*: many projects were implemented without an EIA; procedures were not well-defined and led to conflicts of interest for environmental authorities who were paid to conduct the studies and also to assess them; and the quality of assessments when conducted was limited, often failing to specify appropriate mitigation measures.

The *2002 Law on EIA* addresses the shortcomings of the 1989 Environmental Protection Law. It specifies that EIAs should include: i) an identification and analysis of potential environmental impacts; ii) possible measures to prevent or control the identified impacts; iii) an assessment of the feasibility and costs of the possible measures; and iv) conclusions on the project from an environmental perspective. The 2002 law also specifies the procedures for conducting, reviewing and implementing environmental assessments, and it requires public authorities to hold *public hearings* involving potentially affected parties. The latter requirement was subsequently addressed in SEPA's Measures on Public Participation in the Environmental Impact Assessment Process, which took effect in March 2006. These measures clarify the rights and responsibilities of the various parties with an interest in the EIA and the forms of public participation (e.g. surveys, consultations, seminars, debates and hearings). The measures have generally been welcomed by environmental NGOs.

In 2004, some 310 000 out of a total of 321 000 projects were subjected to EIA, including those requiring a full Environmental Impact Report, a less detailed Environmental Impact Form, or a basic Environmental Impact Registration Form (Figure 7.1). A joint investigation by SEPA and the Ministry of Land and Resources showed that only 30 to 40% of the *mining construction projects* went through the required EIA procedures, and the percentage was as low as 6 to 7% in certain provinces. During a campaign carried out by SEPA in December 2004, the construction of some 30 large projects, most involving hydro or thermal power plants, was suspended as they failed to satisfy EIA requirements.

Recently, SEPA has started collecting *examples of good EIA practices* for a variety of project types to stimulate better assessments and more consistent standards. SEPA also publishes on its website the number of projects that have been rejected under the EIA process.

Overall, the 2002 EIA law provides a much *stronger basis for environmental impact assessment of projects*, and SEPA has taken several decisive steps to support its implementation. However, much will depend on how it is implemented. This will

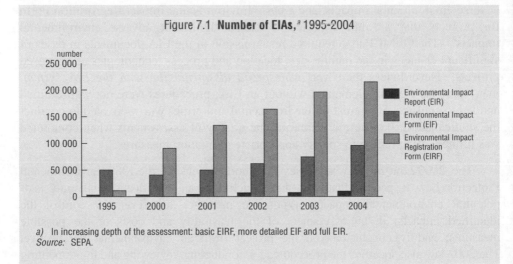

Figure 7.1 **Number of EIAs,** [a] 1995-2004

a) In increasing depth of the assessment: basic EIRF, more detailed EIF and full EIR.
Source: SEPA.

also depend on local governments, and particularly on how they identify projects for assessment, how they apply the provisions for public participation, how they influence the quality of the assessments, and what actions they take in response to the assessments.

1.4 Sustainable development and market-based integration

Sectoral subsidies

Agriculture support in China is relatively low, totalling 7% of gross farm receipts between 2000 and 2003 (OECD, 2005b). This was much lower than the OECD average (31%), and far below support levels in Japan (58%) and in the Republic of Korea (64%). Nevertheless, the cost to the Chinese economy is high as it represented 3.8% of GDP in 2003. This is much higher than the OECD average, and is among the highest for major agricultural producers world-wide. *Support to producers* is mostly provided in the form of market price support and input subsidies, which tend to be the most trade distorting and environmentally damaging forms of agricultural support, as well as providing the lowest income transfer to farmers. VAT is levied on fertilisers and pesticides at the reduced rate of 13% rather than the standard 17%. The prices of these products are also indirectly subsidised by administrative measures and low electricity charges. *Agriculture accounts for more than 70% of water consumption in*

China. Irrigation water is under-priced, which contributes to its inefficient use. Currently only about 45% of the water flowing through China's irrigation systems is effectively used. This compounds China's water scarcity.

Energy prices, notably for products and coal have been increasingly market-based since the deregulation of prices in 1993. By 1996, two-thirds of coal was priced by the market. As a result, coal prices have risen significantly. Annual government-organised coal meetings and fairs bring together suppliers and consumers to agree upon prices, transport and customer allocation. The increase in coal prices has generated conflicts with the electricity sector, where prices are still largely controlled by the NDRC. Ongoing reforms in the energy and power sectors aim to increase efficiency and lower prices by increasing competition among generators.

Oil prices are currently set by the government. Rising international oil prices in 2005 heightened awareness of the country's dependence on oil imports. Nevertheless, domestic oil prices have lagged behind international price rises in the most parts of the country. There are subsidies on oil products. Increases in retail prices were allowed to fall behind increases in crude oil costs to soften the impact of rising oil prices on producers. The gap between the costs for crude oil and receipts for crude products has resulted in losses to oil refineries amounting to CNY 4.19 billion.

Transportation is the fastest growing economic sector in China, generating increasing levels of air pollution and demanding more land. Road construction was prioritised in the review period. China has established a system of road fees, overseen by the Ministry of Construction and local governments, which are used for maintenance and construction. Highways, including new expressways, are generating significant revenues from tolls, and this has created opportunities for the private sector to play an increasingly important role in highway construction. Nevertheless, highway construction has so far largely relied on public spending, which rose to USD 25 billion annually in 1998 and 1999 through a combination of direct government loans and grants to implementing agencies at the provincial level. Road fuel prices and taxes are low by international standards, suggesting there is *scope to internalise environmental externalities* associated with road construction (Figure 7.2).

Environment-related taxes

China has a number environment-related taxes (Table 7.2). In the category of *consumption taxes,* unleaded gasoline is taxed at a lower rate than leaded gasoline, and diesel is taxed at half the rate of unleaded fuel. A registration tax is levied on the sale of motor vehicles, motorcycles and motor cars. Road motor vehicles that meet stipulated low-pollution standards are exempt from 30% of the excise tax. A tax is

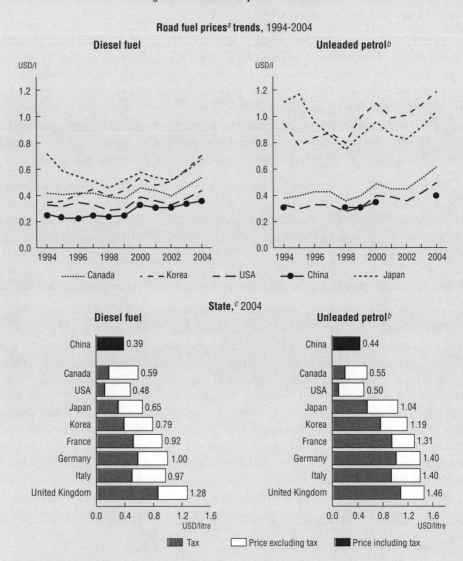

Figure 7.2 **Road fuel prices and taxes**

Road fuel prices[a] trends, 1994-2004

Diesel fuel   **Unleaded petrol[b]**

.... Canada   – – Korea   — — USA   ● China   - - - - Japan

State,[c] 2004

Diesel fuel

China	0.39
Canada	0.59
USA	0.48
Japan	0.65
Korea	0.79
France	0.92
Germany	1.00
Italy	0.97
United Kingdom	1.28

Unleaded petrol[b]

China	0.44
Canada	0.55
USA	0.50
Japan	1.04
Korea	1.19
France	1.31
Germany	1.40
Italy	1.40
United Kingdom	1.46

■ Tax   □ Price excluding tax   ■ Price including tax

a) At constant 2000 prices.
b) Unleaded premium (RON 95); Canada, United States, Japan, Korea and Germany: unleaded regular.
c) In USD at current prices and exchange rates.
Source: IEA-OECD.

Table 7.2 **Environment-related taxes**

Tax	Taxable item	Tax rate (amount per unit)	Revenue 2002[a]
Fuel-related consumption tax	Gasoline		191.87
	unleaded	0.20 CNY/litre	
	leaded	0.28 CNY/litre	
	Diesel	0.1 CNY/litre	
Transport-related consumption tax	Motor vehicle tyres	10%	93.95
	Motorcycles	10%	
	Motor cars	8%, 5%, 3%	
Vehicle and vessel usage tax	Vehicle		28.93
	passenger vehicle	60-320/passenger vehicle	
	cargo vehicle	16-60/tonne	
	motorcycle	32-80/motor cycle	
	non-motorised vehicle	1.2-32/non motorised vehicle	
	Vessel		
	motorised vessel	1.2-5/net tonne	
	non motorised vessel	0.6-1.4/net tonne	
Resource tax	Crude oil	8-30 CNY/tonne	75.08
	Natural gas	2-15 CNY/100 m^3	
	Coal	0.3-5 CNY/tonne	
	Other non-metal resources	0.5-20/tonne or m^3	
	Ferrous metal ores	2-30 CNY/tonne	
	Non-ferrous metal ores	0.4-30 CNY/tonne	
	Salt		
	solid salt	10-60 CNY/tonne	
	liquid salt	2-10 CNY/tonne	
Urban and township land-use tax	Large cities	0.5-10.00/m^2	76.83
	Medium-sized cities	0.4-8.00/m^2	
	Small cities	0.3-6.00/m^2	
	Mining districts	0.2-4.00/m^2	
Farmland occupation tax	1 mu^b or less of farmland/person	2-10 CNY/m^2	57.33
	1-2 mu^b farmland/person	1.6-8 CNY/m^2	
	2-3 mu^b farmland/person	1.3-6.5 CNY/m^2	
	> 3 mu^b farmland/person	1-5 CNY/m^2	
City maintenance and construction tax	City area	7%	470.82
	Country and township area	5%	
	Other area	1%	
Pollution and water charges[c]	n.a.	n.a.	84.74
Total revenue	n.a.	n.a.	608.73
% of total taxation[d]	n.a.	n.a.	*3.34*

a) In hundreds of millions of CNY.
b) 15 mu = 1 ha.
c) Not a tax per se.
d) Including pollution and water charges.
Source: CCICED, Task Force on Environmental and Natural Resources Pricing and Taxation.

applied to road motor vehicles and vessels. A pro-rata tax is applied to the extraction of oil, gas, coal, metals and salt. Land use taxes are applied in urban and rural areas, as well as a city maintenance and construction tax. These revenues contribute to investments in urban infrastructure.

In March 2006, several *additional environment-related taxes* were introduced. A differentiated tax for motor vehicles was introduced according to engine size. The tax on vehicles with engines larger than 2 litres was increased to between 9 and 20%, while the tax on vehicles with engines smaller than 1.5 litres was reduced to 3%; the tax on vehicles with engines between 1.5 and 2 litres remained at 5%. A 5% consumption tax was introduced for disposable wooden chopsticks and for wood floor panels. The scope for oil products subject to taxation was also broadened (CCICED, 2005).

In the last two decades, China has implemented *fundamental tax reforms*. While there is no pressure to increase taxes for budgetary reasons, the transition to a more market-based economy suggests that further reforms will be needed so that the tax structure causes fewer distortions and is more economically efficient. Adjusting environment-related taxes could support this objective while also helping to achieve environmental objectives. They could be imposed in a revenue-neutral way by off-setting increases in environment-related taxes with decreases in other taxes that cause distortions, such as those related to production.

In OECD countries, transport fuels and motor vehicles often account, respectively, for about two-thirds and one-quarter of revenues generated by environment-related taxes. In China they account for one-third and one-sixth. In addition, environment-related taxes account for about 3% of China's total tax revenues, which is low compared to OECD countries. Moreover, the price of road fuel in China is low by international standards (Figure 7.2). This suggests there is *scope for increasing environment-related taxes in China, particularly on transport fuels*. China might follow the approach used in a number of other countries and establish a *"green tax" commission*, i.e. a high-level inter-ministerial group, to analyse options and make recommendations. Such a group would also need to examine the potentially adverse impact of environment-related taxes on poorer parts of the population, particularly in the rural sector.

1.5   Environmental expenditure

PAC investment expenditure

In the period 1996-2000, *environmental investment expenditure* in China averaged 0.8% of GDP, reaching 1.1% in 2000. Between 2001 and 2005 it continued to grow, reaching 1.4% of GDP in 2004 (CNY 191 billion) and surpassing the 1.3%

target set in the 10th FYP.[8] This increase, combined with the increase in the GDP itself, means that environmental investment expenditure in China more than doubled over the period 1999-2005 (Table 7.3).

However, these percentages need to be revised downward in light of the January 2006 adjustment of GDP. So in 2004, environmental investment expenditure in China represented *1.2% of the revised GDP* using the Chinese definition.[9] The pollution abatement and control (PAC) investment expenditure are estimated here to represent in 2004 some *0.6% of the revised GDP*.

In 2004, *investments related to urban environmental infrastructure* accounted for about 60% of environmental investment (using the Chinese definition), or CNY 114 billion. Between 1999 and 2004, investments in urban environmental infrastructure increased by 117% at constant prices (Table 7.3). This mainly reflects investment in waste water treatment and waste management infrastructure, in the context of rapid urbanisation.

Table 7.3 **Environmental investment expenditure,**[a] 1999-2004

(billion CNY)

	1999	2000	2001	2002	2003	2004[d]	2004[e]
Total	83	102	111	137	162	191	191
of which:							
Urban environmental infrastructure[b]	48	52	60	79	107	114	114
Industrial pollution treatment	15	24	17	19	22	31	31
Pollution abatement through "3S" projects[c]	19	26	34	39	33	46	46
Total as % of GDP	1.0[d]	1.1[d]	1.1[d]	1.3[d]	1.4[d]	1.4[d]	1.2[e]
Total as % of GFCF	2.8	3.1	3.0	3.1	2.9	2.7	..

a) Using Chinese definition of "environmental investment expenditure". Measured at current prices. Includes (as opposed to OECD PAC expenditure definition) elements such as investment expenditure in energy efficiency, in fuel switching and in urban amenities.
b) Includes activities related to: household sewerage and sewage treatment plants, collection and disposal of household waste, central heating facilities and landscaping.
c) Refers to "3S" investment schemes for industrial pollution control covering new equipment, adaptation and technical improvement measures.
d) Using GDP figures from NBSC, prior to the January 2006 adjustment of GDP.
e) Using GDP figures following the January 2006 adjustment.
Source: SEPA, NBSC.

In 2004, *investments in industrial pollution abatement and control* accounted for 40% of environmental investment (using the Chinese definition), or CNY 77 billion, more than doubling over 1999-2004 at current prices (Table 7.3). National environmental projects (e.g. projects implemented within the framework of the "three simultaneities" programme) represented about 25% of total environmental investment. This largely reflects industrial investment in energy efficiency, air pollution abatement and waste water treatment, in the context of rapid economic growth.

Over the period 2001-05 (corresponding to the 10th FYP), environmental investment expenditure in China was *allocated to air (40%), water (38.5%) and waste (12.9%)*, and was financed by both public funding sources (57%) and private funding sources (43%).

PAC total expenditure

PAC current expenditure represented 0.3% of GDP (i.e. more than CNY 50 billion) in 2004. Over the years, it has grown slowly following the cumulated growth of PAC investment. It seems to vary within the country, as some facilities simply are not operated due to lack of funds for operation and maintenance. Information on environmental expenditure should be collected using international standards; this could encourage more emphasis on efficiency.

PAC total expenditure (i.e. both investment and current expenditure, using the OECD definition) are here estimated at *0.9% of the revised GDP in 2004*. This is low for a country in transition with major pollution challenges.

For the *period 2006-10*, Chinese authorities have announced that CNY 1 400 billion (or USD 175 billion), or about 1.5% of GDP (i.e. of the expected GDP for the period), will be spent for environmental investment (i.e. investment by different levels of government and by the private sector), addressing mainly air and water pollution and waste management. This may not be sufficient to respond to the government's stated environmental objectives.

Financing expenditure

To achieve China's environmental objectives, and capture related economic, health and social benefits, *it will be a challenge to continue to increasing the environmental share of public expenditure at the same rate* (Box 7.2). China will have to diversify its sources of finance for the environment and increase the *efficiency* with which they are used. The use of environmental funds, during a transition period, might be considered for specific environmental priorities.

Box 7.2 **Financing public infrastructure: co-operation between central and local governments**

Local governments are generally responsible for projects within their boundaries. Full central government financing is not uncommon in the poorer western regions. Local governments are not authorised to incur debt to finance investments, so they have been largely dependent on transfers from central government. However, *sub-national credit* has grown rapidly in recent years, mostly in the form of national bonds and bank loans: the central government has raised revenues for infrastructure construction by issuing national bonds, and allocates them to local governments in the form of grants and loans. Policy banks, such as the China Development Bank, have also played an important role in financing urban environmental infrastructure, particularly in the central and western regions. Commercial banks have played a much smaller role.

Capital spending in China has been growing at the rate of 20% in recent years. Public investment in water (including dams and water transfer infrastructure) and environment accounted for about *25% of government capital spending* in 2003, with much of it concentrated on three rivers (the Hai, Huai and Liao) and three lakes (Dianchi, Chao and Tai).

However, *efficiency in public investment* in infrastructure could be enhanced. Many public infrastructure projects are not completed, not put into operation, or are operated inefficiently, particularly those financed by bonds. Different agencies are responsible for capital and current expenditures. NDRC is responsible for investment expenditure, particularly investment expenditure identified in the FYPs, and the Ministry of Finance is responsible for current expenditure. Co-ordination between these two types of expenditures is not adequate, and the success of projects is further impeded by a third agency having responsibility for allocating personnel. Local governments have incentives to promote high profile projects even if their economic returns are questionable.

In recent years China has experimented with *build-operate-transfer* schemes (Chapter 4). However, the participation of the private sector will not guarantee more efficient public capital spending. Rather, governments, particularly those at local level, should be held accountable for ensuring that projects have an adequate ex ante rate of return and are implemented in a cost-effective fashion. Ultimately there needs to be a better correspondence between expenditure and revenue-raising responsibilities at the local level.

Source: OECD (2006b).

Diversifying sources will imply a fuller implementation of the polluter and user pays principles. More systematic implementation of environmental policies should induce enterprises to invest more heavily in pollution prevention and control. Consumers in the richer provinces are increasingly willing and able to pay for environment-related infrastructure. Public authorities, however, will have a continued role to play in supporting capital investments and in ensuring that the populations in poorer provinces have adequate access to environment-related services. More equitable arrangements for sharing investment costs between richer downstream and poorer upstream provinces ("eco-compensation") are also needed.[10]

China should also try to obtain increased funding from international sources. International financial assistance provided with environmental expenditure about CNY 4 billion in 2004 (largely but not only for pollution abatement and control) in grants and loans (Chapter 9). This represented less than *2% of PAC total expenditure in China.* This is not high, in light of the potential benefits that could result from environmental investments in China, for both China and the international community. FDI could also add to ODA sources of funding.

2. Environmental Policy Implementation

2.1 Environmental planning

Within the context of Five-Year Plans for National Economic and Social Development, there is a number of sectoral plans, including *Five-Year Environment Plans* (FYEPs). They are further supported by individual specific plans (e.g. for hazardous waste management, water management in key rivers and lakes, reduction of air pollution in "two control zones", development of the environmental industry, nature conservation). These plans provide the framework for the FYEPs prepared by the sub-national governments and their EPBs.

The 8th FYP (1991-95) included various provisions to strengthen environmental management, particularly following the 1989 Environmental Protection Law. The *9th FYP (1996-2000)* identified a specific set of environmental objectives to be achieved and distinguished pollution prevention and control, and natural resource management. It contained i) a national programme to control "total" (i.e. volume and concentration) pollution discharges, and ii) a "Trans-Century Green" programme including over 800 water pollution abatement projects. It also called for stricter inspection of industrial pollution control. China subsequently started a number of environmental initiatives, such as the "one control and double attainments" programme.[11]

The *10th FYP (2001-05)* set new targets, and envisaged a number of institutional and regulatory measures (e.g. strengthening decision-making for integration of environmental matters into economic development, strengthening the capacity of environmental management institutions, strengthening environmental regulations and the use of economic instruments). Its implementation was estimated to require an investment (from all financing sources) of CNY 700 billion (or 1.3% of GDP[12] and 3.6% of total fixed investment). It identified a package of 1 137 projects (*"China Green Project Plan – Phase II"*). It was estimated that these projects would require CNY 262 billion, about 37% of the estimated investment for the 10th FYP. The preliminary *assessment of the achievements in the 10th FYP period* shows that performance is mixed. The *11th environmental FYEP (2006-10)*, is under preparation. Environmental targets have been included in the 11th FYP (Box 7.3). A strategic environmental outlook (up to 2020) is under development.

Box 7.3 Targets in the 11th Five-Year Programme, 2006-10

- Energy intensity to be reduced by 20%.
- Water consumption per unit of industrial value-added to be reduced by 30%.
- Water for irrigation in agriculture maintained at current levels.
- Recycling of industrial solid waste to be increased by 60%.
- Area of farmland to be retained at 120 million hectares.
- Total discharge of major pollutants reduced by 10%.
- Forest coverage to reach 20%.
- Control of greenhouse gases to "generate good results".

The *development of FYPs* has been based on extensive analytical work, (including a review of experience from the previous planning period). The plans contain a number of quantitative, time-bound targets. The plans are linked to project programming that is designed to meet these targets. NDRC further links the programming and budgeting processes. Together, the set of FYPs, FYEPs and the specific plans, provides a *very good planning framework* for pursuing environmental progress in China, and has evolved positively, together with the reform in Chinese governance. However, there are a number of ways in which these plans could be improved, in particular by more economic analysis to establish objectives and greater use of environmental performance indicators.

2.2   Inter-governmental institutional arrangements

National and sub-national levels

While the Environmental and Resources Protection Committee (ERPC) of the National People's Congress is responsible for developing, reviewing and enacting environmental laws, the *State Environmental Protection Administration* is the national-level administrative body responsible for developing environmental policies and programmes[13] (Chapter 2). In addition, various issues related to the environment are managed by a number of *other ministries and agencies* of the State Council (Table 2.6).

Primary responsibility for *implementation of environmental policy is at the sub-national level.* China has about 2 000 Environmental Protection Bureaus (approximately 60 000 employees) at the provincial, prefectoral and county level. The EPBs vary in size, with an average provincial EPB employing 13.3 employees (2002), (ranging from 6.4 in Qinghai to 24.1 in Henan). The EPBs' main responsibilities include: overseeing EIA, permitting and the "3S" for projects; monitoring factory emissions; assessing fees for pollution discharges; and initiating legal action against firms that fail to meet environmental requirements. The EPBs are also responsible for environmental reporting, environmental education and awareness-raising activities. *Other administrative units of the sub-national governments* are engaged in environmental policy implementation (Chapter 2). The environmental protection committees of the various tiers of the People's Congresses approve local environmental regulations, review environmental work carried out by lower-level administrations, and consider environmental complaints from citizens. They also contribute to co-ordinating the EPBs' work with that of other government organs.

Assessment

Reducing the *implementation gap at the sub-national level* is one of the major challenges to improving environmental performance in China. This is true in many countries, but particularly so in China. Chinese authorities have identified this problem, most recently with the 3 December 2005 State Council Decision on Implementing the Scientific Concept of Development and Strengthening Environmental Protection. This Decision contains a commitment to environmental progress and a number of measures. Nevertheless, important structural obstacles to *effective environmental policy implementation* at the local level remain.

First, the EPBs receive guidance from SEPA, but they are institutionally and financially *subordinate to sub-national governments*, which have tended to prioritise economic development over environmental considerations. The EPBs have been administratively weaker than many of their sectoral counterparts. However, SEPA is now involved in the selection of the heads of local EPBs, and all provincial EPBs have been established as independent institutions since 2000; in some provinces, local EPBs are required to report directly to provincial rather than sub-provincial levels of government. In Jiangsu, the EPB has been upgraded to a department, the same level as the major economic and sectoral agencies. This could be a model for other provinces. Another development has been the initiation of pilot projects with regional environmental agencies that can help to focus attention on trans-provincial environmental issues. The government recently announced the establishment of nine regional environmental supervision bureaus.

Second, the *performance of local government leaders* has been evaluated using criteria that emphasise GDP growth, with little if any consideration of environmental performance. The State Council has recognised the need to include environment in local leaders' performance assessment criteria. Some experimentation has taken place in Qingdao city, and work is underway to develop a "green GDP". However, the experience of OECD countries over a number of decades suggests that: i) while "greening" economic accounts and developing environmental accounts are meaningful developments, the production of a "green GDP" summary indicator has encountered a range of insurmountable obstacles; ii) while small sets of environmental performance indicators have been developed usefully, aggregate environmental indices are not advisable or reliable.

Third, as in many countries, the *process of decentralisation in China* has resulted in local governments' acquiring responsibilities without necessarily having the means to carry them out (unfunded mandates). The Chinese government has also encouraged local authorities and enterprises to find ways to finance their operations. This has created a number of perverse incentives and conflicts of interest. Until recently, EPBs depended on revenues from the pollution levy system to finance their operational costs. The EPBs were involved in both conducting and assessing EIAs, and environmental monitoring stations could earn revenues from enterprises, but not from EPBs. These conflicts of interest appear now to have been addressed. Nevertheless, EPBs need to identify alternative ways to finance their activities. Charging for services associated with permitting, as practised in some OECD countries, could be one approach.

Fourth, the *accountability of local leaders* and heads of EPBs is largely limited to local People's Congresses which, in the past, have also tended to prioritise economic development over environment. However, there are signs that in the richer provinces an increasingly affluent middle class is demanding a better environment, and that this message is reaching local leaders. In some of these provinces, a better environment is also seen as an attraction for some types of foreign investors and for activities like tourism. There has also been some experimentation with approaches to seek more direct feedback from the public.

2.3   Regulatory instruments

The legal and regulatory framework

The 1989 *Environmental Protection Law* is the main legal basis for environmental management in China. It is currently being revised to adjust the environmental regulatory framework in light of the rapid pace of economic development and the new provisions of the 11th FYP. Since 1989, at least 23 laws dealing with pollution control and natural resource (including energy resource) conservation have been enacted by the National People's Congress (Table 2.5), and more are under development. The legal statutes are supported by more than forty State Council regulations, approximately 500 standards, and more than 600 other legal or norm-creating documents, which set rules and specify tools for implementing the legislation. In addition, some estimates suggest that more than 1 000 environmental laws have been passed at provincial and municipal level. Local governments may set more or less stringent standards than those already covered by the national law (depending on the assimilative capacity of the local environment). Such regulations must be forwarded to SEPA for review and published in a national register.

China now has a comprehensive and modern set of laws to address environmental issues, along with regulatory and economic instruments to implement them. Emission, discharge and ambient (quality) standards are central to the control of emissions from enterprises. Other regulatory instruments are used such as discharge permits and EIAs. Specific instruments have also been applied in urban environmental management, including: "quantitative assessment" for urban environmental control and improvement,[14] the "goal-responsibility system" of environmental protection,[15] environmental model cities, and air pollution indexes.

Over the last decade, the regulatory focus has shifted away from *reducing pollution concentrations toward reducing total pollution loads*. This has been achieved by: reforming emission and ambient quality standards and combining concentration and load discharge criteria; banning the use of high-sulphur coal; and closing down selected particularly polluting production processes (the "15 small types of industry"[16]). There has also been a shift from "end-of pipe" pollution control to pollution prevention by more integrated and comprehensive approaches, such as cleaner production and the "circular economy". Penalties for infringing environmental laws have become more stringent and in some cases are now punishable under civil as well as criminal law.

However, the volume and complexity of China's laws and regulations make it difficult to provide clear signals to the regulated community, and it can be difficult to access certain legal documents. An *environmental code* compiling all of China's environmental acts would help to increase the transparency and readability of the regulatory framework. In addition, the environmental laws and regulations would generally benefit from a sharper focus on reducing the most important human health risks and protecting key natural resources. Compared with a number of OECD countries, China's also has a few regulatory gaps, such as for soil pollution prevention and the management of chemical substances. The current polluter registration and pollution permitting system is fragmented and not backed up by legally-binding provisions and procedures. The more systematic use of an *integrated permitting system* for large enterprises and those with significant environmental impacts, and a simplified system for smaller units, could help strengthen pollution prevention and control.

Environmental enforcement and compliance assurance

Chinese leaders have identified inadequate enforcement as one of the key factors in China's deteriorating environmental situation. The 9th, 10th and 11th FYPs emphasise the need to strengthen *environmental enforcement and compliance assurance*. Responsibility for compliance assurance lies principally at the sub-national level, while SEPA is responsible for providing guidance to national and local enforcement staff for investigating non-compliance and taking enforcement actions. SEPA is involved, however, in compliance assurance actions concerning companies under the direct supervision of the State Council, and it also supports the EPBs in carrying out enforcement campaigns.

In 2003, the *Supervision and Enforcement Bureau* was established as part of SEPA. In keeping with State Council regulations, the bureau consists of 45 enforcement officers and is responsible for: investigating and supervising violations of environmental laws and ecological degradation; co-ordinating resolution of "transboundary" (inter-provincial) environmental disputes; inspecting polluted sites for possible evidence concerning violations of environmental laws; and helping SEPA to formulate policies and standards applicable to the enforcement of environmental regulations.

In 2004, China had more than 3 000 *environmental inspection agencies* with about 50 000 people involved at the state, province, city and county levels (SEPA, 2000-04). Several provinces and centrally administered municipalities, such as Henan, Hubei, Beijing and Tianjing, have established independent institutions, *separate from the environmental protection agencies,* to take charge of *inspections.* However, their capacities are inadequate; even Shanghai, which has received national recognition for its environmental leadership, has just 100 environmental inspectors with responsibility for more than 20 000 factories. SEPA recognised that the dual leadership over local EPBs has compromised the stringency of environmental enforcement. It thus announced the establishment of nine *regional centres* with enforcement co-ordination functions, bringing local enforcement staff directly under SEPA control. The staff size and funding mechanisms are still being discussed.

The *strategy for environmental enforcement* has relied on three key elements: inspections by local EPBs, joint inspections (campaigns) carried out by the central government and EPBs, and mobilising the population, including media and NGOs, to help promote compliance with environmental requirements.

The level of compliance with national pollution standards, permits (violations can include improper action during permit registration and releasing pollutants in excess of allowable limits) and payment of fees is checked through *environmental inspections*. Private enterprises are inspected by the EPB of the jurisdiction where they reside. State-owned enterprises are assigned an administrative status and are inspected by an environmental protection agency with an equivalent administrative status. Many inspections are in fact stimulated by public complaints about pollution incidents.

In cases of *non-compliance*,[17] inspectors have recourse to a variety of instruments, including warning letters, fines, withdrawal of licenses or permits, "pollution control within deadlines" and, in case of persistent non-compliance, shutting down the facility. However, the enforcement resources are spread very thin, capacities are weak and are linked to income levels in the province concerned, and efforts are not sufficiently targeted. Recently, incentives have been introduced to promote technological renovation, phasing-out of outdated technologies and products, and cleaner production, in exchange for extending the deadline for sanctions.

In recent years, more than 2 million inspections have been conducted annually in China, and 80 000 to 120 000 violations penalised (Figure 7.3). The Environmental Protection Law contains provisions for *imposing fines* in cases of refusing an on-site inspection, resorting to fraud during inspections, refusing to file a report or submitting a false pollution report, exceeding national or local discharge standards, or failing to pay fees for exceeding pollutant discharge limits. Different levels of EPBs have differing levels of responsibility and authority to impose penalties. For example, county level EPBs can impose fines up to CNY 10 000, city EPBs up to CNY 50 000, and provincial and municipal EPBs up to CNY 200 000. Severe pollution of a water body can lead to a fine of up to CNY 200 000, and a polluting discharge that causes a water pollution accident can be fined up to 20% of the direct economic damage (up to a maximum of CNY 200 000) or, in case of large economic damage, 30% of the damage (up to a maximum of CNY 1 000 000). Similar provisions are contained in the Air Pollution Law. The revenue from fines goes to the central government budget.

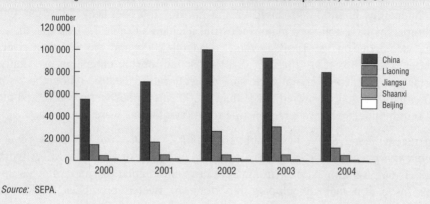

Figure 7.3 **Sanctions for environmental non-compliance, 2000-04**

Source: SEPA.

EPBs can gain court assistance to collect fees and fines. The level of the fines is subject to local government *discretion*, but the decision-making must be "open and fair". The severity of the penalty is determined taking account of such factors as: i) the extent to which regulations were violated; ii) the number of violations; and iii) the responsible party's response (e.g. whether voluntary corrective action was taken). Fines are also levied by the NDRC in cases of serious breaches of the law.

The 1989 Environmental Protection Law, as well as China's water, air and solid waste laws, provides for the application of *criminal sanctions* in the case of particularly egregious cases. The police are charged with investigating cases of environmental crime, together with the prosecutor's office. The first case involving criminal sanctions for an environmental offence involved the Yuncheng City People's Court in Shanxi province in 1998, which convicted a manager of a paper factory for purposeful pollution. He was sentenced to two years' imprisonment and was fined CNY 50 000. To date, a number of high profile cases of environmental crime have been brought to court, but this avenue has been used infrequently due to uncertainly over the legal responsibilities and long judiciary procedures. SEPA is currently examining possible amendments to the Criminal Law to address these issues.

In addition to routine inspections, SEPA has organised special *environmental enforcement campaigns* together with the local EPBs. In the late 1990s, campaigns were launched to clean up the Bohai sea, the Huai river and the Tai lake. Between 1995 and 2000, more than 84 000 heavily polluting plants, usually from township and village industrial enterprises, were shut down along the Huai river. These included oil refineries, cement plants, thermal power plants and metallurgical mills. Campaigns in 2002-04 focused on the mining and chemical industries. Often such campaigns have followed major industrial accidents (Table 7.4), as was the case after the accidents on the Tuojiang river in March 2004 and the Songhua river in November 2005 (Box 8.3). In 2005, SEPA carried out a campaign to improve industrial environmental compliance and protect public health. Altogether, some 30 000 violations were penalised, more than 2 600 enterprises were closed, and more than 300 individuals who were responsible for the violations were penalised.

In recent years, SEPA has sought to *mobilise public opinion to improve the environmental performance of enterprises*. In May 2002, for example, SEPA worked on a television programme on environmental laws and regulations and followed up by sending inspection teams, including reporters, to twelve provinces, regions and municipalities to review the performance of thousands of factories and projects. Environmental performance rating of enterprises and information disclosure schemes are also being used to publicise the environmental performance of enterprises. These schemes have also been taken into account by banks *when assessing loans to enterprises* (Jiangsu province considers itself a leader in this method). In several provinces, environmental telephone "hot lines" enable citizens to report on cases involving possible non-compliance with environmental requirements.

Assessment

Chinese legislation provides a *comprehensive array of tools to enforce environmental laws and to promote compliance* with them. SEPA and other departments under the State Council have worked hard to promote the effective use of these tools with the limited means at their disposal. In some of the richer, eastern provinces, there are positive examples of implementation of environmental laws.

Even though environmental authorities have imposed a number of *sanctions for non-compliance*, a wide gap exists between what EPBs are authorised to do and what they actually do when enterprises violate environmental rules. In many cases, approved and installed pollution control equipment (e.g. to reduce water and air emissions) is put into operation only when inspectors are expected, as the polluters are more interested in saving operation costs or, as in the case of waste water, the communities cannot afford to operate sewage treatment plants. A significant proportion of small and medium-sized enterprises, including in rural areas and in the service sector in urban areas, are not inspected due to lack of capacity, the constraints of "pragmatic" enforcement (considerable discretion is applied by the EPBs in determining how they will enforce environmental requirements), or conflicts of interest between the economic and environmental parts of the administration. Pragmatism is reflected in the EPBs' reliance

Table 7.4 **Selected accidents[a] involving hazardous substances,[b] 1993-2005**

	Date	Location	Origin of accident	Products involved	Number of deaths	Number of injured
1993	11.01	Heilongjiang	Explosion	Natural gas	70	..
	01.03	Guizhou	Poisoning	Arsenic	42	..
	26.06	Henan	Explosion	Chemicals	27	32
	05.08	Guangdong	Explosion	Acid	84	510
	06.08	Guangdong	Explosion	Chemical gas	12	168
	29.08	Fujian	Explosion	Fireworks	27	..
	19.09	..	Fire	Gas	81	19
	28.09	Hebei	Gas leak	Gas	..	500
	11.10	Heilongjiang	Explosion	Natural gas	70	..
	19.11	Beijing	Fire, toy factory	Plastic	84	38
	25.11	Dulin	Explosion and fire	Fireworks	26	..
	26.11	Hunan	Explosion	Chemicals	61	18
	05.12	Hebei	Explosion	Chemical spill	25	..
	12.12	Hebei	Explosion	Chemical spill	6	585
	13.12	Fujian	Fire	Chemicals	61	12
1994	17.06	Guangdong	Fire	Chemicals	76	160
	02.08	Beijing	Explosion	Dynamite	73	99
1996	01.01	Guizhou	Explosion	Chemicals	..	407
	31.01	Hunan	Explosion	Explosives	125	400
	01.07	Sichuan	Explosion	Gas	36	52
1997	21.09	Fujian	Fire	Gas	32	4
1998	24.01	Beijing	Road accident, explosion	Fireworks	40	100
	26.01	Shanxi	Poisoning	Alcohol frelate	27	700
	25.11	Beijing	Industrial accident	Chemicals	35	..
2000	10.03	Jiangxi	Explosion	Fireworks	33	10
	22.04	Shandong	Fire	Chemicals	38	20
	30.06	Guangdong	Explosion	Chemicals	36	160
	05.08	Jiangxi	Explosion	Chemicals	27	26
2001	15.11	Zhejiang	Gas leak, aniline	Gas	..	700
	30.12	Huangmao	Explosion	Fireworks	34	61
2003	28.07	Hebei	Explosion	Fireworks	35	91
	23.12	Sichuan	Explosion	Gas deposit	243	10 000
	30.12	Liaoning	Explosion	Fireworks plant	35	20
2004	20.04	Nanchang	Gas leak	Gas	..	282
	15.01	Heilongjiang	Poisoning	Chlorine	..	134
	15.06	Fujian	Gas leak	Phosgene	1	300
	22.06	Jiangxi	Food poisoning	..	..	258
	04.10	Beijing	Explosion	Chemicals	34	55
	25.10	Fujian	Food poisoning	Chemicals	..	160

Table 7.4 **Selected accidents*a* involving hazardous substances,*b* 1993-2005** *(cont.)*

	Date	Location	Origin of accident	Products involved	Number of deaths	Number of injured
2005	11.01	Shanxi	Explosion	Chemicals	25	9
	21.06	Shanxi	Explosion	Chemicals	..	285
	13.11	Jilin	Explosion	Benzene, nitrobenzene	5	86
	22.12	Sichuan	Explosion	Chemicals	42	336
	23.12	Sichuan	Gas explosion	Gas	44	70

a) Inclusion criteria: Only disasters involving 25 or more deaths, 125 or more injured, or 10 000 or more evacuated or deprived of water supply.
b) Excluding: Oil spills at sea from ships, mining accidents, voluntary destruction of ships or airplanes, damage caused by defective products.
Source: SIGMA, UNEP, CRED International Disaster Database.

on their relationships, or guanxi,[18] with regulated enterprises, i.e. in developing mutual understanding, providing technical and financial assistance, and negotiating reasonable compliance deadlines. The "pragmatic" approach has been applied with some success in China, but EPB staff frequently stop short of revoking permits for serious violations of their conditions, or choose not to fine enterprises for non-compliance in order to maintain harmonious relations.

Local governments sometimes collude with enterprises to circumvent costly environmental requirements that might affect their competitiveness or affect projects favoured by local leaders. For example, enterprises in China are able to escape the supervision of local EPBs by asking local officials to sign permit documents without the approval of environmental administrations. Some local governments set up "umbrella" schemes, prohibiting the environmental enforcement authorities to inspect, and to impose and collect fees and fines from, firms that are seriously polluting, but are considered important to the local economy (as they provide tax revenue or employment).[19] Such interference renders environmental enforcement ineffective. It also leads to an administrative culture in which career prospects are better served by not taking responsibility for problems (and not reporting them to superiors) than by actively trying to resolve them.

The ultimate check on this system is when the pollution leads to demonstrable impacts on human health and the environment, as was the case in several industrial accidents (e.g. the Tuojiang and Songhua river accidents). Such accidents provide strong arguments for strengthen the *evaluation of environmental performance of provincial and local governments*. Doing so would be all the more effective if it were supplemented by greater "downward accountability" towards citizens, for instance, by enhancing the transparency of administrative actions.

2.4   Economic instruments

China uses a *range of economic instruments* as part of its environmental policies. Pollution charges have been used for over twenty years and produce significant financial resources. Other instruments such as user charges, tradable permits and deposit-refund systems have also been applied (OECD, 1997). The legal framework fully recognises their importance.

Pollution charges

The legal basis for China's *pollution levy system* was established in the 1989 Environmental Protection Law and in the laws on air, water, waste and noise. Originally, only discharges that exceeded pollution concentration standards were subject to a charge. In 2003, the State Council promulgated an ordinance on collecting and managing the pollution discharge fee, which directed revenues to the Ministry of Finance or the Department of Finance at the sub-national level. The ordinance bases the pollution charge on the concentration and volume of the pollutant, irrespective of national standards. Polluters have a 20-day grace period to pay the monthly or quarterly charge. In addition to paying fees for pollutants exceeding standards, enterprises violating payment requirements may have to pay four other kinds of penalty charges, referred to as "four small pieces".[20]

In 2004, the total revenue from the pollution levy system was CNY 9.42 billion, collected from nearly 740 000 enterprises (Figure 7.4).[21] This is equivalent to 5% of China's total PAC investment expenditure. However, the *full incentive effect* could be enhanced by applying higher rates and some adjustments to the system. Polluters are required to report increased discharges, and rebates are possible when pollution reductions are verified. Although the EPBs issue notices of discharge fees, the amount is usually negotiated rather than calculated using formulas detailed in regulations. The charge can be reduced or even eliminated at the discretion of local regulators after appropriate inspections. The charge may also be postponed if the polluter cannot afford to pay it, although reductions or exemptions are not allowed in such cases. Such discretion introduces considerable variation in regional enforcement practices, which would be more effective if harmonised. Moreover, pollution charges are still significantly *lower than the cost of pollution reduction*, despite recent rate increases. For example, SO_2 rates were increased from CNY 0.21/kg to CNY 0.42/kg

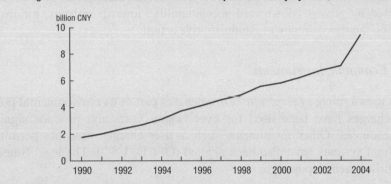

Figure 7.4 **Revenue collected under the pollution levy system,**[a] 1990-2004

a) Based on charges applying to industrial sources and levied on water discharges as well as air emissions,solid waste,
    noise and radioactive substances.
Source: SEPA.

in 2004 and to CNY 0.63/kg in 2005. A new charge of CNY 0.6/kg of NO_x was introduced in 2005. The fee collection rate is still low, estimated to bring in, on average, 50% of the charges imposed (varying between 10% in Western provinces to 80% in coastal areas).

Of the total *revenue produced by the pollution levy system*, 10% goes to the central government and 90% remains at the local level. Since 2003, the revenue from collected charges has been transferred to the ministry or departments of finance. The resources are still earmarked for environmental improvement, but they are no longer used to defray the running expenses of the inspection agency. Instead, they may be used for general environmental protection, the purchase of monitoring equipment, or new technology. The funds are redistributed (in the form of grants or soft loans) for pollution control projects. SEPA and the Ministry of Finance allocate these funds on the basis of proposals provided by the provinces.

User charges

User charges have been gradually applied on household and industry use of environmental services and natural resources. For environmental services, the charges are usually limited to operating costs and do not cover capital investment.

In Langfang, the *price of water for farmers* using their own wells is CNY 0.30/ m^3; in Nanjing it is CNY 0.13/m^3. The *price of water* for industrial production in

Northern China, including treatment and transport, amounts to CNY $5/m^3$. The price for water for industrial purposes in Xi'an is CNY $2.4/m^3$ and does not include taking water from the reservoirs, transporting it over 80 km and purifying it, which costs an estimated additional CNY $1.5/m^3$. The estimates assess the true cost close to CNY $12/m^3$. However, there is much opportunity for on-site recycling of water. *Charges for household water use* is around CNY $2/m^3$, ranging from CNY $2.1/m^3$ in Langfang (which includes CNY 0.40 for sewage disposal) to CNY $1.95/m^3$ in Xi'an (of which CNY 0.35 is for waste water treatment). In Langfang, the use of water in all households is metered. Water prices depend on social factors, although the price structure is not progressive as yet. There are public hearings when prices are set.

In China, a *waste water treatment charge* is paid by all customers connected to a centralised water supply system, often irrespective of whether their waste water is being collected and treated. Development planning commissions at the city level play a role in setting tariffs for projects financed by central government transfers. Waste water charges vary from CNY $0.9/m^3$ in Beijing and Shanghai, to CNY $0.7/m^3$ in Guangzhou, CNY $0.5/m^3$ in Xi'an, and as low as CNY $0.12/m^3$ in some cities in Sichuan province. Some households are also charged a sewage network fee; this may amount to CNY 0.15/t in Jilin, but in most cases there are no separate sewerage charges, as users' waste water charges are also expected to cover the costs of waste water collection. The enterprise responsible for water supply collects charges for waste water treatment.[22] However, many cities do not levy charges for waste water treatment services, and even if such a charge is in place, the revenues still fall far short of what is needed to cover even basic operation and maintenance of waste water collection and treatment facilities.

The fee structure for *municipal solid waste collection* is complex, but is about CNY 6 per household per month. The municipal government pays the fees for those who cannot afford them. Industry fees are different and depend, for example, on who pays to transport garbage to a landfill. Fee collection rates vary between 10% in the Western provinces to 80% in coastal areas.

Emissions trading

During the 1990s, SEPA conducted a number of studies and pilot projects on *emissions trading*, mostly in connection with SO_2. In 2002, in order to assess the feasibility of a nation-wide emissions trading scheme, SEPA organised pilot applications in seven provinces. This work has helped to identify some of the conditions necessary to establish emissions trading in China. In addition, two power plants in Jiangsu province reached an agreement to trade SO_2 allowances to meet total emission limits. SEPA's work demonstrates the potential for using emissions trading to help reduce SO_2 emissions in a cost-effective way. In addition to designing appropriate administrative arrangements, it may also be necessary to consider redesigning other instruments for total emission control, in particular emission/ discharge permits, to help achieve emission reductions in the most cost-effective way.

In order to protect environmentally vulnerable areas, especially in Western China, the Chinese authorities established comprehensive *ecological compensation (eco-compensation) systems*. For example, the government implemented a "Grain for Green" programme (also known as "Sloped Land Conversion Programme") in 1999. The purpose was to set aside sloped cropland so as to increase forest cover and prevent soil erosion. Where possible, farmers set aside all or part of certain types of land and plant seedlings; in return, the government provides free seedlings and compensates the farmers with in-kind grain allocations and cash payments. While there is some concern about implementation in the long run, the "Grain for Green" programme has so far had a positive effect on the welfare (net income per capita) of most farmers in the programme areas and has helped to increase the amount of forested areas.

Further plans include setting up *tradable quotas* for protected areas in Guangdong province. Prefectures short of reserves and protected areas will have to "buy" quotas from regions with more than the provincial average (6.78%) of total land area under protection. This is expected to benefit poor, mountainous areas, which tend to have more protected areas and will be able to "sell" quotas to more developed, urban areas, thus generating funds for ecological maintenance. Similar plans exist in relation to water resources, to encourage inter-regional, inter-industrial and upstream-downstream trading. This is expected to increase the flow of funds from heavily populated water-using areas and industries to more remote areas with a water surplus. Pilot projects in Zhejiang, Ningxia and Inner Mongolia have already proved successful and will be replicated in other regions.

Other economic instruments

The deposit and refund system for *packaging* has been applied on an informal basis in Shanghai, and SEPA is currently investigating the feasibility of applying it to packaging and hazardous waste items. In some provinces, *monetary deposits* (performance bonds) were collected during the construction of enterprises to guarantee the application of the "three simultaneities". If the construction satisfied the requirements, the deposit was returned to the investor; if not, it was retained by the administration as a fine. A similar approach has been used in the recovery of *mining sites*. The Law on Mineral Resources required mining enterprises to undertake measures to recover the land from their operations, such as planting trees or grasses. Hebei province, for example, requires a deposit of CNY 500-1 000 per mu (15 mu = 1 ha) to guarantee land recovery.

2.5 Environmental management in industry

Voluntary approaches

Voluntary approaches are not an important instrument in Chinese environmental policy. SEPA has a programme of recognising national, environmentally-friendly enterprises, but, by early 2006, fewer than 200 enterprises had been named. Some publicly-traded companies are voluntarily reporting on environmental performance. A Chinese branch of the World Business Council for Sustainable Development has been established, largely with participation of multinational enterprises. In general, the environmental awareness and performance of domestic Chinese enterprises is weak. Exposure to foreign markets with well-established environmental regimes, and commerce with multinational companies based in China that have strong corporate environmental policies, are probably the main drivers for improvement at this time.

Environmental management systems

In 1997, the State Bureau of Technical and Quality Supervision adapted the *ISO 14000 series* into equivalent national standards. A national approval scheme for ISO 14001 certification was introduced along with a system of examination of certification entities by a national accreditation body and national registration of auditors. By 2004, the number of certified companies in China reached over 8 000, the second highest number in the world (after Japan). Given that China ranked tenth in the world in 2001, with roughly 1 000 certified companies, the growth rate has been impressive. This growth has been supported by the Chinese government's move to offer tax incentives (mostly in the form of VAT reductions or rebates) to encourage companies to achieve certification.

SEPA's Environmental Certification Centre, established in 2003 with a staff of 40 qualified ISO 14001 auditors, is a financially autonomous body that processes applications for ISO 14001 certifications. The applications come primarily from fast-growing, small and medium-sized Chinese companies that participate in global markets. The standard screening fee for ISO 14001 certification is CNY 30 000. China also has over 100 other *certification entities*, including those working on ISO 9000 certification. Most of these units were established with government support. About 300 experts based at universities and research institutions serve as auditors.

Since the introduction of the "*cleaner production*" (CP) concept in the 1980s, SEPA's main focus has been plant-level demonstration projects, training, capacity building and policy advice. After an initial period of separate and locally-driven initiatives, SEPA issued the 1997 Recommendation on Promoting CP in China, requiring local environmental protection agencies to integrate CP into environmental management policies. A separate, comprehensive Law on Promotion of Cleaner Production came into effect in January 2003.

China's CP activities have been supported by more than 40 *cleaner production centres* set up across the country. They include one national centre, four sectoral centres (covering the petrochemical, chemical, metallurgy and aircraft manufacturing sectors), and 11 local centres (in Beijing, Shanghai, Tianjin, Hohhot, Shaanxi, Heilongjiang, Shandong, Jiangxi, Liaoning, Inner Mongolia and Xinjiang). The National Cleaner Production Centre has held over 600 cleaner production training courses serving more than 20 000 trainees, of which about 2 000 are certified CP trainers.

Several studies show that CP practices have both reduced *pollution and increased the production efficiency of enterprises* in China. These practices contributed to a 20% reduction in emissions while generating economic returns of CNY 500 million annually (Wang, 2003). In addition, some of China's environmental management regulations and processes, such as environmental impact assessment and the "three simultaneities" system, also encourage enterprises to introduce CP to improve their environmental management.

Eco-labelling of products

The China Certification Committee for Environmental Labelling of Products (CCEL) was formed in 1994 by SEPA, the State Bureau of Technical Supervision. The CCEL involves professionals from various disciplines that develop criteria for and oversee the operation of *environmental labelling*. The CCEL is recognized by SEPA as the only authority to conduct third-party certification and to award China Environmental Labels. Annual inspections and random sample checks are carried out to ensure that good standards are maintained. Between 1994 and 2005, assessments were conducted for 800 enterprises, and 12 000 products were awarded various environmental labels.

2.6   Land use planning and regulation

Land and soil are scarce in China. While China's territory is vast, its population of 1.3 billion (21% of the world's population) has the smallest per capita land base in the world. China's growth and development have placed high pressures on the land and soil resources, mainly through soil pollution, soil erosion and the rapid conversion of land into industrial and urban uses. Pollution affects 10 million of the 120 million hectares of cultivated land while 2 million hectares are irrigated with polluted water and 130 000 hectares are used for waste disposal. Worldwide, China is considered to be one of the countries that suffers most from erosion. From 1996 to 2004, arable land conversion led to a loss of 8 million hectares; in 2005, the loss of arable land was reduced to 360 000 hectares due to stricter restrictions on the acquisition of farmland for construction purposes by the central government. Estimates made for the year 1999 of the economic costs of land and soil degradation reach CNY 62 billion (direct costs) and CNY 270 billion (indirect costs), or 4% of China's GDP. Land and soil are key resources for agriculture (Box 7.4).

The two major ministries involved in land management are the *Ministry of Land and Resources* and the *Ministry of Construction*. The Ministry of Land and Resources is responsible for both urban and rural land throughout China, the administration of land ownership and land use rights, the land register, land use planning and the protection of land resources, and the conveyance of land use rights (LURs) in co-ordination with other relevant departments. The Ministry of Construction is responsible for urban planning and the provision of major infrastructure. In *rural areas*, the administration at the county level is responsible for overall land use planning within its respective jurisdiction, including issuing "land contract certificates" to farmers and ratifying the conversion of farmland to non-agricultural uses. It is only at the county level and above that land can be approved for *conversion to non-agricultural use*, with the approval level required increasing progressively depending on the area of land being considered for conversion. In some areas, townships may influence village land policies, including village-wide land reallocations, which can lead to conflicting situations between village leaders and farmers especially when the land is leased to external investors without consensus from local farmers and without proper compensation for lost access to land.

Land ownership and farmland

Farmland is *de facto owned by village collectives, which extend land lease contracts to individual farm households*. Households have most of the rights associated with property ownership: they can use, sub-lease and transfer land, but cannot sell it. This is the result of policy changes initiated in 1978. With adoption of the *"household production responsibility system" (HPRS) in 1978*, farmers, as members of the collectives, have been allocated land use rights for a specified period with the condition that they fulfill grain delivery quotas to the state (or pay with a cash equivalent), pay various fees and charges to the collectives, and "maintain the integrity" of the contracted land. The HPRS broke up commune production teams (established in the mid-1950s) and made the farm household China's basic unit of agricultural production. To maintain egalitarian access to land, households have generally been allocated use rights to agricultural land on a per capita basis, without consideration of the amount of land they owned before the collectivisation of the 1950s (in contrast to land reforms in transition economies of Central and Eastern Europe).

Box 7.4 Agriculture and the environment

Food security has been an overriding objective of agricultural policy in China. Historically this has been achieved by expanding agricultural land. However, although China's land area is vast, cultivated land is scarce and accounts for only 13.5% of the land surface. Since the mid-1980s, cultivated land has decreased at the rate of 0.5 million hectares per year due to industrial and urban development and road building. Increased output has been achieved through greater use of fertiliser, pesticide and mechanical inputs. As a result, China, with 10% of the world's farmland, has been able to feed 21% of the world's population. Despite constantly increasing absolute levels of food production, agriculture's share of GDP fell from just under 30% in 1990 to 15% in 2003, largely because of the faster rate of growth in other sectors. Nevertheless, agriculture still accounts for 40% of employment. Further improvements in agricultural productivity will depend on a substantial reallocation of land resources away from land-intensive crops (such as grains and oil seeds) to more labour-intensive, higher value-added products (such as fruits and vegetables), a continuing trade in products consistent with China's comparative advantage and better husbandry of natural resources.

Despite the remarkable achievement in feeding China's large population, there are increasing environmental constraints on further increases in agricultural productivity. Salinisation affects 8.5% of the land area. Acid rain affects 25% of Chinese territory. Soil erosion occurs on 38% of the total land area, and almost two-thirds of agricultural land. Nearly 9% of the total land area suffers from desertification, with a further 14% at risk. Water scarcity is exacerbated by inefficient use of water in the agricultural sector (agriculture accounts for 70% of total water consumption) and by extensive pollution associated with fertilisers, pesticides and sharply increasing levels of livestock waste. Per capita water resources are between one-quarter and one-third of the world average. Soil erosion and deforestation are linked to serious flooding. One estimate of the costs associated with flooding in the 1990s amounted to CNY 80 billion (about USD 9.5 billion).

The Chinese government began to place a high priority on the environmental impacts from agriculture in the late 1990s, after food became relatively abundant. In addition to the investments in agri-environmental measures (e.g. "grain for green" policy), a variety of laws have been enacted concerning soil and water conservation, wildlife protection and land protection. Various monitoring programmes relate to desertification, soil and water retention and biodiversity. Demonstration projects, supported by the Ministry of Finance and the Ministry of Water Resources, have been implemented to reduce the environmental impacts of agricultural practices in the context of river basin management. Training has been provided to farmers to optimise the application of chemical inputs. Co-operation projects have been conducted with donors to support sustainable agriculture and use of land, water, grassland, forest and coastal resources.

The first round of lease contracts was granted at the beginning of the 1980s within the framework of the HPRS. A new round of leases was launched in 1998. By providing for longer-term leases, typically 30 years, the 1998 Law on Land Administration has created *favorable conditions for voluntary transfers of land use rights among farmers*, thereby allowing for enlargement of farm size. The 2002 Law on Rural Land Contract seeks better protection of farmers' land use rights: the contracted land cannot be taken back and reallocated by the collective during the period of the contract. The law also specifies that migrants settling in towns (but not in cities) retain their land use rights in their original village and that farmers should be sufficiently compensated for land transferred to other uses. Article 10 of the Constitution was amended in 2004 to require *compensation* for land taken over by the State.

Land legislation and planning

Control of land management by the national government is based on the 1998 *Law on Land Administration*. As Chinese authorities have become increasingly concerned with the loss of agricultural land, the law contains strong provisions for its *protection*. It also aims to balance the demand among uses and land use planning to achieve an appropriate density and to contribute to *environmental protection*.

Comprehensive and *long-term strategic land use plans* are prepared at all administrative levels. *Topic plans* focus on land use and development sectors such as tourism, water conservation and land reclamation. *Detailed plans* are project and site specific. In addition, the 1998 law provides a detailed legal framework for land management: *land use plans* must be prepared at each level of government based upon land quotas set by central government that each lower-tier local government has to allocate and meet. The law applies a system of state control over the main use of land, which can be classified as "agriculture", "construction" and "unused". Land use plans, once approved, are to be "strictly carried out". Changes to the plans can only be made with authorisation of the provincial governments and the State Council, and not by local government (Pieke, 2002).

Implementation of the Law on Land Administration has not been fully effective. In 2002, the Ministry of Land and Resources registered over 100 000 violations of it. There is also often a mismatch between measured and actual land, its intended and practical use and reporting. In addition, inadequate *compensation for farmers* whose land has been converted to non-agricultural uses has been an important source of *social conflict* in some parts of rural China in recent years.

Functional zones

The 11th FYP introduced the concept of *functional zones* associated with environmental and natural resource concerns. The purpose is to ensure that local governments respond to *different sets of incentives* according to different functional zones (other than the dominant objective of promoting GDP and growth), emphasizing better environmental performance and taking account of environmental carrying capacity. The four categories of functional zones are: *optimised development zones* (which include areas where land development density is already high and environmental resource carrying capacity starts to decline); *key development zones* (which include areas where resource environmental carrying capacity is relatively strong and the economic and population concentration condition is relatively good); *restricted development zones* (which include areas where environmental resource carrying capacity is relatively weak and a large-scale, concentrated economic and population condition is not desired); and *prohibited development zones* (which include legally established nature reserves).

Notes

1. This implies an annual growth rate of 7%, which the 2005 OECD Economic Survey of China judged to be feasible.

2. Away from the planned economy and production targets, the 11th FYP: i) sets broader objectives and goals (which open up opportunities for market forces to allocate resources more efficiently); and ii) programmes a set of public investment projects.

3. The Commission of Environmental Protection of the State Council played an important role in environmental decision-making in the 1990s. It consisted of key people from 31 ministries and commissions and several representatives of large enterprises and the media. The commission co-ordinated the ministries' environmental efforts and helped resolve controversies in proposed laws related to the environment. During the 1998 reorganisation, the commission was dismantled.

4. Leading groups are influential bodies that countries representatives of the Communist Party and of the administration at the level of the State Council.

5. The system of "three simultaneities" is also called also "three synchronisations", "three simultaneous steps", or "3S".

6. Standards for classifying the potential environmental impacts of a particular construction project, together with a detailed list of examples, are contained in SEPA's Catalogue for Construction Project EIA Classification Management, which is updated periodically.

7. In addition, if the proposed projects are not constructed within five years of EIA approval, the procedures have to be reassessed by SEPA or an EPB before construction can proceed. The EIA procedure is simplified considerably if the project is undertaken within an industrial park (economic development zone) or other government plan for which environmental impact assessments have already been conducted and approved.

8. At the same time, it accounted for 2.7% of gross fixed capital formation in 2004, which fell short of the target of 3.6% established in the 10th FYP.

9. The Chinese definition of environmental investment expenditure and the OECD definition of PAC investment expenditure both: i) cover public and private investment expenditure; and ii) exclude water supply and nature protection expenditure. However, the Chinese definition differs by including a number of additional elements (e.g. investment expenditure in energy efficiency, in fuel switching, in urban amenities). This results in Chinese estimates of environmental investment expenditure being significantly higher, than the PAC investment expenditure as defined by the OECD.

10. The design of eco-compensation schemes in China has been analysed by a special CCICED Task Force set up in 2005. Results are expected by the end of 2006. A direct compensation for ecosystem services was administered through the Forest Ecosystem Compensation Programme (Chapter 6).

11. The "one control" required that total emission loads of major pollutants in all regions be kept within nationally specified levels (1995 standards). The "double attainments" required that emissions from all industrial sources meet both national and local standards by the target date.

12. With GDP calculated using NBSC methods, preceding the January 2006 adjustment.

13. The structure and position of the national environmental agency has been evolving over the last two decades. The first version of China's top environmental body, the Environmental Protection Bureau, was a unit under the State Council set up in 1974 with a staff of 20. Subsequent reorganisations of the governmental system in China, particularly in 1998, led to the creation of a SEPA placed directly under the State Council. Even though SEPA does not have a permanent seat in the State Council, its head has the status of Minister, and participates in State Council meetings when environmental matters are discussed. Currently, SEPA's Head reports directly to the Vice Premier in charge of environmental protection.

14. By use of environmental indicators.

15. Through this system provincial governors, city mayors and county magistrates sign contracts specifying environmental goals to be achieved within their jurisdiction.

16. Mostly township and village industrial enterprises (TVEs), in particular small paper mills and dying and metal coating enterprises.

17. The following *offences* are the most common in order of frequency: failure to comply with the EIA or "three synchronisations" requirements; failure to pay a pollution levy; operating pollution control equipment without the necessary permits; engaging in hazardous waste disposal without the required permit; and releases to air and water as a result of industrial accidents.

18. Guanxi, which has long been an element of Chinese life, is based on a blend of exchanges and mutual relationships that "create feelings of responsibility and obligation on the one hand and indebtedness on the other". In general, guanxi is maintained over long periods.

19. Even worse are cases where county governments revoked EPB decisions to fine an enterprise or did not permit the local EPB to apply for a court order to execute an administrative fine.

20. These include: i) a late payment fee of 0.1% per day; ii) an annual 5% increase in the charge rate after paying the charge for two years; iii) a double charge for enterprises built after 1979 that violate standards, close down existing treatment facilities without approval of the local EPB, or fail to comply with administrative orders requiring pollution control by a fixed day; and iv) a fine to compensate for economic losses or adverse human health effects caused by waste release.

21. This included approximately 46% collected from charges for exceeding concentration standards, 32% collected as "small four pieces", 18% from charges for SO_2 emissions, and 4% from charges on waste water volume discharges. Sixty per cent of the charges on exceeding concentration standards came from exceeding waste water standards, 25% came from exceeding air standards, 12% from noise and 3% from solid waste standards.

22. This arrangement is expected to increase collection rates as well as to decrease the costs of billing and collection, because the willingness to pay for drinking water is typically higher than willingness to pay for waste water treatment.

Selected Sources

The government documents, OECD documents and other documents used as sources for this chapter included the following. Also see list of websites at the end of this report.

CCICED (China Council for International Co-operation on Environment and Development) (2005), "Final Report of the Task Force on Environmental and Natural Resources", *Pricing and Taxation*, Beijing.

CCICED (2004a), *An Environmental Impact Assessment of China's WTO Accession: An Analysis of Six Sectors*, Beijing.

CCICED (2004b), *Financing Mechanisms for Environmental Protection in China*, Beijing.

China National Cleaner Production Centre (2002), *Five-Year Summary Report (November 1997-March 2002) Policy and Regulations*, China – Canada Co-operation Project on Cleaner Production Expert Study Group, Beijing.

Day, Kristen, A. (ed.) (2005), *China's Environment and the Challenge of Sustainable Development*, Columbia University, New York City.

Deutsche Bank Research (2006), *Environmental Sector in China*, BDR, Frankfurt.

Development Research Center (2004), *National Energy Strategy and Policy Report 2020*, China's Economic Science Press, Beijing.

Economy, Elizabeth (2004), *Rivers Run Black: The Environmental Challenge to China's Future*, Cornell University, Ithaca, New York.

Feng, Zhiming, Baoqin Liu and Yanzhao Yang (封志明， 刘宝勤，杨艳昭) (2005), 中国耕地资源数量变化的趋势分析与数据重建: 1949- 2003 (Analysis on the Trend of Quantitative Changes of Chinese Cultivated Land and Data Reconstruction: 1949-2003), 中国科学院地理科学与资源研究所 (Geographical Science and Resources Research Institute, Chinese Academy of Sciences), 自然资源学报，第20卷第1期, 2005年1月 (Journal of Natural Resources, January 2005).

Hansen, Stein *et al.* (2002), *Green taxation for poor in China*, Nordic Consulting Group ECON-Report.

Ho, Mun and Chris Nielsen (eds.) (2006), *Clearing the Air: Assessing the Health and Economic Damages of Air Pollution in China*, MIT Press, Cambridge.

IEA (2006), *World Energy Outlook*, OECD/IEA, Paris.

Jin, Jian Min (2004), *The Future of Chinese Eco-Industry* (in Japanese), Fujitsu Research Institute, Tokyo.

Ma, Xiaoying and Leonard, Ortolano (2000), *Environmental Regulation in China: Institutions, Enforcement and Compliance,* Rowman and Littlefield Publishing Group, Lanham, Maryland.

Meng, Wei, Luo Han and Lu Lianhong (2006), *The Strategic Environmental Assessment of the Collection of Fuel Oil Tax in China*, SEPA, Environmental Protection, Beijing.

NBSC (National Bureau of Statistics of China) (2006), China Statistical Yearbook on Environment, China Statistics Press, Beijing.

NDRC (2006), *Guidelines of the Eleventh Five-Year Plan for National Economic and Social Development*, Xinhua News, Beijing.

NDRC (2005), *China Medium and Long Term Energy Conservation Plan*, China Environmental Science Press, Beijing.

Nickesen, Alfred and Mitchel, Stanfield (2000), *Toll Road Securisation in China, Roads and Highways*, Topic Note RH 3, World Bank, Washington DC.

OECD (2006a), *Investment Policy Reviews: China*, OECD, Paris.

OECD (2006b), *Challenges for China's Public Expenditure Policies – Towards Greater Equity and Effectiveness*, OECD, Paris.

OECD (2005a), *OECD Economic Surveys: China*, OECD, Paris.

OECD (2005b), *OECD Review of Agriculture Policies: China*, OECD, Paris.

OECD (2005c), *China in the Global Economy: Governance in China*, OECD, Paris.

OECD (2005d), *Financing Strategy of the Urban Waste Water Sector in Selected Municipalities of the Sichuan Province in China*, Project Report, OECD, Paris.

OECD (1997), *Applying Market-Based Instruments to Environmental Policies in China and in OECD Countries*, OECD, Paris.

Pieke, Frank (2002), *The Politics of Rural Land Use Planning in China*, Max Planck Institute for Social Anthropology Working Paper No. 40, Halle.

SEPA (State Environmental Protection Administration) (2006), "China Green Natural Accounting Study Report 2004", SEPA/NBSC, Beijing.

SEPA (2005), Policy and Regulation Department, International Cooperation Center, Jiangsu Province Environmental Protection Bureau, and Nanjing University Environment College (国家环境保护总局法规司，国家环境保护总局对外合作中心，江苏省环保厅，南京大学环境学院), 企业环境行为评价及信息公开项目总报告 (Report on Enterprises Environmental Performance Assessment and Projects on Public Access to Information) 2005年12月.

SEPA (2000-04), China Environment Yearbooks 2000-2004, China Environmental Science Press, Beijing.

Shankar, Raja and Anwar, Shah (2001), *Bridging the Economic Divide Within Nations: A Scorecard on the Performance of Regional Development Policies in Regional Income Disparities*, World Bank, Washington DC.

State Council Information Office (2006), *White Paper: Environmental Protection in China (1996-2005)*, Xinhua.

State Council of the People's Republic of China (2005), *Decision on Implementing the Scientific Concept of Development and Strengthening Environmental Protection*, Beijing.

State Council of the People's Republic of China (2003), *Programme of Action for Sustainable Development in China in the Early 21st Century*, Beijing.

State Council of the People's Republic of China (1994), *China's Agenda 21: White Paper on Population, Environment and Development in China in the 21st Century*, Beijing.

Stockholm Environment Institute (2002), *China Human Development Report 2002: Making Green Development a Choice,* Oxford University Press, Stockholm.

Wang, Jinnan and Chazhong, Ze (2003) *Social Environmental Management System in China,* Chinese Academy of Environmental Planning, Beijing.

Warford, Jeremy and Yining, Li (2002), *Economics of the Environment in China,* Environmental Economics Working Group, CCICED, Beijing.

Wen, Jiabao (2006), *Report on the work of the environment,* 10th National People's Congress, March 2006, Beijing.

World Bank (2001), *China: Air, Land and Water Environmental Priorities for a New Millennium,* World Bank, Washington DC.

World Bank (1997), *Clear Water, Blue Skies: China's Environment in the New Century,* World Bank, Washington DC.

World Energy Council (2005), *China's Energy Supply: Many Paths – One goal,* World Energy Council, Berlin.

Xu, Ming and Tian-zhu, Zhang (2005), "Material Input Analysis of China Economic System", *China Environmental Science,* 2005, Tsinghua University, Beijing.

Zhao, Jimin and Marc, W. Melaina (2006), *Transition to hydrogen-based transportation in China: Lessons learned from alternative fuel vehicle programs in the United States and China,* Energy Policy 34, Beijing.

Zhou, Wei and Joseph, S. Szyliowicz (eds.) (2005), *Towards a Sustainable Future. Energy, Environment and Transportation in China.* China Council for International Co-operation on Environment and Development, China Transportation Press, Beijing.

ENVIRONMENTAL-SOCIAL INTERFACE

Features

- Social disparities and the environment
- The Great Western Development Strategy
- Health and the environment
- Environmental information
- Environmental NGOs
- The Jilin accident and the Songhua river pollution

Recommendations

The following recommendations are part of the overall conclusions and recommendations of the environmental performance review of China:

- further improve health and living standards, particularly in less developed areas, by reducing the share of people without *access to sound environmental services* (safe water, basic sanitation, electricity); taking account of affordability constraints, give higher priority to water infrastructure in development strategies (e.g. for the poorer central and western China);

- consolidate and strengthen information on health and the environment and develop a *national health-environment plan* of action; implement the most cost-effective measures; promote pollution release and transfer reporting by enterprises; build capacity to report on exposures of specific population groups to environmental health risks (e.g. occupational health, health impacts near polluting facilities, children's health);

- continue to improve environmental information by developing and using *indicators of environmental performance*, environment-related *economic information* and analysis, and environmental accounting tools such as *material flows accounts*; expand the coverage of environmental information (e.g. to diffuse pollution, toxic substances, hazardous waste); continue to improve consumer protection and public *access to environmental information*;

- further expand *environmental education* and awareness, particularly among young people;

- continue efforts to work with *NGOs and the public* to achieve environmental policy goals; strengthen co-operation and partnerships with *enterprises* and corporate social responsibility.

Conclusions

China's economic growth has helped raise living standards and has contributed to significantly reduce poverty. In recent years, government policies have emphasised *economic growth with due attention to social and environmental concerns*: environmental issues associated with rapid urbanisation and development of coastal regions, with poverty, and with development challenges in less-advanced western parts of the country are being addressed. Considerable progress has been made since the mid-1990s in the development of *environmental information*, access to this information, and participation on environmental issues. China produces each year comprehensive environmental statistics and environmental reports. The media and the rise of committed and outspoken environmental NGOs reinforce the demand for environmental progress. Progress can also be seen in environmental education and awareness-raising through primary education.

However, the rapid economic growth has led to *very wide and increasing disparities* between the rich and the poor, urban and rural communities, and coastal and inland provinces. While some aspects of the urban environment have improved in China's mega and large cities, additional *demands for environmental services* (e.g. water supply, water sanitation, solid waste management) are resulting from the large population migration from western and central China to coastal China. At the same time, the needs for environmental services of the expanding towns and townships and of the rural poor, particularly in the central and western regions, are also growing. To reduce industry relocation and environment-related distortions to competitiveness and trade within China, *national environmental standards* (i.e. product, emission and quality standards) should be implemented by all provinces effectively and efficiently, minimising transition periods when transitions are necessary. Concerning *health*, pollution is contributing to an increase in *respiratory diseases*, cancer and birth defects. Environmental and health information should be strengthened to support priority setting and to generate related economic and health benefits. Concerning *environmental information*, improvements could be made with respect to indicators of environmental performance, environment-related economic information, environmental and material flows accounts, the coverage of environmental information, and monitoring. *Environmental education* should be further strengthened (e.g. at university level) and expanded, particularly for young people. Environmental awareness should be increased in Chinese enterprises.

◆   ◆   ◆

1.   Social Disparities and the Environment

1.1   Social disparities

China has made remarkable progress in the last ten years in economic growth, population stabilisation and poverty reduction. However, China's economic advances have outpaced social and environmental advances. Although Chinese authorities, including those at the highest level, wish to build a "harmonious society" balancing economic growth and social and environmental progress, China has experienced *increasing inter-provincial,*[1] *income and, particularly, urban-rural disparities* (Figure 8.1). These *disparities* and the related social issues are *both large and important* and are associated with considerable environmental policy challenges. Average GDP per capita in China is still very low compared to OECD average

Figure 8.1 **Regional disparities,**[a] 1985-2004

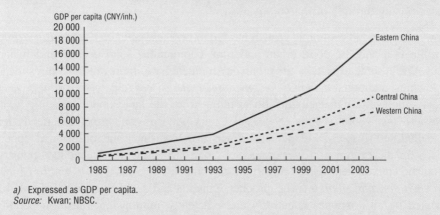

a) Expressed as GDP per capita.
Source: Kwan; NBSC.

(USD 5 975 compared to USD 25 300 at 2000 prices and PPP), and China ranks beyond 100th in the world on this measure (OECD, 2005). In nominal terms, income per capita is close to 6% of the OECD average.

Inter-provincial disparities

Inter-provincial disparities are particularly important among: i) the more advanced coastal provinces in the east; ii) the less-developed central provinces, and iii) the poorer western provinces. GDP per capita ratios range from 1 to 7 (i.e. from Guizhou to Shanghai) and household income per capita from 1 to 10 (i.e. from rural Guizhou to urban Shanghai) (Table 8.1).

Based on GDP per capita, Shanghai, Beijing and Tianjin would be classified by the World Bank as upper-middle income economies (comparable to several OECD countries), while most Chinese provinces and autonomous regions would be classified as lower-middle income economies, and several western provinces would be classified as low income economies (Figure 8.2). China has been described as *"one China, four worlds"*, to reflect that: 2.2% of the population live in cities such as Shanghai, Beijing and Tianjin, under conditions similar to those found in some OECD countries; 22% live in coastal regions such as Guangdong, Zhejiang, Jiangsu and Liaoning under conditions similar to upper-middle income countries; 26% live in Hebei, Jilin, and large parts of central China under lower-middle income country conditions; and the remaining 50% live in central and western China under low income country conditions.

Table 8.1 **GDP and household income,** by province, 2004

Provinces[a]	GDP[b] (billion CNY)[c]	GDP[b] per capita (CNY/inh.)[c]	Urban household income per capita (CNY/inh.)[c]	Rural household income per capita (CNY/inh.)[c,d]
Beijing	428	28 689	15 638	6 170
Tianjin	293	28 632	11 467	5 020
Hebei	877	12 878	7 951	3 171
Shanxi	304	9 123	7 903	**2 590**
Inner Mongolia	271	11 376	8 123	**2 606**
Liaoning	687	16 297	8 008	3 307
Jilin	296	10 920	7 841	**3 000**
Heilongjiang	530	13 893	7 471	**3 005**
Shanghai	745	42 768	16 683	7 066
Jiangsu	1 540	20 723	10 482	4 754
Zhejiang	1 124	23 820	14 546	5 944
Anhui	481	7 449	7 511	**2 499**
Fujian	605	17 241	11 175	4 089
Jiangxi	350	8 160	7 560	**2 787**
Shandong	1 549	16 874	9 438	3 507
Henan	882	9 072	7 705	**2 553**
Hubei	631	10 489	8 023	**2 890**
Hunan	561	8 379	8 617	**2 838**
Guangdong	1 604	19 315	13 628	4 366
Guangxi	332	6 791	8 690	**2 305**
Hainan	77	9 405	7 736	**2 818**
Chongqing	267	8 537	9 221	**2 510**
Sichuan	656	7 514	7 710	**2 519**
Guizhou	159	4 078	7 322	**1 722**
Yunnan	296	6 703	8 871	**1 864**
Tibet	*21*	7 720	9 106	**1 861**
Shaanxi	288	7 783	7 492	**1 867**
Gansu	156	5 952	7 377	**1 852**
Qinghai	47	8 641	7 320	**1 958**
Ningxia	46	7 829	7 218	**2 320**
Xinjiang	220	11 208	7 503	**2 245**

a) Provinces, centrally administered municipalities or autonomous regions.
b) GDP before the 2006 revision by NBSC.
c) Maximum and minimum values are colored in each column.
d) Per capita average household income below USD 1 per day are in bold (CNY/USD = 8.27).
Source: NBSC.

Urban-rural disparities

Demographic trends in China indicate that while the current *urban* population is about 430 million (43% of the total), it will reach 850 million by 2015. Indeed, the annual urban growth rate has been approximately 1% since the early 1980s.

Figure 8.2 **Inter-provincial[a] disparities in an international perspective, 2004**

Per capita GDP[b]

High-income economies[c]

Italy
Singapore   Hong Kong
New Zealand   Macao
Greece   Portugal

Upper-middle-income economies[c]

Chile   Shanghai

South Africa   Beijing

Russian Federation   Tianjin

Lower-middle-income economies[c]

Brazil   Thailand   Guangdong   Jiangsu   Zhejiang

                                            Liaoning   Shandong   Fujian

Morocco   Hebei   Heilongjiang

Egypt   Huebi   Jilin   Inner Mongolia   Xinjiang

Indonesia   Shanxi   Hainan   Henan

Sri Lanka   Qinghai   Chongqing   Jiangxi   Hunan

                                  Anhui   Sichuan   Tibet   Shaanxi   Ningxia

----

Low-income economies[c]

Cameroun   India   Guangxi   Gansu   Yunnan

Guinea   Kenya   Guizhou

a) Provinces, centrally administered municipalities or autonomous regions.
b) Per capita GDP for Chinese provinces at current exchange rate (CNY/USD = 8.27); values for 2004 not including the
   GDP correction made for China as a whole by NBSC in January 2006.
c) According to World Bank classification.
Source: NBSC; World Bank.

Approximately 13 million people migrate from rural to urban areas each year (Hårsman, *et al.*, 2005), mainly from the northwest to the southeast, despite the household registration system (hukou) which plays a role in restricting migration. The future of the *hukou reform* is uncertain. The main *pull* factor is the attractiveness of urban areas (most of which are located in the eastern provinces) which offers economic opportunities (e.g. migrant workers earn 10 to 20 times more in urban employment than what they earned in farming). *Push factors* in rural areas are the increasing number of landless farmers (e.g. 40 to 50 million farmers may have lost their land as a result of large-scale expropriation of farmland for infrastructure construction and development of residential areas) and environmental degradation (e.g. lack of environmental infrastructure, increasing pressures from pollution). Redressing the in-balance between urban and rural income has become a priority for the Chinese government.

Poverty prevails in rural areas, where the average rural household income in most provinces is below USD 1 per capita per day. When China began its reform process in 1979, more than 650 million people were living on less than USD 1 per day. Between 1990 and 2001, this number declined greatly, from 375 million (33% of the population) to 211 million (16.6% of the population). Recent World Bank data suggest that the number had fallen to 135 million by the end of 2004. However, the World Bank estimates that some 500 million Chinese still live on less than USD 2 per day. According to Chinese statistics, the rural population living in absolute poverty (with an annual income below CNY 637) declined from 85 million in 1990 to 29 million in 2003, with more than 90% of the decline taking place between 1990 and 1996. However, the number of rural poor increased by 800 000 between 2002 and 2003, the first increase since the economic reforms began in 1979. More than 300 million rural residents have no access to safe drinking water. The large labour flows from rural areas to eastern cities also generate *urban poverty*. In the late 1990s and through the early 2000s, urban household incomes reached about 3.5 times those of rural households, with urban workers transferring part of their earnings to family members who remained in the rural areas.

Development policies

The *objectives of China's development policies* are specified in the National Economic and Social Development Five-Year Plans (FYPs). Over the two last decades, they aimed to maintain high and consistent economic growth, raise people's living standards and eradicate poverty. The 9th FYP (and Long-Range Objectives to 2010) established economic development as an "absolute principle", driven by deep reform of the economy and ensured through political and social stability. The objectives included doubling the 2000 GDP by 2010 and maintaining the population size at 1.3 billion in 2000 and 1.4 billion in 2010. The 10th FYP emphasised economic growth, addressing problems from rapid urbanisation and providing support to less-developed western parts of the country. The 11th FYP calls for a better balance of economic, social and environmental goals. This is also expressed in the different concepts of development advanced by Chinese leaders (Chapter 7).

China's "Great Western Development Strategy", adopted in 1999, covers *12 western provinces* and autonomous regions, representing 70% of the country's area and close to 30% of the population (Box 8.1).

In October 2003, the government announced a programme to rejuvenate the *north-eastern provinces* (i.e. Liaoning, Jilin and Heilongjiang). The programme focuses on state-owned heavy industries concentrated in these provinces (70% of total industrial assets) and largely relies on *structural adjustments* such as change of ownership, management innovations, incentives to attract foreign and domestic capital, and creation of new labour-intensive industries to absorb surplus labour.

In March 2004, the government announced an initiative to "promote the fast rise of the *central region*" (i.e. Hubei, Shanxi, Henan, Hunan, Jiangxi and Anhui). National programme documents related to this new initiative have not been released yet. These inland provinces represent 10.7% of China's territory, 28.1% of the population, and 23% of the country's GDP. The programme focuses on developing agriculture and "constructing the new social rural villages", as well as on the central region's comparative advantages (e.g. energy, mineral products, natural products).

1.2   Related environmental challenges

Providing environmental services to urban and rural populations

While some aspects of the urban environment have improved in China's *mega and large cities*, providing environmental infrastructure and services (e.g. water supply, waste water collection and treatment, solid waste collection and treatment) presents a *double challenge*: i) addressing the investment backlog, and ii) addressing the influx of new urban residents (Table 8.2). Environmental infrastructure and services are also needed for *small and medium-sized cities*. Small cities or townships (e.g. population between 5 000 and 10 000) are already growing fast, and the number of medium-sized cities (e.g. more than 100 000 inhabitants) is expected to increase from 630 (in 2001) to over 1 000 in 2015. The provision of environmental services in *rural areas* is no less important, as it will not only bring the direct benefits of water supply and sanitation but also the indirect benefits associated with improved health and education. The issue of *affordability* of environmental services is clearly more acute in rural areas than in large urban centres.

The westward relocation of polluting plants

In spite of the government's encouragement to develop cleaner production, and bans on relocation of polluting plants, *relocation of polluting industrial installations continues*. Most of China's major cities have adopted local policies to move polluting

Box 8.1 **The Great Western Development Strategy**

Developing the West

In 1999, the government launched the "Great Western Development Strategy" (Xibu Da Kaifa) to foster the development of the *less-advanced western areas of China*, which include the provinces of Gansu, Guizhou, Ningxia, Qinghai, Shaanxi, Sichuan, Tibet, Xinjiang and Yunnan, as well as Chongqing municipality and the autonomous regions of Inner Mongolia and Guangxi. Together these represent an area covering more than 70% of the country, a population close to 30%, and a GDP of 17% (Table 2.2). They also include most of the country's minorities. These western areas are rich in minerals, energy (including hydropower), land resources and tourism, but they face a number of obstacles to development (e.g. insufficient infrastructure, limited investment, weak education system). The environment has often been damaged as a result of "third front" industrialisation in the 1960s, and extensive development of small township and village enterprises through the 1990s.

Development objectives

The Strategy aims at a long term development effort to establish a "new western China" by the middle of the 21st century. After the intitial stage (2000-05), it has entered in the development stage (2006-15). During the 11th FYP, the Great Western Development Strategy focuses on: *environmental protection and reforestation*; *investment in infrastructure*; promotion of high tech industries *education* and human resources; regional *economic development*; *framework conditions* (e.g. improving the investment context, strengthening the legal system, reforming the economic system and state-owned enterprises to attract foreign direct investment.

Infrastructure *mega-projects* have been placed at the core of the Strategy, including the south-north water diversion, a west-east natural gas transfer, a west-east power transmission and construction of the Qinghai-Tibet railway.

Implementation and funding

The task of implementing the Strategy has been allocated to the national agencies responsible for identified priorities as well as to provincial and other territorial governments. *National co-ordination* is done by the Leading Group for the Western Region Development, located in the National Development and Reform Commission.

In addition to the support provided by the national government for some national-level projects, a *mix of funding* from public sources (national, provincial and lower levels of government) and private domestic and international sources supports other projects. For instance, provincial and other local authorities are responsible for finding investors for projects that conform to the Strategy objectives. State-owned enterprises were expected to go to the market to raise equity financing. Some projects are funded by bank loans, equity financing, and bonds. The wealthier eastern/coastal

Box 8.1 **The Great Western Development Strategy** *(cont.)*

provinces were asked to contribute to develop new markets and bring advanced management and production styles to western enterprises. In the first year of the Strategy, Shanghai had already signed 200 co-operative contracts with a total value of over CNY 10 billion. Foreign investment and World Bank loans were sought to supplement and stimulate domestic funding. Preferential financing was also used via tax exemptions, preferential interest rates or repayment schedules.

While overall *expenditure* statements for the Strategy have not been published, about CNY 500 to 850 billion were spent over five years on major infrastructure development projects, and between CNY 720 billion and CNY 2 380 were provided to minority groups to support priorities identified in the Strategy. Expenditure has been growing at a rate of 21.7% annually, according to the State Ethnic Affairs Commission.

Environment

Projects under the Strategy's first objective are specifically environmental. In addition, projects under the second objective are expected to have an environmental impact assessment (EIA).

China also faces a risk of environmental damage from i) unsustainable exploitation of natural resources and ii) the relocation of polluting factories towards the western areas. Responding to such concerns, in 2000 SEPA issued a "Notice of Prohibition to Move Pollution to Western Areas" to provincial EPBs and Economic and Trade Committees. The notice prohibited the transfer or construction of virtually all outdated installations in terms of pollution and resource use. The notice also prohibited the transfer of hazardous waste to western areas. However, a number of cases of relocation of heavily polluting factories have been recorded.

plants outside of the city. The Beijing Development and Reform Commission recently announced that by 2008 all polluting plants would be relocated in the rural areas outside of the fifth ring-road. Shanghai moved more than 700 factories outside the city limits during the seven years up to 2004. In Guangdong province, all plants in the category of "fifteen small industries" (i.e. highly polluting small heavy industries, for which there are special restrictions) have been moved to remote mountainous areas within the province. A number of heavy polluting plants were also moved out of Shanghai to nearby southern Jiangsu province in the 1980s, and then were moved again to less developed areas of northern Jiangsu and to Anhui province, as local implementation of environmental standards became stricter in southern Jiangsu.

Table 8.2 **Environmental indicators,** by province, 2004

Provinces[a]	Intensity of water use[b] (%)[e]	Water use by agriculture[c] (%)[e]	SO$_x$ emissions per GDP (kg/CNY 1 000)[e]	Particulate emissions[d] per GDP (kg/CNY 1 000)[e]	Municipal waste production per capita (kg/inh.)[e]	Hazardous waste production (1 000 t)[e]	Total energy consumption per GDP (tce/CNY 1 000)[e]
Beijing	152.3	37.5	0.4	0.2	329	40	0.11[f]
Tianjin	153.5	54.3	0.8	0.3	177	60	0.13
Hebei	126.7	75.1	1.6	0.8	109	130	0.14[f]
Shanxi	59.9	58.9	4.7	3.6	178	80	0.31
Inner Mongolia	39.1	87.1	4.3	2.4	138	390	0.24
Liaoning	45.4	65.8	1.2	0.8	185	470	0.17
Jilin	30.6	67.0	1.0	1.1	211	80	0.18
Heilongjiang	39.8	71.8	0.7	1.0	278	120	0.14
Shanghai	472.9	15.9	0.6	0.2	350	360	0.09
Jiangsu	257.6	54.9	0.8	0.3	110	860	0.09
Zhejiang	30.7	51.6	0.7	0.2	149	200	0.08
Anhui	41.8	58.1	1.0	0.5	72	90	0.13
Fujian	25.7	56.4	0.5	0.2	83	70	0.07
Jiangxi	19.7	63.2	1.5	0.6	60	40	0.10
Shandong	61.1	71.8	1.2	0.4	135	650	0.10
Henan	49.3	62.1	1.4	0.9	70	100	..
Hubei	26.1	54.3	1.1	0.5	148	130	0.13
Hunan	19.7	62.5	1.6	0.9	73	300	0.13
Guangdong	39.1	51.7	0.7	0.2	188	580	0.07
Guangxi	18.0	72.2	2.8	1.6	47	1 260	0.13
Hainan	27.1	81.7	0.3	0.1	100	0	0.10
Chongqing	12.0	30.1	3.0	0.8	76	430	0.11
Sichuan	8.5	57.6	1.9	1.3	66	170	0.16
Guizhou	9.5	55.1	8.3	2.0	52	1 560	0.38
Yunnan	6.9	74.6	1.6	0.6	45	210	0.17
Tibet	0.6	91.6	0.0	0.1	139	0	..
Shaanxi	24.2	65.8	2.8	1.3	95	40	0.17
Gansu	70.6	79.4	3.1	1.0	112	500	0.25
Qinghai	5.0	72.4	1.6	1.6	107	630	0.29
Ningxia	750.9	92.7	6.4	2.1	230	0	0.50
Xinjiang	58.1	91.9	2.2	1.2	176	390	0.22

a) Provinces, centrally administered municipalities or autonomous regions.
b) Water resources correspond to water resources internal to the province only.
c) Per cent of total water used; water use by agriculture includes uses for irrigation of farms and forests, animal husbandry and fishing.
d) Particulate emissions refer to volume i) in smoke emitted in fuel burning by industrial plants, and ii) emitted by fuel burning from all activities and facilities other than industrial activities. They are calculated on the basis of coal consumption.
e) Maximum and minimum values are colored in each column.
f) Data for 2003.
Source: NBSC.

Relocation of polluting plants generally follows one of three routes: from city core to city periphery; from city core to small and medium-sized towns in the same province or region; or from eastern cities to central and western emerging towns. China's less developed areas lack the infrastructure and skilled labour to attract and develop high-tech production, and their *competitive advantage lies in attracting labour-intensive, traditional industries* (e.g. small-scale pulp and paper, cement, fertiliser, chemicals, coal mining, energy production, metallurgy), which are often associated with low resource efficiency and heavy pollution. While relocation helps to reduce pollution in large city centres and to provide development and employment opportunities for smaller towns and rural areas, this process has caused extreme pollution problems, especially in the Huai river basin. *Internalisation of pollution damages* is needed to avoid having local authorities "compete" to attract industrial investment by subsidising polluting activities.

Environmental disparities

China has important *disparities in environmental conditions, environmental infrastructure and environmental governance* capacity (Table 8.2). In part, these result from differences in densities of population and activities, natural resource endowment, level of economic development, economic structure and institutional history. However, China's "environmental disparities" also reflect an inefficient use (and over-use) of natural resources, due to i) highly distorted price signals (e.g. water and energy prices), and ii) inefficient production and consumption processes that result in pollution damages that are unaccounted for by enterprises or citizens.

1.3   Perspectives

To achieve environmental effectiveness and *economic efficiency in managing natural resources and pollution*, resources (e.g. water services) should be priced at cost and pollution-related damages (e.g. industrial pollution) "internalised", respectively in line with the user-pays and polluter-pays principles. Because of the wide disparities described above, *differentiated transitions* will be needed according to levels of development. China's current legislation takes account of this concerning the provision of water services, industrial emission charges and regional development efforts of Chinese authorities (e.g. for western and central China).

Water services

While full-cost pricing of water could rapidly be established in coastal China, with appropriate attention to access to water for the poor, it will have to be introduced more progressively in central China, and with some further delay in western China.

The process should ensure the establishment of *full-scale water supply and sanitation infrastructure* over the time frame of a five-year plan in coastal China, of two or three plans in central China, and of a generation in western China. The aims of *economic convergence and environmental convergence* among the different regions of China should go hand by hand.

In a developing country like China, the speed of the *environmental transition* would in practice depend both on governance capacity (of provincial and lower levels of government) and on funding capacity. The share of private funding (from water charges), public funding (from governments), and grants and loans (from domestic sources as well as international sources) will vary over space and time. The priority given to water infrastructure in public funding by the national, provincial and lower levels of government should be high (together with other essential infrastructure needs), as it is now recognised that achieving water development goals contributes to achieving other development goals (e.g. health, education, labour) and brings multiple economic benefits. Spatially, private funding would be higher in coastal regions and public funding higher in western regions. Over time, public funding and borrowing would gradually decrease at a speed largely dependent on economic growth in the relevant provinces. Increased funding needs to be matched by reforms in the governance of the water sector, including the establishment of water utilities as autonomous, but fully accountable institutions.

Industrial pollution

To achieve environmental effectiveness and economic efficiency, the establishment of an internal market in China, without pollution-related distortions, will require: i) *uniform environmental standards on products* (e.g. automobiles, chemicals); and ii) *implement the polluter pays principle* for enterprises associated with industrial and energy production. This implementation could rely on a mix of instruments such as: pollution and land-use regulations, economic instruments (e.g. taxes, charges, emissions trading) and voluntary approaches (e.g. branch agreements, industrial certification, environmental reporting by enterprises). The aim should be to eliminate as rapidly as possible differences in the implementation of the PPP over space.

In practice, the *environmental transition* will require pollution abatement and control investment (e.g. cleaner fuels, desulfurisation equipment, waste water treatment plants) associated with the evolution of the industrial structure (closing obsolete plants, investing in competitive plants, ensuring adaptation periods for intermediate plants) over time and space. Related environmental progress will require effective *governance and implementation* as well as the withdrawal of various forms of subsidies, and attention to competitiveness and distributive issues. Enterprises should adopt corporate responsible approaches (i.e. measure and take into account the actual and potential economic, social and environmental impacts of their decisions). Such efforts should bring multiple benefits related not only to pollution abatement and health but also, for example, to international environmental progress (e.g. ozone layer, climate change) and domestic goals (e.g. more efficient use of water and energy).

2. Health and the Environment

The *impact of pollution on human health* is increasingly recognised in China. For example, the Ministry of Health has linked environmental pollution to a 25% increase in birth defects recorded between 2001 and 2003. China's rapid industrialisation and urbanisation are linked to an increase in the incidence of disease and chronic illness including cancer related to exposure to air and water pollution. The related economic damage is also large. The ongoing transfer of some polluting production from the eastern part of the country to the central and western parts adds health risks to already disadvantaged areas.

The *limited availability of information on environmental health* (e.g. concerning drinking water, health impacts from environmental pollution, exposure of local populations) limits the capacity of Chinese authorities and citizens to act in preventive or curative ways. The latest nationwide survey on drinking water and water-related deseases was conducted in the period 1983-88. It is suggested here that China should develop a *white paper* on health and environment, a *national health and environment plan*, and *reporting mechanisms* for enterprises and authorities related to the exposure of specific population groups to pollution and related-health risks.

2.1 Air pollution

China has more than 270 million urban residents living in areas with an *air quality* below the ambient air quality standard (Chapter 3). Epidemiological surveys in China have established a link between air pollution and respiratory diseases (e.g. linking 1.5 million cases of bronchitis, 23 000 deaths from respectively respiratory disease and 13 000 deaths from heart disease to air pollution) (Stockholm Environment Institute and UNDP, 2002).

Overall, *damages linked to air pollution in China were estimated at several per cent of GDP* depending on the valuation method used (e.g. contingent valuation method, human resource method). Health damage from air pollution was estimated at 1.8% of GDP with an uncertainty ranging between 0.65 and 4.7% of GDP. Under a business-as-usual scenario, health effects by 2020 in China were predicted to include: 600 000 premature deaths in urban areas, 9 million person-years of work lost due to pollution-related illness, 20 million cases of respiratory illness per year, 5.5 million cases of chronic bronchitis and health damage; an overall cost of health damage reaching 13% of GDP (Ho, 2006; World Bank, 1997).

2.2 Water pollution

A majority (75%) of the water flowing through China's urban areas is unsuitable for drinking or fishing, and 30% of the rivers monitored are worse than grade V (i.e. highly polluted) (Chapter 4). The Minister of Water Resources stated (on 22 March 2005, World Water Day) that *300 million people drink contaminated water on a daily basis*. Of these, 190 million suffer from related illnesses and more than 30 000 children die annually from diarrhoea due to polluted water. Along China's most polluted rivers, towns and villages record high rates of cancer, diminished IQ and miscarriages (Economy, 2004b).

Achievement of the water-related *Millenium Development Goals* of halving, by 2015, the number of people without access to safe drinking water and the number of people without access to basic sanitation would bring significant health and economic benefits and should be pursued actively. The Ministry of Health recently stated that at least 300 million rural residents have no access to safe drinking water.

2.3 Occupational health

A recent survey in town and village enterprises from 30 counties and urban districts across 15 provinces showed that *occupational harm* occurs in 83% of the surveyed enterprises that 60% of the enterprises do not take protection measures, and over 90% of the surveyed workplaces with dust do not meet the national health standard. While exposure to dust can be improved through ventilation and protection measures as well as basic worker safety education, even easy-to-implement measures are often not applied (Zhao, 2006).

In some small coal mines, the concentration of dust is 130 times higher than the national standard, and 35% of the workers develop *coal workers' pneumoconiosis* (especially silicosis). In addition to suffering from the disease, victims face difficulties in getting financial, medical and legal support (Zhang, 2001). As the disease is incurable, and as the number of people (especially young people) affected is high, related economic and health damages are high, and add to the human losses in mining accidents (Table 2.4).

Migrant workers are the most affected by occupational health hazards. Occupational diseases are seen as expanding from urban industrial districts to rural areas, from the eastern part of the country to the central and western parts, and from large and medium-sized enterprises to medium and small enterprises.

3. Environmental Democracy

3.1 Provision of environmental information

Environmental data and reporting

China's *Environmental Protection Law* stipulates that "the competent departments of environmental protection administration under the State Council, and governments of provinces, autonomous regions and municipalities directly under the central government, shall regularly issue bulletins on environmental situations".[2] Similar provisions appear in China's sectoral laws, such as the Law on Prevention and Control of Air Pollution, the Law on Prevention and Control of Water Pollution, the Marine Environmental Protection Law and the Law on Prevention and Control from Environmental Noise.

In line with these requirements, national authorities publish a *range of environmental reports*, bulletins, brochures and news releases. In particular, SEPA publishes annually a compendium of environmental data (in Chinese), a state of the environment report (since 1991 in both Chinese and English, and an environment yearbook (in Chinese and English) (SEPA, 2006a, b, c). Together, these three reports provide solid and comprehensive national information. Only a few other countries have the equivalent publications yearly.

In addition to SEPA's yearly compendium of environmental data, the *National Bureau of Statistics* publishes in its annual China Statistical Yearbook a section on environmental data, which builds on SEPA's environmental data as well as on data on water resources, urban waste, land use, forest resources and marine disasters, provided respectively by the Ministry of Water Resources, the Ministry of Construction, the Ministry of Land and Resources, the State Forestry Administration and the State Oceanic Administration. These administrations publish their own bulletins of statistics and activities.

The national *state of the environment* reports present a factual, compact and comprehensive view of the environment (covering water, the coastal and marine environment, atmospheric and acoustic conditions, as well as waste, radiation, land resources, forests, grassland, biodiversity, climate change, and natural disasters), provide information on a range of selected topics (including international, economic and implementation issues relating to the environment), and contain environmental indicators. The structure of the reports follows the OECD "Pressure-State-Response" model since 1998.

The *environment yearbooks* are published in co-operation with the China Forum of Environmental Journalists and provide comprehensive coverage of events and actions taken during the year, including by the legislative branch (the National

People's Congress and its Environment and Resources Protection Committee), the executive branch at national level (different ministries and agencies), and the executive branch at provincial level. It also covers international co-operation.

Governments of *provinces, autonomous regions and municipalities* directly under the central government issue their own annual state of the environment reports as well as information on the environment in a variety of forms. The state of the environment reports prepared by the Environmental Protection Bureaus in co-operation with other sectoral bureaus, cover environmental pollution, the ambient environment, and environmental protection measures. Some provinces publish their state of the environment reports on the Internet.

In addition to the national and provincial state of the environment reports, reports on the environmental situation in key environmental management regions and river basins are produced.

Nevertheless, China has room for progress in providing environmental information. First, it could further develop and use *environmental performance indicators*, including at provincial or sub-provincial levels, drawing on international experience (OECD, 2003). Second, *economic information* related to the environment should be further developed (e.g. environmental expenditure, green taxation, economic assessments, water and energy prices). Third, the greening of economic accounts and *environmental accounting*, such as *material flow accounts* and natural resource accounts, should be further developed without too much emphasis on aggregate indices. And fourth, environmental information should also be extended to cover a wider range of pollutants (e.g. diffuse pollution, toxic substances).

Environmental monitoring

Since the mid 1990s, SEPA has worked to strengthen *environmental monitoring* to support three objectives: to provide technical support for the implementation of environmental policy and management; to enable technical supervision of environmental policy enforcement; and to provide technical support for social and economic infrastructure development. FYPs provide guidance for development of the monitoring system. In the late 1990s, a nationwide system was established at four administrative levels: i) SEPA's China National Environmental Monitoring Centre (CNEMC); ii) 38 provincial environmental monitoring centres; iii) 391 municipal environmental monitoring centres; and iv) 2 229 county-level environmental monitoring stations (OECD, 1999).

Each *monitoring organisation* at the sub-national level is subordinated to the upper level monitoring station and to the same level environmental protection administration. Some cities must also report to the CNEMC. In 2002, the 2 229 air monitoring stations across China included 474 automated air-monitoring systems in 179 cities. Routine monitoring of air, surface water and noise is carried out by 80-85% of city-level stations and 56% of county-level stations; 41% of city-level stations have begun monitoring soil conditions; 39% of stations have begun bio-monitoring; 22 stations have begun monitoring nature protection. Over 20 cities have established automatic air quality monitoring systems comprising up to 100 automatic substations. A number of cities provide their inhabitants with daily and weekly air quality reporting based on this monitoring activity and the Air Pollution Index (Box 8.2).

In addition, environment-related monitoring is also done by a *range of other departments*. For example, the Ministry of Water Resources carry out some monitoring related to water bodies, and the Ministry of Health on drinking water quality. The departments responsible for the management of agricultural, forestry, ocean, hydrological and other resources, and industrial, military and railway administrations, operate several thousand stations.

Box 8.2 Weekly and daily air quality reporting

Weekly and daily *reports on urban air quality* are released to the public. These reports are based on routine monitoring of traditional pollutants and on the national Standard on Ambient Air Quality. They are expressed as an *air pollution index* (Box 3.2).

In 1997, the 10th session of the Third Environment Commission of the State Council decided that a *weekly report on urban air quality* (covering such pollutants as SO_2, NO_x and total suspended particulates) should be established in 47 key Chinese cities. The city of Nanjing was the first to provide weekly air quality reports in newspapers and on TV. Currently, China Environmental News releases environmental quality reports for these key cities every Saturday. China Environmental News also publishes the air quality indices of key cities on the Internet.

In 1999, after two years of weekly bulletins on urban air quality in key cities, the Chinese authorities decided to issue *daily air quality reports in 42 key cities.*, and there are plans to add five additional coastal cities to this daily reporting scheme. The classification criteria and pollutants were also adjusted in the daily air quality reporting. The weekly and daily reports have been presented via TV, newspapers, 168 telephone information stations, information highways and street displays.

Urban air quality weekly and daily reports have played a positive role in increasing the *environmental awareness* of citizens, industry and local governments, in linking air quality and health concerns, and in facilitating the comparison of air quality among Chinese cities.

Many industries carry out *self-monitoring* of their emissions. Industry is required by SEPA and local Environmental Protection Bureaus (EPBs) to file a pollutant discharge registration every year. Some local EPBs have increased the frequency of industry self-reporting to once a quarter and work with enterprise managers to enhance industry's self-monitoring capacity. Large enterprises usually hire state-owned monitoring stations or private monitoring service providers (certified with the State Quality and Technology Supervision Administration) to take measurements, and then report the data to their local EPBs.[3] In Jiangsu province, for example, some enterprises are financially capable of *continuously monitoring their operations*. The monitoring equipment is licensed and verified every year by the official monitoring station to ensure it functions properly. The firms' environmental specialists are licensed by the Jiangsu provincial EPB to verify and endorse their company's self-reporting forms. The official monitoring station conducts regular and surprise inspections on these monitoring service providers for quality assurance. Small and medium-sized enterprises often lack self-monitoring capacity and there are concerns related to the validity of self-monitored data.

Efforts have been made to *co-ordinate the various monitoring networks*. Some progress has been made, but it has not been possible to establish a single, nationally-unified environmental information collection and transmission system. The reasons are technical (e.g. the large volume of data generated, technical limitations in data collection, lack of resources) and institutional (e.g. co-ordinating data collection, storage, processing, evaluation and reporting across the range of institutions). Concerns have also been expressed that the incentives for conforming with national policy objectives may also affect the reliability of data transmitted to higher levels, and that monitoring data may not cover some new areas of concern. All of these factors limit the extent to which China can use environmental monitoring data to support policy development and implementation. Environmental monitoring should progress further on *both the supply side and the demand side*. In other words, progress is needed in improving the quality of environmental data and the efficiency of its production and in linking them more clearly to decision-making.

3.2   Access to environmental information

Information technology has played an important role in China in facilitating public access to environmental information and promoting public awareness of environmental issues. Since the launch of the "Government Online" project in 1998, many reports have been posted on the Internet. Various organisations and individuals have established more than 2 000 environment-related websites.

Recently, central and local governments have been using a more systematic mechanism for disclosure of environmental information to influence the environmental behaviour of enterprises. China has several examples of the successful application of an environmental *performance rating and information disclosure* scheme, which provides a basis for communication between different stakeholders including the government, industry and the public. However, information about *pollution discharges from industries* (e.g. Pollutant Release and Transfer Register),[4] be it continuous or accidental, is not accessible to the public on a routine basis. The Jilin chemical accident illustrates not only the environmental management issues faced by industrial facilities, but also the environmental information and communication issues faced by government authorities (e.g. ensuring reporting on actual discharges, preventing local officials from covering up pollutant releases, and avoiding delays for the public to receive proper information) (Box 8.3).

3.3   Access to courts

China's legal framework provides opportunities for the public to take environmental disputes to court and to *take legal action against polluters*. A number of legal actions have also been taken against local governments that fail to act against non-compliance. Cases involving illegal pollution discharges from a factory can be tried through the Civil Law, with the aim of stopping the discharges and receiving compensation for damage. Administrative units such as a local EPB (or its employees) can be sued under Administrative Law for deficiency in fulfilling administrative responsibilities (e.g. refusal to provide information, absence of active enforcement of environmental laws and regulations).

The Civil Law also contains provisions on *collective litigation*, which have proven effective in providing legal protection to the victims in environmental civil lawsuits. Many environmental civil lawsuits are currently handled through this approach. However, only public prosecutors can bring a lawsuit to court in the public interest. The Centre for Legal Assistance to Pollution Victims (at the University of Politics and Law in Beijing) has been active in providing legal assistance in many cases, including the widely publicised case in Pingnan county, Fujian province, where 1 721 villagers filed a successful lawsuit against a large chemical plant.

In practice, pollution victims may have difficulty *providing courts with evidence of the damage claimed*, for instance if samples of polluted air or water were not collected at the time when the polluting discharge occurred. Pollution victims may also have difficulty *taking cases to court*; for example, farmers may not be able to afford the cost of filing a lawsuit, hiring lawyers and conducting appraisals. Another potential obstacle is that local governments sometimes back the polluting enterprise.

Box 8.3 **The Jilin accident and the Songhua river pollution**

The accident and its immediate consequences

On 13 November 2005, an explosion at a petrochemical plant in Jilin city (Jilin province) killed six people and injured 70. It also generated an 80-km spill of *benzene compounds into the Songhua river,* *which flowed* towards Harbin city (Heilongjiang province), which has 3.8 million inhabitants (Table 7.4). The spill then reached Russia's Amur river which provides water to the city of Khabarovsk, with 500 000 residents.

On 21 November, the *water supply in Harbin city* was cut off. Infrastructure maintenance was cited by local government as the cause. On 22 November the media reported that water could have been contaminated after the explosion, and authorities acknowledged that 100 tonnes of toxic chemicals (benzene and nitrobenzene) had polluted the river, significantly exceeding permissible levels. Subsequently, local schools were closed, about 10 000 people were evacuated as a precautionary measure, and residents were warned not to use water from the river. The water supply in the city of Harbin was shut down for four days as a safety measure (and resumed afterwards). Bottled water and water brought by tanks was distributed to the population. SEPA issued emergency monitoring instructions to its Environmental Protection Bureau in Heilongjiang province five days after the accident. The Jilin provincial authorities ordered water to be released from a dam into the Songhua river in an attempt to dilute the pollution levels within Jilin's borders.

In December 2005, the *State Council* stated that SEPA "underestimated the possible impacts of the incident" and the Environment Minister resigned. A package of USD 1.2 billion was committed to combat pollution in the Songhua river basin. For Harbin itself, a new water supply infrastructure will be completed in 2007, reducing its reliance on the Songhua river is raw waters. The general manager of the petrochemical plant in Jilin was dismissed by its parent company. Two workshop managers at the Jilin plant were also dismissed as the investigation pointed to human error in handling the industrial operations. *Co-operation with Russian authorities* took place at the technical level and at the highest level of both governments. Chinese authorities offered substantial support to the local Russian populations affected. On 24 January 2006, monitoring data showed that fish from the Songhua river was safe for consumption.

The UNEP assessment and SEPA response

An assessment of the accident's impact was carried out by UNEP (after a field mission on 9-16 December 2005) at the request of the Chinese authorities. It concluded that: i) during the initial response phase, government *communication and information-sharing* with the general public was not adequate to ensure appropriate responses by the affected population; and ii) in order to reduce the risk of such accidents, analysis of the *internal risk management practices used in enterprises* should be undertaken through a random sample of industries. The analysis also showed that further investigations were needed to clarify whether existing early warning systems and contingency plans were sufficient.

Box 8.3 **The Jilin accident and the Songhua river pollution** *(cont.)*

In the four months following the Jilin accident, SEPA enhanced its *communication and information on pollution accidents*, reporting a further 73 pollution accidents affecting China's river systems. These accidents included: a cadmium spill along the Beijiang river in South China's Guangdong province that also threatened the local drinking and agricultural water supplies; chemical spills along Northeast China's Hun river; central China's Xiangjiang river in Hunan province; a diesel spill along the Yellow river in Henan province; and an oil spill in Ganjiang river in Jiangxi province.

Also early in 2006, *SEPA inspected more than 100 industrial sites* and ordered corrective actions at 20 chemical and petrochemical enterprises that were found to pose serious safety threats. SEPA stopped or postponed approval of projects at 44 sites with a total investment of CNY 149.5 billion, because their locations were considered unsafe. Twelve of the 20 projects that inspectors found to pose hazards were located along China's two main riverways, the Yellow and Yangtze.

The Songhua accident has received wide domestic and international *press coverage*, and has been added to the worldwide list of well-known industrial accidents. It contributed to raising the priority for environment on the government agenda.

In 60 to 70% of cases the pollution victims are not successful. The National People's Congress has put forward proposals to open up opportunities for the public to sue in the public interest.

Chinese courts do not currently play a major role in addressing environmental disputes for several reasons. First, in China, disputing parties often prefer to resolve their differences using *informal negotiations* in which compromises are made to reach consensus. Third parties often facilitate conflict resolution by means of mediation and conciliation. This approach comes from a Confucian tradition which emphasises moral values and moral instructions (not fear of legal sanctions) as a basis for guiding behaviour and maintaining social co-operation. Second, even when courts are used, they decide their cases not only on the basis of environmental law, but also by relying on official policy, the views of local governments, and their own sense of justice and fairness in contractual dealings. Third, China does not have enough *judges trained in environmental law*. Nevertheless, *examples of legal cases brought by the public* are a positive sign of a change that could strengthen the influence of the public on the environmental behaviour of industry and local officials. Greater involvement by the court system could strengthen enforcement of environmental law. More efforts are needed to inform the public of its rights in bringing cases of serious environmental pollution (for example, those resulting from deliberate actions by individuals or enterprises) to court.

3.4   Public participation

Many surveys and public opinion polls have consistently shown a *positive attitude towards the environment* in China. Recent surveys show that environmental concerns are at the top of a long list of concerns. However, *actual behaviour* does not always match attitudinal survey responses: for example, a willingness to pay for environmentally-friendly products or to change consumption patterns does not always follow declared intentions.

China's Environmental Protection Law and other sectoral laws require the authorities to allow individuals to participate in decision-making processes, and regulations describe a number of ways for citizens to participate in decision-making. *In practice*, however, the degree of participation varies greatly and depends on factors such as the type of decision being made and the time and budget available to receive public input.

A tradition allowing Chinese citizens to make *complaints* to government authorities has existed for centuries. Citizens can express dissatisfaction to *special offices* at various levels of government. Many people send signed or unsigned letters of complaint. Concerning environmental matters, citizens can direct their concerns to: i) EPBs, as many of these have "complaint divisions" to hear public concerns; ii) mayors' offices, as many cities have a vice mayor responsible for environmental protection, with staff to deal with citizens' complaints; and iii) their elected representatives, particularly local People's Congresses and their Environmental Protection Committees.

Many cities have established *hotlines* for residents to report environmental problems. For example, Dalian city (Liaoning province) has installed a 24-hour telephone hotline to receive citizens' complaints about environmental pollution. The city also has a radio talk-show that gives people an opportunity to discuss their environmental problems. Frequently, citizens first complain to the factory causing the problems and turn to government authorities or the media only if the factory is not responsive.

In addition, China has more *interactive mechanisms*, which allow the public to express its opinion about governmental projects and policies and to influence them. These include *environmental impact assessment* (EIA) mechanisms and informal instruments (e.g. consultations through town meetings, hearings or advisory panels).

3.5   Environmental education

Environmental and sustainable development education, which used to be an ad hoc, informal advocacy activity, is now included in *formal education*. Experiments with environmental education started in 1973, following a statement from the State

Council that "knowledge of environmental science should be popularised in primary and secondary schools". In the 1980s, environmental education was expanded to a number of schools countrywide. After 1992 and the UN Conference on Environment and Development (UNCED), environmental education was extended to cover sustainable development, including the "harmonious" interaction between environmental, economic and social development.

Chinese authorities defined an *environmental education* perspective for 1996-2010 in a national compendium of actions on environmental education. In 1997, China launched the "National Green Education Programme", based on guidelines developed by NEPA, the Central Bureau for Publicity, and the Ministry of Education. Subsequently national environmental education guidelines were issued. The result was the creation of almost 17 000 *"primary, secondary and tertiary green schools"* (2.5% of the schools in China) in 2004 and a dozen centres to train teachers in environment and sustainable development, to support the programme (Wang, 2004). It is expected that this number will reach 76 000 by 2008 (over 10% of the schools). While the high level of education reaches 15% of the adult population (versus 65.6% on average in OECD countries, 73% in the Republic of Korea, 84% in Japan), universities have produced increasing numbers of graduates in a wide array of environmental disciplines over the last ten years. Concerning *environmental law*, out of 400 universities with law schools, 10% do research on or teach environmental law.

Chinese authorities co-operate with *international partners* on promoting environmental education (e.g. with WWF, with Japan and the Republic of Korea in the tripartite Conference of Environment Ministers).

3.6   Environmental NGOs

Since 1994, the year when the first environmental non-governmental organisation (NGO) was founded in China, the country has seen a *large increase* in the number of environmental NGOs. In the early 2000s, the total number of student environmental organisations had reached 184, while the number of non-student environmental NGOs was 73 (Yang, 2005). The *current number* of environmental NGOs is estimated at 3 000, according to the 2005 database of the All China Environment Federation (ACEF). Government-initiated groups account for nearly 50%, student clubs for 40%, Chinese civic organisations for over 7% and Chinese branches of international NGOs for 3%. However, other estimates refer to about 2 000 registered environmental NGOs, with perhaps another 2 000 registered as for-profit business entities, or not registered at all. Some experts estimate that there are about 100 000 environmental groups in China.[5]

Scope of activities

Environmental NGOs in China have evolved from focusing on environmental education and biodiversity protection to debating a wide range of environmental issues, including large-scale infrastructure projects, lack of enforcement of environmental requirements, and misappropriation of funds.

The *Chinese government* has generally adopted a positive attitude toward environmental NGOs. The 9th and 10th FYPs encouraged society and its representatives to address environmental degradation concerns and to promote environmental education and awareness. For example, in 2001, 20 NGOs joined in the drafting of Beijing's plan for a "Green Olympics" in 2008. In 2002, more than 30 Chinese NGOs attended the World Summit on Sustainable Development in Johannesburg. In 2004, environmental NGOs launched in Beijing the "26 °C campaign" to save energy during hot summer periods (Box 8.4). SEPA, in particular, supports environmental NGO activity and works very closely with them both formally and informally. At the local level, however, EPBs remain ambivalent about environmental NGOs and their potential role in assessing environmental performance.

China also has many grassroots environmental NGOs, including environmental clubs and student groups which promote environmental consciousness, public participation in decision-making and sustainable development through a *mix of traditional and new collective actions*. These groups' traditional actions include public lectures, workshops, conferences, forum discussions, field trips, publication of newsletters and multimedia documents, and new forms of "electronic action" such as online discussions, online mailing lists and Internet petitions. Typically, the groups encourage learning, co-operation and participation. In addition, new collective actions led by some NGOs include supporting citizens through legal action, notably to protect pollution victims and induce polluting industries to operate within the environmental legislative and regulatory framework. For example, the Centre for Legal Assistance to Pollution Victims operates a telephone hotline about environmental legal issues and has taken more than 30 cases to court on behalf of pollution victims, winning about half.

Since the late 1990s, *international NGOs* have begun to launch projects or set up offices in China. Seven of them have offices in Beijing, including the World Wildlife Fund for Nature, Friends of the Earth, Greenpeace and the Jane Goodall Institute (Yang, 2005). These offices promote global environmental issues and often form partnerships with Chinese NGOs (e.g. co-organisation of conferences, seminars and lectures; financial support to Chinese NGOs for education and awareness-rising). The government views foreign donations to environmental NGOs like other foreign direct investments: as a positive contribution to the Chinese economy.

Box 8.4 **The Beijing 26 °C Campaign**

Summers in Beijing are hot and humid, and the use of air conditioners is widespread. As a result, air conditioners account for *40 to 50% of total power consumption* during the three summer months. Air conditioners are typically set at a temperature of around 22-24 °C or even lower, thus using unnecessarily high amounts of electricity and causing discomfort and illness.

Since 2004, *six environmental organisations* in Beijing, namely Global Village of Beijing, WWF China, China Association for NGO Cooperation, Friends of Nature, Institute for Environment and Development, and Green Earth Volunteers have carried out the "26 Degree Campaign", similar to the 28 Degree "Cool Biz" Campaign run in Japan, where employees are permitted to wear light clothing in the office during summer and are asked to put on an extra layer in winter, when thermostats are set at 18 °C.

The 26 Degree *Campaign*, which is supported by the government, urges companies, shopping malls, offices, embassies and other institutions to set their air conditioners at 26 °C in order to save electricity and protect the environment. A range of hotels, offices and embassies joined the campaign in the first few weeks after its launch. As with other campaigns, it is often hard to keep up the momentum over time, and without monitoring and publication of results in the media, it is difficult to say how successful the campaign has been.

Source: Website Global Village Beijing.

Organisational forms

Two 1998 *regulations* provide the general framework for the registration and management of all social organisations and non-profit organisations in China. Officially, NGOs must have a government sponsor, to whom they report their membership, funding sources and activities. In a given administrative area, there should not be more than one organisation for any specific theme of work. In practice, government-sponsored NGOs cover many themes. NGOs are not permitted to have branch organisations in various provinces. NGO leaders say that the degree to which all of these restrictions are enforced varies according to the sponsor.

China has *five main types* of environmental NGOs: i) those established or sponsored by government agencies with most of their activities responding to agency requests for information, research and analysis related to environmental policy and laws; ii) registered non-profit public or private organisations; iii) registered for-profit business enterprises;[6] iv) university research centres and student associations affiliated with institutions of higher education (often registered with campus Youth Leagues);[7] and v) unregistered voluntary groups functioning as NGOs, some operating mainly through the Internet.[8]

Many of the most renowned environmental NGOs are based in Beijing. However, they undertake *activities throughout the country*, including significant activities in Tibet, Yunnan and Sichuan. Many smaller, locally-based NGOs have also sprung up to address local concerns such as biodiversity protection, dam construction and water pollution. As these often struggle with the government-mandated registration process, funding and membership requirements, the Beijing-based NGOs often try to support their development through training in grant writing, in developing materials and programmes, and even through financial support. As many environmental NGOs face funding and staff challenges, they partly depend on funding from international sources. Except for some government sponsored environmental NGOs, most have a *small professional staff* (e.g. not more than 20),[9] and rely heavily on volunteers and members. These are predominantly college students and recent graduates.

Notes

1. i.e. disparities among provinces, centrally administered municipalities and autonomous regions.
2. This is consistent with the 1978 OECD Council Recommendation on Reporting on the State of the Environment, and with the 1998 OECD Council Recommendation on Environmental Information.
3. Some monitoring stations prefer contracts from industry rather than from EPBs because they earn revenue from the former which can supplement their income. This creates a conflict of interest and reduces the monitoring capacity available to the government.
4. See the 1996 OECD Council Recommendation on Implementing Pollutant Release and Transfer Registers.
5. Estimates of the total number of NGOs (i.e. environmental or not) in China vary significantly. The Ministry of Civil Affairs reports 280 000 registered NGOs in 2005 (including some 6 000 Chinese branches of foreign or international NGOs), up from a total of 4 800 NGOs in 1988. The World Bank reports 300 000 to 700 000 NGOs (Mooney, 2006, Siarnacki, 2006).
6. Since the NGO registration process is still complex, many green associations opt for the easier route of registering as for-profit business enterprises, even though this requires them to pay tax.
7. Many organisations call themselves university student environmental groups to avoid the NGO registration process.
8. Unregistered organisations are in principle illegal; however, as evidenced by Green Earth Volunteers and Han Hai Sha, they not only operate publicly but enjoy considerable recognition.
9. For example, Global Village of Beijing has 11 staff members; Friends of Nature and the Institute of Environment and Development each have 12.

Selected Sources

The government documents, OECD documents and other documents used as sources for this chapter included the following. Also see list of websites at the end of this report.

CCICED (China Council for International Co-operation on Environment and Development) Lead Expert Group (2003), *Sustainable Industrialisation in China and a Well-off Society*, Discussion Paper for the Second Annual Meeting of the Third Phase of the CCICED, NDRC, Beijing.

Chai, Xilong, Linghui Kong *et al.* (柴西龙, 孔令辉, et al) (2005), "建设项目环境影响评价公众参与模式研究." (Study on the Public Participation Mode in Environmental Impact Assessment of Construction Projects), 中国人口·资源与环境 15(6) (China Population, Resources and Environment).

Chen, Shaohua and Martin Ravallion (2004), *China's (Uneven) Progress Against Poverty*, World Bank Research Working Paper 3408, Washington DC.

Chen, Tinghui (陈廷辉) (2003), 民间环保组织在环境保护中的作用 (NGO's Function in Environmental Protection), 中山大学法学院 (Sun Yatsen University School of Law), 中山大学学报论丛, 2003年第23卷第4期 (Sun Yatsen University Forum, 2003, Vol. 23, Issue #4).

China Development Research Foundation, UNDP (United Nations Development Programme) (China) (2005), *China Human Development Report 2005: Towards Human Development with Equity*, UNDP, New York.

Ding, Y. *et al.* (2005), *Migration and Job Creation in China's Sustainable Urbanization Strategy*, Background Paper for the Fourth Annual Meeting of the Third Phase of the CCICED, NDRC, Beijing.

Economy, Elizabeth (2004a), *Economic Boom, Environmental Bust,* Council on Foreign Relations www.cfr.org/publication/7548/economic_boom_environmental_bust.html.

Economy, Elizabeth (2004b), *The River Runs Black: the Environmental Challenge to China's Future*, Cornell University Press, Ithaca, New York.

Fan, C. Cindy (2005), *Modeling Interprovincial Migration in China, 1985-2000*, Eurasian Geography and Economics, Vol. 46, No. 3, March 2005, Bellwether Publishing.

Ho, S. Mun and P. Chris Nielsen (2006), *Clearing the air: assessing the health and economic damages of air pollution in China*, MIT Press, Cambridge.

Kwan, Chi Hung (2005), *Regional Disparities Have Gone Beyond Acceptable Limits – The Path to an All-Round Well-off Society Remains Distant in China in Transition*, www.rieti.go.jp/en/china/ 05112901.html.

Lee, F. Yok-shiu (2005), "Public Environmental Consciousness in China: Early Empirical Evidence", in K.A. Day (ed.), *China's Environment and the Challenge of Sustainable Development*, M.E. Sharpe, Inc., Armonk, New York.

Li, Fugui and Bing Xiong (李富贵, 熊兵) (2005), "环境信息公开及在中国的实践." (Public Access to Environmental Information and Its Practice in China), 中国人口·资源与环境 15(4): 22-25 (China Population, Resources and Environment).

Li, Lizhen and Lan Ma *et al*. (李丽珍, 马兰, et al) (2005), "环境影响评价中的公众参与现状及其改进建议." (Present Situation and Suggestions for Improvement in Public Participation in Environmental Impact Assessment), 云南环境科学 24 (Additional Issue, Yunnan Environment Sciences).

Liu, Jianguo and Jared Diamond (2005), *China's Environment in a Globalizing World*, in Nature 435, 30 June 2005, pp. 1179-1186.

Liu, Yingling (2005), *SEPA Releases New Measure on Public Participation in Environmental Impact Assessment Process*, in Worldwatch Institute, China Watch, www.worldwatch.org/node/3886.

Ministry of Health, People's Republic of China (中华人民共和国卫生部) (2005), "妇幼保健与社区卫生工作简讯(2005年第11期)." (Briefing on Work Related to Women and Children's Healthcare and Community Health, 2005), Beijing.

National People's Congress Website (中国人大网) (2002), "什么是环境诉讼? 环境诉讼有什么特点?." (What Is an Environmental Lawsuit? What Are the Characteristics of Environmental Lawsuits?), Beijing.

NBSC (National Bureau of Statistics of China) (2006), China Statistical Yearbook on Environment, China Statistics Press, Beijing.

OECD (2003), *Proceedings of the OECD-China Seminar on Environmental Indicators*, OECD, Paris.

OECD (1999), *Proceedings of the OECD-China Seminar on Environmental Monitoring*, Beijing, 12-14 April 1999, OECD, Paris.

Pan, Yue (潘岳) (2004), 环境保护与社会公平 (Environmental Protection and Social Justice), 环境教育 (Environmental Education), Beijing.

Siarnacki, Anne (2006), *The Evolving Role of Environmental NGOs in China*, NGO Watch Website, www.ngowatch.org/articles.php?id=311.

Smil, Vaclav (1995), "Environmental Crisis: Regional Conflicts and Ways of Cooperation", in G. Bächler and K.R. Spillmann, Environmental Crisis: Regional Conflicts and Ways of Cooperation, Center for Security Studies (CSS), Zurich.

SEPA (State Environmental Protection Administration) (2006a), Annual Statistics: Report on Environment in China, China Environmental Science Press, Beijing.

SEPA (2006b), China Environment Yearbook, China Environment Yearbook Press, Beijing.

SEPA (2006c), State of the Environment Report, SEPA, Beijing.

SEPA (2005), Policy and Regulation Department, International Cooperation Center, Jiangsu Province Environmental Protection Bureau, and Nanjing University Environment College (国家环境保护总局法规司, 国家环境保护总局对外合作中心, 江苏省环保厅, 南京大学环境学院) (2005), 企业环境行为评价及信息公开项目总报告 (Report on Enterprises Environmental Performance Assessment and Projects on Public Access to Information), SEPA, Beijing.

State Council of the People's Republic of China EPA (1994), *China's Agenda 21: White Paper on China's Population, Environment, and Development in the 21st Century*, China Environmental Science Press, Beijing.

Stockholm Environment Institute and UNDP (China) (2002), *China Human Development Report 2002: Making Green Development a Choice*, Oxford University Press, Hong Kong.

Tao, Ruijun, Xuemei Wu (陶汝君，吴雪梅)（2005），环境保护与公众参与 (Environmental Protection and Public Participation), 眉山市东坡区环境监察大队，四川眉山；四川省环保产业发展中心，成都 (Dongshan District Environmental Inspection Team, Meishan, Sichuan; Sichuan Province Environmental Protection Development Center, Chengdu), 四川环境，2005年第24卷第5期 (Sichuan Environment, 2005, Vol. 24, Issue #5）.

Wang, Hongqi (2004), *Education for Sustainable Development in Primary and Middle School in China*, School of Environment, Beijing Normal University, Beijing.

Wang, Jin (汪劲) (2005), "中国环境法学研究的现状与问题—对1998－2003年中国环境法研究状况的调查报告." (Present Situation and Problems in the Study of Chinese Environmental Laws – Investigation Report on the Study Situation for Chinese Environmental Laws from 1998 to 2003), www.riel.whu.edu.cn/show.asp?ID=3219.

World Bank (1997), *Clear Water, Blue Skies: China's Environment in the New Century,* World Bank, Washington DC.

World Health Organization and UNDP (2001), *Environment and People's Health in China*, www.wpro.who.int/NR/rdonlyres/FD5E0957-DC76-41F2-B207-21113406AE55/0/CHNEnvironmentalHealth.pdf.

Wu, Kai, Haiping Lei and Xikun Li (吴凯，雷海平，李希昆) (2005), 论我国公众参与环境保护法律机制的缺陷与完善 (The Defection and Improvement of the Legal Mechanism Regarding Public Participation in Environmental Protection), 昆明理工大学法学院，云南昆明 (Kunming University of Science and Technology School of Law, Kunming, Yunnan), 昆明理工大学学报，第5卷第3期，2005年9月 （Journal of Kunming University of Science and Technology, Vol. 5, Issue #3, September 2005), Kunming.

Xiong, Ciguang (熊慈光: 91) (2005), 论环境侵权民事诉讼的新发展 (Regarding New Developments in Environment-Related Acts of Tort Civil Lawsuits), 河南省政法管理干部学院学报 (Henan Province Political and Legal Administration Cadre College News).

Yang, Yongheng, Angang Hu and Zhang Ning (2005), *Regional Disparities and Imbalance of China's Human Development: A Historical Perspective of "One China, Four Worlds"*, Centre for China Study, Tsinghua University.

Yang, Guobin (2005), *Environmental NGOs and Institutional Dynamics in China*, in The China Quarterly 2005, Cambridge University Press, Cambridge.

Zhang, Heping (张和平) (2001), 特写: 首例群体矽肺病人身损害赔偿特大官司扫描 (Close-Up: A Glance at the First Major Personal Injury Compensation Lawsuit for Group of Coal Workers' Pneumoconiosis Patients).

INTERNATIONAL CO-OPERATION

Features

- China's engagement in international environmental co-operation
- The desertification challenge
- Transboundary air pollution
- Marine issues
- Progress in combating stratospheric ozone depletion
- Climate change issues
- Trade and environment

Recommendations

The following recommendations are part of the overall conclusions and recommendations of the environmental performance review of China:

- continue China's *active engagement in international environmental co-operation*, seeking to improve the effective and efficient use of i) domestic resources, and ii) international support mechanisms (e.g. the World Bank's Clean Development Fund, the Multilateral Fund under the Montreal Protocol, and the Global Environment Facility);

- strengthen *monitoring, inspection and enforcement capabilities* in support of the implementation of international commitments (e.g. on trade in endangered species, in forest products, in hazardous waste and in ozone-depleting substances, as well as on sound chemicals management, ocean dumping and fisheries management);

- improve governmental oversight and environmental performance in the *overseas operations of Chinese corporations* (in the spirit of the OECD guidelines for multinational enterprises);

- develop *partnerships with foreign enterprises* to contribute to environmental progress through provision of training, technical support and cleaner technology; ensure environmental requirements are not relaxed to attract *foreign direct investments*;

- continue to assign high priority to domestic and regional *anti-desertification efforts*;

- intensify domestic and international co-operation to reduce *transboundary air pollution* in Northeast Asia by, inter alia , introducing cleaner coal technology, improving energy efficiency and fuel switching;

- ensure that the interim and final targets for the phase-out of *ozone-depleting substances* under the Montreal Protocol continue to be achieved on schedule;

- prepare a *coherent national plan on climate change* which draws together the array of climate-related activities currently underway and planned to improve their collective efficiency and impact;

- strengthen efforts to protect and improve *water quality in coastal waters and adjacent regional seas* from land-based pollution sources, and upgrade environmental management regulations and government oversight in the aquaculture industry;

- integrate environmental considerations systematically into China's growing *development co-operation* programme.

Conclusions

The last decade has seen a *dramatic increase in China's engagement with other countries* in addressing environmental challenges. This reflects a growing recognition across the spectrum of Chinese institutions of the important economic, social and ecological stakes that China has in meeting these challenges, and also of its shared interests with the international community. China is now an active, constructive participant in a broad array of regional and global environmental conventions, institutions and programmes, and is *drawing heavily on international financial institutions and special mechanisms* (e.g. the Montreal Protocol's Multilateral Fund) to augment its own resources and ensure that China's international commitments are met. Since 1995, it has reduced its production and consumption of ozone-depleting substances more than any other country; established comprehensive and ambitious policies and legal regimes in the areas of marine pollution and fisheries management; provided international leadership in efforts to control transboundary movement of hazardous waste; recognised and taken initial steps to confront its emissions of greenhouse gases; and undertaken a detailed examination of how its *trade and investment policies* can work to support environmental management goals.

China, however, remains the *second largest contributor of greenhouse gases*, and is still the world's largest producer and consumer of ozone-depleting substances. Its largely coal-fired economy is a major source of acid rain and other transboundary air pollutants in Northeast Asia, and is a significant contributor to global-scale air pollution, including mercury. Its *coastal waters and regional seas* are suffering from an increasing burden of land-based pollution in many areas; and the environmental management and food-sanitation regimes for China's rapidly expanding marine aquaculture industry need strengthening. A lack of strong *monitoring, inspection and enforcement capabilities* and associated penalties are limiting the effectiveness of otherwise sound policies, laws and regulations established to further China's domestic objectives and international commitments in the areas of marine fisheries, coastal water quality, hazardous waste transport, and the control of *illegal trade in endangered species, forest products and ozone-depleting chemicals*. Stronger efforts are needed by the government to ensure that *Chinese corporations operating overseas*, particularly in such environmentally-sensitive industries as forest products and mining, are positive contributors to China's stated goal of building an international reputation for sound environmental management and sustainable development. *Funding limitations* and *inadequate institutional co-ordination* are constraining the pace at which China is able to carry out an ambitious international environmental agenda that includes a range of difficult challenges (e.g. desertification control, greenhouse gas reduction, marine management). To achieve success, increased financial efforts from China as well as major technical support and targeted financial assistance to China from *OECD countries and international financial institutions* will be required.

◆ ◆ ◆

1. Policy Objectives, Institutions and Mechanisms

1.1 Policy objectives

China's *9th and 10th Five-Year Plans* (FYPs) for Economic and Social Development (covering 1995-2005) each includes a general call for *expanded international co-operation* on environmental protection. Over the decade, *specific policy objectives* were articulated in a variety of pronouncements by senior Chinese officials, e.g., in international forums such as the UNEP Governing Council, the UN Commission on Sustainable Development and the 2002 World Summit on Environment and Development; in major national reports such as China's *Agenda 21*; and in action plans developed in response to China's accession to multilateral environmental conventions, e.g. on biological diversity and persistent organic pollutants. Collectively, they reveal a consistency of perspective and intention in this area with the following principal components:

- a *growing awareness* of the role international environmental co-operation can play in furthering China's political, economic and social development agendas;
- a commitment to enhance bilateral, regional and multilateral environmental co-operation;
- a desire to *import advanced environmental technology* and to develop a domestic environmental industry;
- a willingness to *devote expanding resources* to an ever-broadening international environmental agenda;
- acknowledgment of *China's contributions* to regional and global-scale environmental threats, and a commitment to be part of the solution through, inter alia , playing a leadership role on selected issues; and
- a commitment to reach out and *assist developing countries in Asia and beyond* to address both shared and unique environmental problems.

China continues to approach its goals and responsibilities on international environmental matters from the *position of a developing country*, having been accorded this status (with associated preferences) in a variety of international conventions (e.g. on waste management, ozone depletion and climate change). It maintains, in this regard, that the international community must implement the principle of the "differentiated responsibility of countries at different levels of economic development" for the financing and implementation of environmental management programmes. China's position is that it will require *major technical and financial support* from the industrialised countries of the OECD if it is to be able to carry out fully its international environmental treaty obligations and aspirations.

1.2 Institutional responsibilities

Principal responsibility for the pursuit of China's international environmental agenda resides with the International Co-operation Department of the *State Environmental Protection Administration* (SEPA). Its activities derive in large part from the need to organise, co-ordinate and conduct work programmes in support of a broad array of bilateral and regional agreements and multilateral conventions. SEPA's responsibilities include oversight of international affairs on nuclear safety.

Over *20 other central ministries and associated state institutes are involved*, both in supporting roles or with their own specialised international activities. These include the:

- Ministry of Foreign Affairs (overall foreign policy guidance and negotiation of government-to-government agreements);
- National Development and Reform Commission (NDRC) (energy, climate change and desertification);
- Ministry of Finance (management of multilateral-derived funds, e.g. from the Global Environment Fund (GEF) and the World Bank Multilateral Fund);
- Ministry of Foreign Trade and Economic Co-operation (environment-trade relationships);
- Ministry of Commerce (trade in toxic chemicals and hazardous waste);
- Ministry of Agriculture, Bureau of the Fisheries (coastal and high seas fisheries management);
- State Oceanic Administration (coastal and marine water quality management and biodiversity conservation);
- Ministry of Communications (emergency response to marine oil spills);
- Ministry of Science and Technology (Agenda 21 and Sustainable Development);
- Ministry of Health (coastal water quality and food safety in aquaculture);

– General Administration of Customs (inspections related to illegal trade in endangered species, hazardous waste and ozone-depleting substances).

Increasingly, *inter-ministerial task groups* have been created to plan, and then implement, China's participation in multilateral environmental conventions (e.g. on desertification and climate change). Research, policy analysis and programme support is also obtained from an array of *government-sponsored and private institutions*, such as the Energy Research Institute, the Centre for Ecological and Environmental Economics, and the Chinese Academy of International Trade and Economic Co-operation. Broad policy analysis and recommendations are provided by the *China Council for International Co-operation on Environment and Development (CCICED)* established in 1992, with high-level representation from both China and OECD country counterparts (Box 9.1).

At the *sub-national level*, Environment Protection Bureaus and municipal and provincial governments, principally those located in coastal regions and where transboundary watersheds exist, play selected roles.

1.3   Co-operation mechanisms

Bilateral agreements

At the end of 2005, China had over 60 bilateral agreements on environmental protection with *42 countries*, including 22 in Europe, 7 in the Americas, 11 in Asia and the Pacific and 2 in Africa (Table 9.1) These range from broad government-to-government agreements, to interdepartmental memoranda of understanding, to joint communiqués. Included are bilateral agreements on nuclear safety with France, United States, Germany, Pakistan, Spain, Republic of Korea, Russia and Canada. In addition, China has fisheries agreements with some 20 countries that have increasingly addressed resource conservation and sustainable fisheries issues rather than allocation of fish stocks.

While many of the mainstream environmental agreements are limited to information exchange, training and periodic meetings, others are *project-oriented and broad-ranging* (Table 9.2). Two examples are China's longstanding environmental co-operation with *Japan* (Box 9.2), and the *Sino-Italian co-operation programme* for environmental protection. The latter includes the establishment of an office (and joint task forces in Beijing and Shanghai) that is overseeing 45 projects (involving monitoring, energy efficiency, natural resources management and urban planning), with funding of some USD 175 million over the 2001-05 period. *Japan, Republic of Korea, Canada, United States, Germany and Norway* have also carried out substantial, project-oriented, environmental co-operation with China over the last decade.

Box 9.1 China Council for international co-operation on environment and development (CCICED)

China's participation in the 1992 UN Conference on Environment and Development stimulated the creation that same year of the CCICED. The council's mission is to serve as a high-level *consultative and advisory body* for China's government, based on studies it undertakes on major issues in the field of environment and development.

The CCICED has a unique *international character* with its membership and associated task groups including senior officials, experts and scholars from other countries and international organisations. Operating under the leadership of the Vice Premier, the CCICED currently has 47 domestic and international members, including ministers or vice-ministers from ministries under China's State Council. The CCICED itself meets annually. In 2005, policy analysis was being conducted by seven task groups on: integrated management of river basins; non-point source pollution by agriculture; pricing of natural resources; taxation; sustainable transportation; agriculture and sustainable rural development; and WTO and the environment. Task force reports and recommendations that are endorsed by the CCICED are delivered to ministries, provinces, autonomous regions, and municipalities governed directly by the State Council.

In addition to serving as a bridge between China's government and the international community, the CCICED has had a significant *influence* on China's domestic environmental policies. The 10th FYP, for example, included a series of key indicators of sustainable development, and a call for the promotion of China's environmental industry, both based on CCICED recommendations.

Source: SEPA.

The *pace of international co-operation* has picked up in recent years. Prior to 1995, China had formal agreements with nine countries, beginning with the US agreement in 1980. Between 1995 and 2000, China engaged another 15 countries, and added a further 18 in the last five years. Significantly, the trend reflects China's intention to *extend co-operation to developing countries* outside of the Asian region, reflected in agreements with Brazil, Iran, Morocco and Egypt since 2002.

Regional mechanisms

The *Tripartite Environment Ministers Meeting*[1] *(TEMM)*, established at the first meeting of a China, Japan and Republic of Korea summit in 1999, is one of China's

---

most important and active forums for environmental co-operation at the regional level. Over the past six years, TEMM projects have been initiated on long-range transboundary air pollution; environmental consciousness-raising and training; freshwater lakes pollution; prevention of land-based marine pollution; co-operation in the environmental industry; dust storm control and ecological conservation in Western China; and acid deposition monitoring.

In 1992, the *Northeast Asian Conference on Environmental Co-operation (NEAC)* was created by China, Japan, Republic of Korea, Russia and Mongolia as a continuing forum for the exchange of views and information on the state of the environment. The following year, the *Northeast Asian Subregional Programme of Environmental Co-operation* (NEASPEC) was launched, leading to a series of joint projects on, inter alia , pollution reduction in coal-fired power plants, sand and dust storm prevention, China-Mongolia transboundary nature reserves, protection and development of the Yellow sea, and implementation of the "circular economy". A Northeast Asia Centre for Environmental Data and Training has been established, headquartered in Seoul.

Specialized mechanisms are also in place for co-operation among coastal states in the region on *marine resources and pollution management*. Multi-state action plans exist for the Yellow, Bohai, East and South China seas, implemented within the framework of the UNEP Regional Seas Programme. In 1994, China played a lead role in the design and adoption of the *Action Plan for the Protection, Management and Development of the Marine and Coastal Environment of the Northwest Pacific Region* (NOWPAP).

Table 9.1 **China's bilateral environmental**[a] **agreement partner countries,** 2005

Asia	Japan, North Korea, Republic of Korea, Mongolia
	India, Iran, Pakistan, Singapore, Sri Lanka, Tajikistan, Thailand, Uzbekistan
Africa	Egypt, Morocco
America	Brazil, Canada, Colombia, Cuba, Peru, United States
Europe	Belgium, Denmark, Finland, France, Germany, Iceland, Italy, Netherlands, Norway, Spain, Sweden, United Kingdom
	Bulgaria, Czech Republic, Hungary, Poland, Romania, Russia, Slovak Republic, Ukraine
Oceania	Australia

a) Fishing agreement partners not all included.
Source: SEPA.

Table 9.2 **China's bilateral environmental programme** (content of selected agreements)

	Completed and on-going co-operation	Signature time of agreements
Japan	Sino-Japanese Friendship Centre for Environmental Protection, demonstration city project, development of urban information system for 100 Chinese cities, restoration of water environment of Taihu lake, circular economy	1994-03-20
Republic of Korea	Investigation on marine environment of the Yellow sea, development and promotion of environmental industries, ecological protection, dust and sandstorm, research and development of environmental technology, transboundary movement of air pollutants	1993-10-28
Mongolia	Natural reserves and conservation of species, desertification	1990-05-06
Sri Lanka	The impact of farm chemicals and organic foods	1998-12-28
Egypt	Comprehensive utilisation of crop residues	2003-09-01
Canada	CCICED, climate change, environmental technology authentication, environmental protection relating to the raising of livestock and poultry, environmental education, conservation of biodiversity, POPs and clean production	1993 1998-01-16 2003-09-31
United States	Environmental and health impact, environmental monitoring, ambient air quality and planning, water pollution and control, POPs, environmental laws and regulations, environmental industry, promotion of and education on environmental protection	1980-02-05 2003-12-09
France	Air quality monitoring and modelling	1997-05-15
Germany	Management of hazardous waste, capacity building in environmental protection, corporate environmental management, organic food, environmental certification, circular economy	1994-09-26
Italy	Sino-Italian Environmental Co-operation Office, ecological conservation, environmental governance and ecological investigation, climate change, sustainable urban transportation, utilisation of renewable energy, energy saving and energy efficiency, sustainable buildings, alternative technology for ozone depletion materials, sustainable and biological agriculture, afforestation, biodiversity, sustainable water management, POPs, air quality monitoring, development of Chinese Centre for the implementation on international conventions, disposal of medical waste, training in environmental management and sustainable development.	2000-10-19
Netherlands	Taihu lake, green ship dismantling, environmental law enforcement training	1998-09-21 2002-02-13
Norway	Acid rain, environmental economy, EIA, green annual report, environmental statistics, environmental information, environmental planning, ISO 14000 certification	1995-11-06
Poland	Information exchange on environmental funds, environmental training	1996-12-02
Sweden	EIA training, clean production, environment and trade	2002-08-31
United Kingdom	Environmental economy training, coastal environmental planning, urban environmental protection	1998-06-17
Russia	Nature reserves, environmental protection of boundary rivers	1994-05-27
Australia	Rehabilitation of mining sites, investigation of environmental industry	1995-04-05 2000-05-11

Source: SEPA.

Box 9.2 Bilateral environmental co-operation between China and Japan

Framework

China and Japan have actively pursued environmental co-operation in East Asia for *nearly three decades*. Not only the national governments, but also other stakeholders (e.g. local governments, private companies, NGOs), have been actively involved in a variety of projects including environmental education, monitoring, installation of pollution control equipment and afforestation.

The *China-Japan joint commission on environmental protection* plays an important role in promoting environmental co-operation as well as exchanging environmental technology and experience between the two governments. Some examples of successful bilateral co-operation at government level are described below.

This bilateral environmental co-operation is, of course, not exclusive of other forms of co-operation, and is complemented by a *number of regional co-operation* initiatives including: the Northeast Asian Conference on Environmental Co-operation (NEAC) created by China, Japan, the Republic of Korea, Russia and Mongolia (in 1992); the Acid Deposition Monitoring Network in East Asia (EANET) (since 1993); the Action Plan for the Protection, Management and Development of the Marine and Coastal Environment of the Northwest Pacific Region (NOWPAP) (in 1994); the Joint Research Project on Long-Range Transboundary Pollution in North East Asia (launched in 1995); and the Tripartite Environment Ministers Meeting (since 1999).

Examples

The *Sino-Japan Friendship Centre for Environmental Protection,* established in 1996, benefited from grant aid from the Japanese government (about USD 100 million) and funding from the Chinese government (about USD 7 million). The Centre is directly affiliated with SEPA. Its purview includes environmental scientific research, technology development, information exchange and personnel training. The centre is now in its third phase, which began in 2002 and whose activities include provision of Japanese experts, training of Chinese professionals, and provision of necessary equipment. While continuously strengthening the co-operation with Japan, the Centre has established relationships with other countries, regions and international organisations, and conducts exchange and co-operation on various environmental issues.

The *China-Japan Environmental Development Model City Scheme* (agreed in 1997) provides for co-operation in air management. Chongqing, Guiyang and Dalian were selected as model cities for projects that aim at preventing air pollution (acid rain), establishing a circular industry/society, and disseminating success stories in China. The projects were funded by Japanese ODA loans (e.g. supply of heating gas, thermal power, steps against factory pollution). In 2003, annual SO_2 emissions in the three model cities decreased respectively by 170 000 tonnes, 130 000 tonnes and 11 000 tonnes, respectively, compared to 1999 emissions.

Box 9.2 **Bilateral environmental co-operation between China and Japan** *(cont.)*

Japanese ODA (grants and loans commitments totaling USD 1.3 billion in 2004) is by far the largest among OECD donor countries, although it is declining. Its environmental share is more than 20% (a high share compared to other OECD donors), and has had a significant impact on environmental improvement in China. The analysis of the effects of 16 projects (launched during 1996-2000) revealed a 190 000 tonnes reduction in SO_2 emissions and a 340 000 tonnes decrease in COD effluents in 2003. Beneficiaries of the projects are estimated to be almost 4 million people in the 10 cities covered by urban gas projects, 0.9 million in the six cities covered by regional heat supply projects, and 13 million in the 28 cities covered by sewerage projects.

The *Japanese Crested Ibis* (Nipponia Nippon) is one of the most endangered bird species. It now only lives in China in the wild. Its total world population is 600, including some population being bred in the Sado Japanese Crested Ibis Conservation Centre in Japan. A couple of ibises were given as present to Japan by China in 1999, as a first step of co-operation. In October 2003, the China-Japan agreement on co-operation in protection of the Japanese Crested Ibis was signed. Thanks to this co-operation, the population of the Japanese Crested Ibis in the Sado Conservation Centre has now reached 98.

CDM is a promising area for the co-operation between the two countries, since they can play key roles in the overall GHG reductions in the world. Private companies are active in designing CDM projects, with support of the two Governments.

In Southeast Asia, following a 2002 proposal from the Association of Southeast Asian Nations (ASEAN), China, Japan and Republic of Korea have been participating in the *Environmental Ministers Meeting between ASEAN and China, Japan and Republic of Korea ("10 + 3")*. This provides a broad-ranging policy dialogue in such fields as natural resources conservation, ocean environment, environmental technology training, clean technology, and transboundary pollution. Further, the *Environmental Co-operation in the Greater Mekong Sub-region* offers a *sub-regional* co-operation mechanism operated by the Asian Development Bank, involving China and its Yunnan province, Vietnam, Laos, Cambodia, Myanmar and Thailand. It has established an environmental working group, launched a programme of periodic meetings of environment ministers of the region, initiated a Greater Mekong "biodiversity protection corridor programme" and developed environmental performance analysis in co-operation with the OECD.

With *Europe*, China and the European Union agreed in 2003 to co-operate on environmental protection, with initial discussions carried out on climate change, water management and biodiversity. Protecting the environment and promoting sustainable development has been included as one of three key areas for co-operation between China and the EU in the "country strategy paper" that sets priorities for EU assistance to China; and a joint biodiversity conservation programme with USD 40 million in funding was launched in 2004.

Co-operation with *Africa* has also been initiated. In 2000, the China-Africa Environmental Protection Co-operation Forum was established, with the initial meeting held in Beijing. With support from UNEP, the *China-Africa human resources development environmental protection training programme* began in 2005, with China providing assistance to enable environmental officials and experts from African countries to participate in the training.

For the last ten years, China and the *OECD* have been co-operating in the environmental field under an agreement which focuses on policy dialogue, information exchange and environmental assessment. The current environmental performance review was undertaken within the framework of this co-operation.[2] China has under consideration possible adoption of OECD procedures on chemical safety; and has been utilising the lists adopted by OECD members for the control of hazardous waste in international trade.

Multilateral mechanisms

At the end of 2005, China was *participating in virtually all of the major multilateral organisations*. Its objectives are: to project its views on environmental issues, influence the content of international agreements, participate in programme design and implementation, and obtain technical and financial support for its domestic environmental management efforts. *UNEP* has been one of the most important partners, since China became a member state in 1973, along with the *GEF* (Box 9.3), *UNDP, UNESCO, WHO, IMO, UNIDO, FAO, the World Bank, the Asian Development Bank (ADB)*, and some dozen *international NGOs, such as IUCN*.

Since 1980, China has signed and acceded to *more than 20 multilateral environmental conventions and associated protocols* (References II.A, II.B), almost all of the major multilateral environmental agreements (MEAs) negotiated over the past decade. Particular prominence has been accorded by Chinese officials to the Basel Convention on the control of transboundary movements of hazardous waste; the Montreal Protocol on ozone-depleting substances; the Rotterdam Convention on

Box 9.3 **China and the Global Environment Facility (GEF)**

The GEF provides new and additional *grants and concessional funds to UNDP and World Bank eligible countries* to meet the incremental costs of projects and programmes in six areas: biodiversity, climate change, international waters, land degradation, the ozone layer and persistent organic pollutants. The GEF was formally established in 1994 by the World Bank, UNDP and UNEP. Countries access GEF resources either through one of these three "implementing agencies", or through several "executing agencies", including the Asian Development Bank (ADB). Since 2004, the ADB has assumed a much more prominent role in the design and implementation of GEF projects. Since 1991, the GEF has provided over USD 5 billion in grants and has generated close to USD 20 billion in co-financing from other sources for projects in developing countries and countries with economies in transition. Altogether, GEF has provided financing for some 1 500 projects in 160 countries. The World Bank accounts for over 70% in terms of total funds provided, followed by UNDP, ADB, UNEP and UNIDO.

China joined the GEF in 1994. The *Ministry of Finance*, through its Department of International Co-operation, is the political and operational focal point for GEF in China, and is the funding window for the World Bank and ADB. *SEPA* is the technical support department for GEF projects; and the *NDRC* plays a prominent role by virtue of its mandate to co-ordinate China's climate change activities. A separate GEF Secretary Office was created in 2000 to expand GEF funding and promote project management efficiency.

China has, to date, received more GEF funding than any other country, with the majority devoted to the areas of *climate change and biodiversity*. Through 2003, GEF had allocated close to *USD 500 million to China* for 42 China-based projects, with 70% of it going to 23 climate-related projects (e.g. energy efficiency and renewable energy sources). Protection of biodiversity efforts received 15% of the funding, with the remainder dedicated to international waters and land degradation projects.

Source: World Bank, 2005.

chemicals and pesticides in international trade; the Stockholm Convention on persistent organic pollutants (Box 9.4); the Convention on Biodiversity; the Biosafety Protocol; and the Nuclear Safety Convention.

China has compiled a *good record in responding to its commitments* as signatory to the MEAs. It has consistently ratified expeditiously the agreements it has signed; met deadlines for submission of reports and data; and enacted implementing domestic legislation and programmes in a timely fashion. In many instances, target

Box 9.4 Persistent organic pollutant control

SEPA started to deal with POPs in 2001, when China signed the *Stockholm Convention*. The Convention entered into force in 2004 in the country. China has adopted *five "Strategies" on POPs*: for phase-out of pesticide POPs, for reduction and disposal of PCBs, for reduction and control of unintentionally produced POPs (UP-POPs), for inventory and disposal of POPs, and for waste and contaminated sites. The five Strategies were translated in *China's POPs National Implementation Plan*, which was endorsed by the State Council in spring 2006 and will be presented at the Conference of Parties of the Stockholm Convention by the end of 2006.

The annual total production of *pesticide POPs* in China is estimated at around 110 000 million tonnes. While the production of *toxaphene* and *hexachlorobenzene* (HCB) was stopped in the 1980s and in 2004, respectively, China still produces *chlordane*, *mirex* and *DDT*. DDT is used to produce dicofol, malaria prophylaxis and additives for antifouling paints. The production of DDT (3 236 million tonnes in 2003) and the levels of DDT in the environment have declined since China prohibited its use as a pesticide in 1983. However, DDT can still be found in the atmosphere, farmland soil, waters, sediments, cereals and vegetables, meat, animals and the human body in many areas of China. DDT is mainly exported to Africa and Southeast Asia to be used mostly for malaria control and, to a lesser extent, to protect wood. A single company exports DDT.

The production of *PCBs and PCBs-containing products* has been prohibited in China since 1974. Typical PCB waste are PCB-containing electrical products and material (like soil) and PCB-contaminated equipment. Due to China's low disposal capacity, most PCB-containing products have been kept in storage, awaiting final disposal. As a result, China has abandoned storage sites (without monitoring and/or warning signs), some of which have exceeded the 20-year legal storage limit, and sealed and buried PCB capacitors that are difficult to identify. Some 600 000 to 800 000 out-of-use PCB-containing capacitors (i.e. 30 000 to 40 000 tonnes of PCBs) are scattered throughout China. Some additional 15 000 to 25 000 PCB-containing capacitors are in use in China. Although China never produced PCB-containing transformers, some were imported in the past. Their number is unknown.

UP-POPs (e.g. polychlorinated dibenzo-para-dioxins and polychlorinated dibenzo-furans or PCDD/Fs) and *HCB* are released through waste incineration, ferrous and non-ferrous metal production, mineral production, production and use of chemicals and consumables, disposal and landfill of waste, etc. The 2005 estimate of China's total annual release of PCDD/Fs was 15 kg of toxic equivalent, of which more than half is in soil (e.g. in landfills or from waste incineration) and the rest is in air (e.g. from cement kilns, iron ore sintering). Since no special legislation directly deals with UP-POPs, China's laws on EIA and on the promotion of cleaner production are the bases for controlling the UP-POPs.

commitments (such as those set under the Montreal Protocol) have been met; in other instances, shortages of technical and financial resources have limited China's ability to implement the plans it has set out to pursue MEA objectives.

Overall assessment

Overall, China is *deeply engaged* in the design and conduct of international environmental affairs in today's world, and is viewed widely as an *important actor and responsible partner*. The challenge for the government is to mobilise and manage efficiently the financial and manpower resources needed to carry out a rapidly broadening spectrum of programmes and projects in this area. Over the past five years, the growth in the scope of activities and their complexity *has outpaced the expansion of the government's investment*, including in the international offices of key ministries and departments. The result is a proliferation of many under-funded, under-staffed and delayed efforts, which fail to capture all of the benefits that might otherwise accrue to the country. The options are to mobilise expanded resources and/ or consolidate and prioritise the activities, to ensure that they can deliver maximum contributions to China's environmental agenda, as well as to its economic and social development objectives.

In this regard, consideration should be given to carrying out a *high-level, inter-ministerial review* of China's portfolio of international environmental activities to prepare and articulate a coherent policy and programming framework. The review should reassess priorities, establish necessary funding levels, examine lead and supporting-agency responsibilities, and set out measures to improve interagency co-operation and co-ordination. It would, ideally, be requested by the State Council and spearheaded by the Ministry of Foreign Affairs to ensure its linkages to China's foreign affairs, economic and social policies. Absent this level of review, the international environmental portfolio should *nonetheless be kept under periodic review within the government* to promote greater efficiency and impact of scarce resources.

2. Regional Issues

Over the past four decades, the *rapid growth in population, industrialisation and urbanisation in East Asia* has given rise to myriad environmental problems which most countries of the region share or have in common. The fact that the majority of these countries, including China, belong to the developing world whose first priority is to expand economic growth has limited China's ability to invest in environmental protection. Today, as the environmental problems of East Asia have intensified to the point of adversely affecting GDP goals as well as human health and the well-being of citizens, the priority being assigned to environmental protection and rehabilitation has risen, at a time when China and its neighbours can more readily afford to invest in China's solutions.

Regional-scale problems of greatest interest to China are: degradation of arable lands from the process of desertification; transboundary air pollution, especially acid precipitation; pollution of coastal waters and regional seas by inflows of industrial, agricultural and urban waste, as well as waste dumping and oil spills at sea; depletion of marine fisheries through overfishing and degraded water quality; and conservation of migratory birds and wildlife habitat (Chapter 6).

2.1   Desertification

At the end of 2004, an estimated *27.5% of China's total land area* (2.64 million km^2) was in a desertified condition. Some 400 million people were affected by the extensive soil salination and alkalization and blowing sand and dust was resulting in suspension of communication and transportation, buried villages, delayed or cancelled flights, reduced life of irrigation works and respiratory diseases. The *direct and indirect costs* to China each year are estimated at USD 40 billion (NBCD, 2005a). Regionally, severe economic and human health impacts are being felt as far away as the Republic of Korea; globally, dust derived from deserts and desertified areas in China and Mongolia are being carried around the world by upper atmosphere currents, with deposition detectable in the Arctic and on the western coast of North America.

Institution building and regional co-operation

China has, for decades, assigned anti-desertification a *high priority in its economic and socialdevelopment policies*, and hence has made major investments in this area. It was a strong supporter of the 1977 United Nations Conference on Desertification, and of the follow-up UN Convention to Combat Desertification (UNCCD), which it ratified that same year. Even prior to negotiation of the convention, China had prepared a national ten-year anti-desertification programme. In 1987, a head office for research and training in desertification control was set up by UNEP in Lanzhou (Gansu province). In 1994, China established a *national co-ordinating group to combat desertification* and a *national committee for the implementation of UNCCD* with representation from 18 ministries of the State Council. Three years later, a national bureau to combat desertification was created to co-ordinate, manage and promote anti-desertification efforts.[3]

In 2002, the environment ministers of China, Japan and Republic of Korea agreed to develop a joint *dust and sandstorm (DSS) monitoring network* within the framework of the TEMM. They also decided to promote joint training and education programmes to help build capacity in addressing DSS mitigation, and to seek expanded co-operation with international organisations. Consequently, in 2003, China, Japan, Republic of Korea and Mongolia agreed to conduct a project on prevention and control of dust and sandstorms in Northeast Asia, in collaboration with the ADB, the United Nations Economic and Social Commission for Asia and the Pacific (ESCAP), UNEP and the UNCCD Secretariat, with investments by the ADB of USD 500 000 and by the GEF of USD 1 million. Outputs have included an institutional framework for future regional collaboration; a master plan that sets out a phased approach to establishing and operating a regional monitoring and early warning system; and an investment strategy with recommendations on sustained financing and project priorities.

China's National Action Programme on Desertification

The *goals and broad policy and programme framework* for these institutions are set out in a *National Action Programme* that has been reformulated several times since its development in 1991 (NBCD, 2005b). This programme is financed through four channels: central government allocation; locally-raised funds; discounted-interest loans; and overseas financial support from the World Bank, UNDP, UNEP and the ADB, plus bilateral donors. From 1993-96, Germany assisted China to establish ecological afforestation programmes in desertified areas with grant funding of some USD 23 million. Bilateral agreements on desertification have also been signed with Japan, the Netherlands, Australia, Canada, Belgium, Norway and Sweden. China has drawn on technical support offered by UNEP and the UN Office to Combat Desertification and Drought; and, in 1996, the Secretariat of the UNCCD supported efforts by NGOs in China to assist villages with sand dune control and awareness-raising.

In the mid-1990s, China's *National Action Programme was redesigned* to encompass three time frames. The *goal set for 2000* was to mitigate "*to some extent*" the continuous expansion of desertification and to raise "substantially" the living conditions of the residents of affected areas. By *2010*, the objective is to have the environmental conditions "*improved considerably*" on a regional scale, with people's living standards greatly improved. By *2050*, the aim is to have *all but the most severely desertified areas* brought under effective control, with nature reserves covering more than 90 million hectares. The scope and complexity of the challenge, as acknowledged in the national programme, is evident in the fact that an estimated 10% of China's desertification is due to excessive pasturing, 30% to over-cultivation of the land, 40% to deforestation and woodcutting, and the remainder to improper water use, including poorly planned irrigation schemes and surface water diversions.

To *engage citizens* in afforestation and other anti-desertification activities, the Chinese government has *introduced preferential policies*, including discounted government loans for desertification projects, and deductions or exemptions from taxes on benefits generated by the development and stabilisation of barren and fragile land. The auctioning off of the "four barrens" (barren hills, barren gullies, barren desert and barren land) has been employed, and subsidies and other incentives have been provided to villages and individuals to plant shelter belts and convert sloping farmland into forest or grassland. Enterprises have been encouraged and assisted to relocate to desertified areas for purposes of helping to improve both environmental and economic conditions.

Overall assessment

It is difficult to establish whether China is on track to meet its goals, given widely varying estimates of the trend in desertification over the past decade. Government reports differ quite markedly as to the rate of change, and whether affected areas are shrinking or continuing to expand. The consensus is that the *problem has been mitigated in recent years*, in conformance with the year 2000 goal, as the result of the major investments China has been making. To maintain this momentum, and achieve its year 2010 objective, China will have to continue to *assign high priority to its anti-desertification efforts*, and enlist the aid of *other countries* and multilateral financial and technical institutions. China's support for Mongolia's efforts is also an essential component.

2.2   Transboundary air pollution

Northeast Asia has been identified by UNEP as one of the *global "hot spots" for air pollution*, along with Europe and North America. China is a major source of the regional pollution. It is principally acid precipitation due to China's heavy use of coal for power generation, residential and commercial heating and industrial manufacturing, in fact some 70% of the country's primary energy consumption. Given that China possesses the world's largest reserves, and projects economic and energy growth (Table 2.3), regional transboundary air pollution will increase rapidly unless China can undertake major countermeasures[4] (UNEP, 2000).

Research, monitoring and education

Chinese officials have acknowledged the transboundary air pollution situation since the mid-1980s, and pledged best efforts to address it. At the same time they have stressed the *need for technical and financial support* from the international community if they are to be in a position to mount the remedial measures. In 1996,

China joined with Republic of Korea and Japan to launch a joint research project on long-range transboundary pollution in North East Asia, with the aim of first identifying the sources and movements of long-range air pollution in the region, and then mitigating the damages. The first phase was completed in 2004 with preparation of a system for measuring and modelling the major pollutants (SO_2, NO_x and particulate matter), and agreement on a second phase (2005-07) involving co-operative research to measure additional pollutants and the identification of the most cost-effective methods for emission reduction.

China is also one of 13 East Asian countries participating in the *Acid Deposition Monitoring Network in East Asia (EANET)*, supported by UNEP. Four of nine monitoring stations in the network are located in China (Chongqing, Xi'an, Zhuhai and Xiamen), where measurements are carried out on wet and dry deposition, soil quality and vegetative impact. EANET is the first *region-wide, co-operative scientific network established in East Asia*, and could provide the basis for international political agreements on transboundary air pollution similar to those in Europe and North America. Further, ESCAP and the ADB have financed projects involving China, Mongolia, the Republic of Korea and Japan to establish training and education facilities and materials on environmental protection associated with coal-fired power plants, to improve environmental monitoring, and to develop an action programme for improving the efficiency of particulate matter abatement in existing power plants.

Intentions and actions

Beyond expanded research, monitoring and education, there is an *urgent need to begin to control the sources* of transboundary air pollution. In this regard, in 2004 *China announced stepped-up measures* to reduce the impact of coal-fired power plants on air quality domestically and regionally. These include requiring all new plants to install desulfurisation equipment, increasing the monitoring and inspection of existing plants, and providing incentives for firms to scrub emissions. Previously, power companies were able to pay (relatively small) fees to the government rather than invest in pollution control.

Regarding *mercury*, Chinese officials are considering a designated programme to reduce airborne emissions, and technical discussions have been undertaken with the US Environmental Protection Agency. *Organic pollutants* have also begun to receive attention, in part as the result of momentum and incentives provided through China's accession to the Stockholm Convention on Persistent Organic Pollutants (POPs) (Box 9.5). A work programme on organics has been launched with France and the US, and Chinese experts are working with Italian counterparts on VOCs monitoring.

The *challenge for China* and for other countries in and beyond the region that have direct interest in seeing transboundary air pollution mitigated, is *to co-operate* to: i) improve the overall energy efficiency of the Chinese economy; ii) increase the efficiency of coal use; iii) introduce further coal washing and other clean coal technology as rapidly as possible; and iv) look to changes in fuel mix in the transport sector. The latter could, inter alia , reduce the emissions of nitrous oxides which are an increasing contributor to regional-scale acidification. Many of the related actions would bring multiple benefits concerning not only transboundary air pollution, but also reduced local air pollution and health damages (Chapter 3), as well as reduced emissions of greenhouse gases (GHGs). *Expanded international support* in the form of technology development and transfer, technical assistance and funding will be essential.

2.3   Management of regional seas

The *degradation of the health of the seas* that China shares with its immediate neighbours (i.e. Bohai sea, Yellow sea, East China sea and South China sea) has become a matter of growing concern as the economic, social and ecological impacts are increasingly evident. *Pollution from land-based sources* (inorganic nitrogen, phosphates, oil, chemical oxygen demand (COD) and heavy metals) is a major factor, along with contamination from ocean dumping, marine oil spills, and oil and gas exploration activities (Chapter 4).

Recently, China's State Oceanic Administration reported that the *Yellow sea and South China sea had "generally good" water quality* overall, showing slight improvement over earlier years (SEPA, 2005a). The Bohai sea was in an intermediate, fairly static condition; while water quality in the *East China sea was judged to be poor* and continuing to degrade. Based on readings from 293 monitoring stations, 32% of coastal waters met grade I (good) quality standards; 35.2% met grade II standards; 8.9% met grade III standards; and 23.9% fell into grade IV (or above) standards (indicating high pollution) (Figure 9.1, Table 4.5). In 2003, 2004 and 2005, *red tides* continued to present problems, with 119, 93 and 82 occurrences, up from about 30 to 80 in 2000-02. The total area affected has risen from 10 000 km^2 to 26 000 km^2. Most (60 to 80%) red tides occur in the East China.

The resource-rich regional seas of Northeast Asia and the North Pacific are also among *the most heavily fished, and over-fished, waters of the world*. China is a party to numerous bilateral fisheries agreements with neighbouring countries that are intended to promote sustainable fishing practices. However, there are no region-wide agreements or forums which bring together all fishing countries of the region.

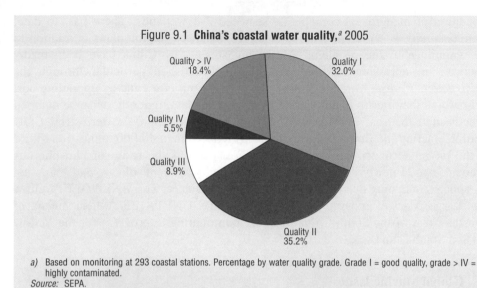

Figure 9.1 **China's coastal water quality,**[a] 2005

a) Based on monitoring at 293 coastal stations. Percentage by water quality grade. Grade I = good quality, grade > IV = highly contaminated.
Source: SEPA.

China's four bordering seas are components of *UNEP's regional seas programmes* for the *Northwest Pacific*, with China, Japan, Republic of Korea and Russia the involved countries; and for *East Asia*, involving China, Australia, Cambodia, Indonesia, the Philippines, Malaysia, Republic of Korea, Singapore and Thailand. Each sea, in turn, is the focus of an *action plan* (covering both the marine and coastal zone environments) negotiated by the associated states, and supported to varying degrees by UNEP and other multilateral institutions (e.g. ADB, GEF).

In 1994, China played a lead role in the design and adoption of NOWPAP, which also involves Republic of Korea, Japan and Russia. The principal activities have focused largely on the Yellow and East China seas. NOWPAP's Data and Information Network Regional Activity Centre is located in Beijing. China and Republic of Korea have conducted joint research since the mid-1990s under the "Yellow Sea environment joint investigation" to assess water quality at 33 points along four designated lines. Under the *Blue Bohai action plan*, some USD 2 billion have been invested over the past decade by three Chinese provinces and one municipality bordering the Bohai sea, with 87 projects underway at the end of 2004. Further, UNEP and the GEF have been supporting efforts by China and South East Asian countries under the *East Asia regional seas programme* to reduce degradation in the *South China sea* and in Thailand bay, with a recent focus on protection of mangroves, coral reefs, seaweed and wetlands. There has, to date, been a *scarcity of resources* applied to the regional seas action plans in relation to the magnitude of the problems. Major investments by China have been made, however, in addressing land-based pollution sources, in particular through the *construction of waste water treatment plants in coastal cities* and by upgrading port and harbour facilities to handle waste from vessels that might otherwise be dumped directly into the sea. For instance, during the period 2000-03, terrestrial COD volume entering the Bohai sea was reduced by more than 40 000 tonnes (one-third of the target set under the Blue Bohai action plan) as the result of phosphorous removal by 21 newly-constructed waste water treatment plants and bans on phosphate-containing detergents in four Bohai provinces. The UNEP/GEF "Yellow Sea Large Marine Ecosystem Project" and the UNDP/GEF "Management of Coastal Areas of the South China Sea" are contributing to progress for the Yellow sea and South China sea.

3. Global Marine Issues

China's economic and social development has historically been closely linked to the *wealth and vitality* of the marine environment. At present, some 30% of China's population (over 350 million people) live within 100 km of the coastline which stretches 18 000 km from the Yalujiang river in the north to the Beilun river in the south. Abutting the Bohai, Yellow, East China and South China seas, China has looked to the marine environment for trade, food, recreation, minerals and raw materials, with a perspective that has evolved from concern with under-exploitation until the mid-1980s to, most recently, rational exploitation and sustainable development.

Chinese officials were active participants in the international discussions of Law of the Sea issues, including environmental management aspects, up to and following China's ratification of the UN Convention in 1992. That same year, a *"China Ocean Agenda 21"* was formulated, setting out a sustainable development strategy for the country's marine environment. Major elements included:

- safeguarding the *new international marine order*, and the rights and duties of states, through consultation and on the basis of international law and principles of fairness;

- strengthening the planning and management of the coastal zone and offshore areas to promote *sustainable use of marine resources* and improve the prosperity of coastal areas;

- promoting the development of *marine industries*, balancing development and protection;

- improving the monitoring, surveillance, *law enforcement and management* of the marine environment, with emphasis on controlling land-based marine pollution and implementing a system for controlling the total quantity of pollutants;

- *strengthening* oceanographic science and technology, research and development;

- establishing a *comprehensive management system* for the marine environment; and

- participating actively in *international co-operation* in the field of marine management, and conscientiously fulfilling the obligations defined in the Law of the Sea Convention.

A decade later (May 2003), the State Council issued a *national plan for ocean economic development* which presented objectives and targets for oceanic ecological development and resource protection. They included: application of the principle of integrating "total volume control" with economic and social development centred around prevention and sustainable development; implementation of a catch quota for fisheries; control and reduction of the intensity of traditional fishing practices; introduction of closed fishing seasons; strict control on contaminants entering the sea, with priority on heavy metals, COD and nutrients; maintenance of water quality in unpolluted areas and improvement in bays and near coastal cities; and establishment of nature preserves, including cetacean monitoring zones.

In support of this array of policy objectives, China has created over the years a comprehensive and effective regime (comprising policies, laws, regulations and programmes) for protecting its coastal waters and seas from damage from coastal construction projects, offshore oil development, vessel pollution, waste dumping and land-based pollution sources. In the process, China has increasingly engaged other countries through a variety of bilateral, regional and multilateral agreements and programmes to buttress and extend its domestic marine management goals, and to enable it to *fulfill its treaty commitments* to the rest of the international community.

China signed and ratified *virtually all of the major conventions and protocols* on marine management; is an active member in over 25 international bodies engaged in marine protection; and has some two dozen bilateral agreements in place covering marine pollution control, ocean science, and fisheries management.

3.1 Marine pollution

The *priority* that China attaches to marine pollution control *has increased sharply over the past decade* as the adverse impact on fisheries, coastal aquaculture and recreation has become more evident. Research and monitoring of the causes and impacts of the pollution has been expanded. Inspection and regulation of vessels, offshore mining and coastal enterprises has been tightened. New sewage treatment plants have been built in coastal cities. Advanced oil-water separation and control technologies have been introduced in the offshore construction and drilling industries; and fines and penalties have been increased for both inadvertent and purposeful pollution.

Pollution in China's offshore environment had lessened in recent years, but is still at high levels (SEPA, 2003). Nitrogen and phosphorous contamination from land-based sources (i.e. from agriculture and municipalities) is of particular concern. The most severely affected marine areas had decreased in 2003 to 142 000 km^2 from 174 000 km^2 in 2002. Of the four major marine zones, the Bohai and East China seas were the most polluted area, with the Yangtze river the largest source of land-based pollution in the East China sea. The Yellow and South China seas were in reasonably good shape (Chapter 4).

Dumping at sea

China is a contracting party to the *1972 Convention on the Prevention of Marine Pollution by Dumping of Waste and Other Matter* (the "London Dumping Convention"), as well as subsequent protocols. It has a good record in providing periodic reports and obtaining dumping permits as required under the convention, and in otherwise taking steps to manage and reduce waste disposed of in the oceans. In 2003, there were *72 designated marine dumping zones*, 35 of which were permanent, with monitoring carried out at 24 of the sites by the State Oceanic Administration, which has overall authority for at-sea dumping. It issued some 600 certificates in 2003 for the dumping of dredged materials, and approved the disposal of 116.7 million tonnes of dredge spoils, bringing revenues of USD 675 000 to the government. The dumping is limited to dredged materials from land as well as from harbour improvement projects. *Radioactive waste dumping has been banned* since 1994, and *incineration of waste at sea* is not permitted.

Given the large and growing volume of dredged materials to deal with as coastal construction and harbour maintenance intensifies, coupled with the highly polluted character of much of the spoils, *protecting the quality of the marine environment near dumping sites* will be a major technical and financial challenge.

Ship waste discharges

China has issued specific regulations on the prevention and control of marine pollution caused by ships, pursuant to its domestic marine management goals and policies, and as a party to various conventions of the International Maritime Organisation (IMO). *Discharge of ship waste in China's waters is banned*, and *port facilities* have been equipped or upgraded to handle ship waste of various types. Surveillance by port and maritime authorities has been strengthened and the penalties for infractions increased.

Nonetheless, the large volume of shipping into and out of China's numerous ports in relation to the limited number of trained monitors and inspectors assigned to environmental surveillance has resulted in continuing reports of *illegal dumping* by vessel operators.

3.2   Oil spills

While China has *escaped major oil spills in recent years*, the threat remains high in the face of the large (and growing) volume of traffic in Chinese waters. The Shanghai Maritime University reported that between 1973 and 2000, there were 29 relatively large accidents involving oil tanker spills, of which 7 involved foreign tankers which paid an average of USD 998 000 in compensation. The 22 spills involving Chinese tankers resulted in penalties in just seven cases, with average payments of USD 184 000. This has highlighted the need for the establishment of a strong and equitable compensation mechanism, a matter that is receiving attention by China in conjunction with other countries in East Asia.

China has an overall state emergency plan for oil pollution management on the high seas, supported by emergency response teams in port areas, consistent with IMO conventions and protocols. Under the leadership of the Ministry of Communications, an updated version is being formulated. Some USD 5 million was invested in 2002-04 to improve oil spill control facilities at several major ports, involving spill management technologies, monitoring, identification of sources, warning and recovery. A model project was implemented by the Ministry of Communications and the Port of Qingdao to establish an emergency spill demonstration centre at the port for the recovery of small-scale spills and the control of larger ones. At the moment, a large gap remains between Chinese capacity to deal with major oil spills and *the international state-of-the-art*. Few of its ports are adequately equipped with emergency facilities, nor is there a sufficient number of trained personnel.

Concern also exists about spills and releases from *offshore oil exploration*, leading to the enactment in 1996 of a "crash programme to combat oil spills during offshore oil exploration and exploitation", and the installation on all drilling rigs and ships of oil barriers and chemical de-oiling agents. Spill recovery ships are available in all of China's offshore oil fields.

Ship dismantling

International concern has risen in recent years about *worker health and safety* as well as *environmental impacts* associated with the *ship-breaking and vessel demolition industry*. Over the years, the global industry has grown rapidly with the rising prices for scrap steel and the pressure from new international regulations to replace single-hull tankers with double-body vessels. The ecological and workplace risks derive from the large amounts of toxic substances and hazardous materials (e.g. asbestos) that can be released into the environment during the demolition and recycling processes, with some waste products often disposed of directly into the sea. Since the 1980s, ship-breaking has increasingly *moved to Asian countries*, driven by the demand for scrap steel, the low labour costs in some Asian countries, and the tighter environmental regulations in Europe and North America (Danish Environmental Protection Agency, 2004).

During the 1994-2002 period, *China ranked third* among countries with ship-breaking activities, with 10% of all commercial vessels demolished worldwide (379 ships), and 13% of world tonnage (4 734 000 light displacement tonnes)[5] (Table 9.3). Chinese officials are keenly aware of the environmental issues and risks

Table 9.3 **Ship dismantling**, 1994-2002

Dismantling location[a]	Vessels (number)	Total tonnage[b] (Ldt)	Vessels (%)	Total tonnage (%)
India	2 245	16 135 949	58	45
Bangladesh	529	7 737 562	14	22
China	379	4 734 533	10	13
Unknown	241	1 255 762	6	4
Pakistan	192	3 521 888	5	10
Turkey	109	379 641	2.8	1.1
Indian Sub. Cont.	84	1 191 793	2	3
Vietnam	29	372 882	1	1
Spain	18	59 439	0.46	0.17
Mexico	18	75 746	0.46	0.21
Taiwan	5	31 272	–	–
Philippines	4	49 035	–	–
Brazil	4	20 041	–	–
Total[c]	3 877	35 681 405	100	100

a) Countries where more than two vessels were dismantled. In addition, two vessels were dismantled in Portugal, United Kingdom, Peru; one vessel was dismantled in Canada, Greece, Italy, Japan, Netherlands, Columbia, Cuba, Egypt, United Arab Emirates, Venezuela.
b) Expressed as light displacement tonnes (Ldt).
c) Total applies to all countries.
Source: Danish Environmental Protection Agency.

associated with ship-breaking. *Regulations on the prevention and control of pollution from ship-breaking* go back to the mid-1980s (with recent revisions) to promote high standards from workplace safety as well as environmental protection. Responsibility for regulation and oversight of the industry is with the Ministry of Construction offshore, and with local Environmental Protection Bureaus for onshore facilities. China is also engaged *in the international discussions* on how to improve health, safety and environmental conditions in the industry that are being carried out within the UNEP Secretariat for the Basel Convention on hazardous waste, the International Labour Organisation (ILO), and the IMO. It played a lead role in a 2002 ILO international tripartite meeting of experts on safety and health in ship building for selected Asia Countries and Turkey, a meeting which produced the first-of-its-kind guidelines on the topic.

While China has made efforts in recent years to foster improved environmental and workplace safety, *there remains the need to*: strengthen existing regulations; improve workplace conditions and the handling of toxic materials through in-house training, contingency planning and upgraded technology; and expand the number and quality of monitoring and inspection of ship-breaking firms, with increased penalties for violations. This is especially timely given reports that *China is planning to expand its ship-breaking industry* which has languished over the last decade. Ships intended for disposal/recycling are included in China's list of imports which are not banned or restricted under the Basel Convention.

3.3 Marine resources

Marine fisheries

China is the *world's leading fishing country*, with an output of close to 46 million tonnes in 2003, constituting close to half of the global catch. Of China's total, some 25 million tonnes (56%) were taken from the marine environment (coastal waters and high seas), including 11.1 million tonnes from marine fish farming and ranching (Table 9.4).

Globally, marine fish catch *levelled off beginning in the late 1990s*, fluctuating between 85 million and 95 million tonnes. Chinese fishing authorities predicted in 1997 that *increases of capture fisheries yield would soon reach zero* given the intensive pressures being placed on major world fisheries, including the rich fishing grounds of East Asia, by more boats, better capture technology and loss of breeding grounds to development and pollution. Over the past decade, captures in China's regional seas have increasingly consisted of trash fish and steadily younger individuals, with the average size of catch getting smaller and smaller. Due to overfishing, the *composition of marine catches has changed substantially*, and marine species of high value have declined sharply. Of the four species with greatest commercial value in China, i.e. the large and small yellow croaker, the hairtail and the squids, only the hairtail is still well-represented in its regional seas, although younger individuals are increasingly present in the catches. Conversely, total production of pelagic fish and crustaceans has increased. In both the East China sea and the Yellow sea, demersal and predatory species with a longer life span and a higher commercial value have been replaced by low value species, primarily small pelagic fish such as the chub mackerel, the black scraper and the Japanese anchovy.

The deteriorating fisheries situation has played a significant role in the *public's growing concern about environmental degradation in China in general*, encouraging national efforts to address problems of land-based marine pollution and oil spills, and the engagement of other countries to protect fisheries resources in coastal waters and the high seas. It has also stimulated the rapid expansion of China's aquaculture industry.

Table 9.4 **Fisheries production in China,** 1970-2003

	Total production (P, 000 mt)	Marine fisheries		F/P (%)	F/C (%)
		Capt. fisheries (C, 000 mt)	Farming/ranching (F, 000 mt)		
1970	3 254	2 168	182	5.6	8.4
1980	4 455	2 772	415	9.3	15.0
1990	13 137	5 790	2 023	15.4	34.9
1991	14 254	6 373	2 255	15.8	35.4
1992	16 579	7 325	2 919	17.6	39.8
1993	19 708	8 169	3 884	19.7	47.5
1994	23 834	9 539	5 070	21.3	53.2
1995	28 418	10 955	6 448	22.7	58.9
1996	31 897	12 419	6 725	21.1	54.2
1997	35 038	13 835	6 949	19.8	50.2
1998	38 025	14 950	7 576	19.9	50.7
1999	40 030	14 955	8 570	21.4	57.3
2000	41 568	14 754	9 411	22.6	63.8
2001	42 579	14 379	10 101	23.7	70.2
2002	44 320	14 305	10 827	24.4	75.7
2003	45 692	14 300	11 149	24.4	78.0

Source: FAO.

Since the 1980's, China has established *co-operative fishing agreements* with more than 20 countries and regions, with emphasis shifting from an initial concern with sharing out the resource to conserving and managing the stocks on a sustainable basis (Box 9.5). China currently: has forward-looking agreements and management programmes with the Republic of Korea, Japan and Vietnam; participates in an array of regional and global fisheries conservation programmes and institutions; and is party to most of the principal multilateral conventions designed to protect and enhance the world's fisheries, e.g. on straddling and highly-migratory fish stocks, conservation of Atlantic Tuna, and fishing vessel compliance with international conservation measures on the high seas.

In 1987, a "Details of the Fisheries Law" was ratified by the State Council which has been the *foundation of the Chinese fishery legal system.* This regulated the development of the fishing industry, including aspects of aquaculture for the first time, and assigned priority to management over exploitation, to culture over capture, and to quality over quantity. To protect the ecology of China's coastal fishing grounds, "Water Quality Standards of Fishing Grounds" were promulgated, and "Regulations on the Supervision and Control of the Environmental Sanitation of Shellfish-Raising" plus other regulations were issued by the concerned departments, including SEPA, the Ministry of Health, the State Oceanic Administration and the Ministry of Construction. Currently, a *complex fisheries management regime is in place*, combining overall policy with implementing laws, regulations, conservation programmes, and well-defined institutional responsibilities at the state and provincial levels. Closed fishing seasons are employed, along with a permit and quota system for certain species, closed fishing areas, marine sanctuaries, banning of harmful fishing gear and methods, and restricting the size of net meshes and the proportion of young fish that can be captured.

The *challenge for China* will be to move ahead vigorously to take the pressure off the fish stocks from overfishing and the loss of coastal habitat. Its commitment to proceed is reflected in a new programme to scrap 30 000 fishing boats and relocate 300 000 fisherman by 2010. Affecting some one million households, this is a painful but necessary step in what must be a comprehensive approach to fisheries rehabilitation and sustainability if more substantial economic and social impacts are to be avoided.

Aquaculture

Marine aquaculture has become a *major economic activity in China* since the 1980s, operating across 11 provinces along the coastline, with state enterprises employing some 1.5 million workers and another 375 000 employed in related sectors, e.g. processing, transport and distribution. The private sector is estimated to involve three to four times these numbers.

Box 9.5 Regional fisheries agreements

The Yellow sea, East China sea and South China sea have for centuries provided rich fishing grounds for *China and its neighbours, Japan, the Republic of Korea and Vietnam.* As fish catch rates began to decline rapidly in the 1980s, pressures grew for regional co-operation to place the remaining stocks under sound, sustainable management. However, historical territorial disputes prevented meaningful co-operation.

The situation changed with the enactment of the *UN Convention on the Law of the Sea* (UNCLOS) in 1994. UNCLOS granted coastal states the right to declare sovereignty and resource control over an "exclusive economic zone" (EEZ) up to 200 nautical miles off their coastlines. China, Japan, the Republic of Korea and Vietnam all quickly ratified the UNCLOS and declared their respective EEZs. In the case of countries bordering semi-closed seas, like those off China, where EEZ claims overlap, the UNCLOS calls for establishing joint resource management areas and provides guidelines for doing so, even where territorial claims are unresolved. It was on this basis that the four Asian neighbours have since made *steady progress in negotiating a network of bilateral agreements* to manage their common fishery resources.

China signed an agreement with Japan in 1997 for co-operative fisheries management in the South China sea, and a Sino-South Korea agreement on Yellow sea fisheries management was negotiated a year later. Two agreements were signed between China and Vietnam, one on fisheries management and the other on boundary delineation. The *fisheries agreements* address three key issues. They reaffirm each country's exclusive rights over fish resources and fishing activities in its own EEZ; they establish general principles for reciprocal fishing access in each other's EEZ; and they create a co-operative management regime for their shared fishery resources.

The biggest difference among the agreements is that the Sino-Vietnam protocol for the Beibu/Tonkin Gulf fisheries includes a permanent *maritime boundary delimitation.* By contrast, no maritime boundary agreements exist between China and the Republic of Korea or Japan. This has enabled a joint fisheries committee to be established under the agreement with Vietnam that is a permanent body with full operational authority, including a dispute settlement mechanism. The main limitation of all the agreements is that they provide for *management of designated areas*, whereas unregulated fishing for migratory and non-migratory species continues in waters where territorial disputes remain unresolved.

Nonetheless, the *fisheries agreements are pioneering efforts*, reflecting political compromises among countries of the region with strikingly different levels of economic development, domestic political systems and foreign policy concerns.

Source: Rosenberg, 2005.

China's total output from marine aquaculture stood at 11.1 million tonnes in 2003, amounting to some *24% of the country's total aquatic production* (Table 9.4). China's goal is to expand considerably aquaculture production (both inland and coastal), given the levelling off of marine capture coupled with the growing global demand for fish and shellfish and what it views as an area of comparative advantage based on China's extensive experience with fish farming and ranching. Although China is the biggest exporter of marine products in the world, its exports make up less than 10% of its current total output. Further, aquaculture was *among the first deregulated industries in China*, and is thus burdened with few distorting institutional and regulatory effects, and may be only marginally influenced by changes in import tariffs.

An October 2004 report by the CCICED examined the *implications of WTO accession for China's aquaculture industry*. It highlighted the potential for significant economic benefits (e.g. expanded breeding areas, increased consumption of products, enhanced attractiveness to foreign capital), as well as the positive social benefits (e.g. reduced poverty, strengthened national food security, promotion of regional development). The CCICED report also, however, pointed to a mixed picture for the environment: reduced overfishing of the regional and high seas, increased emission of nutrient salts, leading to an increase in red tides and changes in the composition and distribution of offshore biotic populations; threats to the remaining mangrove forests; and increased salinisation of soil and underground water due to fresh water withdrawals from farm operation. Beyond this, a key question is how to strengthen food safety in the open market. Soon after WTO accession, China experienced a *sanction on its aquatic product exports to the European Union* for failing to meet EU requirements on drug residues. It was not until the beginning of the decade that Chinese authorities took serious note: i) of the threats that environmental degradation posed for the aquaculture industry; and ii) that environmental pollution from marine aquaculture itself needed urgent attention by public authorities (SEPA, 2003).

To address these challenges and opportunities, China's central and local governments have stepped up their efforts to *build sound regulatory and standards regimes* for the marine (and inland) aquaculture industry, and have redefined their support policies. New laws and regulations have been issued over the past five years to, inter alia , standardise farming activities (including rearing density, nutrient balances and water quality levels) control the use of various feedstuff and drugs; and improve data collection and monitoring/inspection. In 2002, a series of *national standards* were issued covering the quality of marine organisms, the quality of marine sediments, and technical requirements for marine farming inspection. Food safety *certification* is also being pursued vigorously by Chinese authorities, including consultations with counterparts in North America and Europe and elsewhere in the Asia and Pacific region.

4. Global Issues: Stratospheric Ozone Depletion

4.1 Commitments and achievements

China has engaged in a *successful pursuit* of the goals and targets of the *1987 Montreal Protocol on Substances that Deplete the Ozone Layer*. This effort exemplifies the country's broader commitment to address global-scale environmental problems, its willingness and ability to employ an array of market, regulatory and voluntary instruments in doing so, and its ability to make effective use of the international machinery available to it.

From the mid-1980s through most of the next decade, China's production and consumption[6] of ozone-depleting substances (ODS) grew 12% annually, principally chlorofluorocarbons (CFCs) and halons, due to increasing demand for refrigerants, rigid foams and fire extinguishing agents. By 1997, China produced some 95 800 tonnes of ODS and consumed 87 600 tonnes, calculated on an "ozone-depleting potential", or ODP, basis. At that point, *China was the world's largest contributor of ODS*, more than doubling the output of India, in second place. Consumption by the developed countries of North America, Europe and Asia was much reduced as they had been carrying out ODS reduction and phase-out efforts for almost a decade under their Montreal Protocol commitments.

China ratified the Montreal Protocol in 1991 as an *Article 5 developing country*, and subsequently acceded to all subsequent amendments (i.e. London 1990, Copenhagen 1992, Vienna 1995, Montreal 1997 and Beijing 1999). Its commitments included freezing both production and consumption of principal ODS by 1999 at China's average 1995-97 levels, then reducing CFCs, halons and carbon tetrachloride by 50% by 2005 and 85% by 2007, and phasing them out completely by 2010. Methyl bromide, an economically important soil fumigant used in large quantities by China's horticulture industry, must be reduced 20% by 2005 and phased out by 2015, under the 1992 Copenhagen Amendment.

China has, to date, *met each of its targets on schedule*. It is also well on its way to *achieving the key 2010 target* for complete elimination of CFCs, halons and carbon tetrachloride; and is actively pursuing alternatives to methyl bromide production. By 2005, production and consumption of CFCs had decreased by 40% and 55% respectively from China's 1995-97 baseline of 47 000 tonnes ODP, while production and consumption of halons had dropped 85%. World Bank statistics indicate that *one-half of the ODS that have thus far been eliminated globally is the result of China's efforts*, an achievement that earned awards from both the World Bank and UNEP in 2003 (Figure 9.2, Table 9.5).

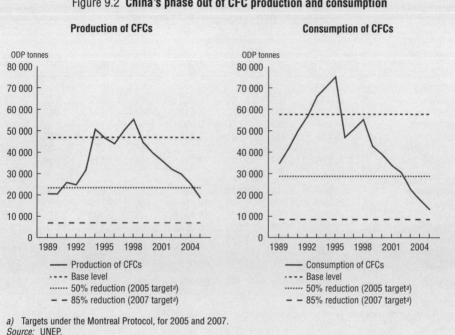

Figure 9.2 **China's phase out of CFC production and consumption**

a) Targets under the Montreal Protocol, for 2005 and 2007.
Source: UNEP.

4.2   Actions taken

Following ratification of the Vienna Convention for the Protection of the Ozone Layer in 1989, China prepared a *national plan on ozone depletion* and moved quickly to *ban new halon and CFC-based aerosol production facilities* in 1990 and 1991, and then in 1993 to prohibit all new CFC production facilities. Also in 1993, China entered into a partnership with the *World Bank's Montreal Protocol Programme* (World Bank, 2004a). This programme is a component of the *Multilateral Fund* established under the Protocol to provide technical and financial assistance to Article 5 developing countries.

As of 2004, some *USD 740 million has been committed by the Multilateral Fund* to help China meet its Montreal Protocol obligations, including funding for multi-year projects that will be released up to the 2010 target date for complete elimination of the production and consumption of most CFCs and halons. The funds were initially

Table 9.5 **CFCs production,**[a] 2002-09

(ODP tonnes)

	2002	2003	2004	2005	2006	2007	2008	2009
Non-Article 5 (1) countries								
Greece	1 440	1 168	1 000	906	900	272	272	272
Italy	9 131	6 000	5 000	3 613	3 600	1 084	1 084	1 084
Netherlands	9 214	2 888	2 500	2 000	..	..	..	..
Spain	6 491	4 948	4 500	2 878	2 850	863	863	863
United States	..	..	..	..	1 200	1 200	800	800
Subtotal	26 276	15 004	13 000	9 397	8 550	3 419	3 019	3 019
Article 5 (1) countries								
Argentina	3 015	3 018	3 020	1 647	1 647	686	686	686
China	32 269	30 000	25 300	18 750	13 500	9 600	7 400	3 200
India	16 855	15 058	13 176	11 294	7 342	3 389	2 259	1 130
Korea (Rep. of)	7 507	7 500	7 500	5 061	5 000	1 518	1 518	1 518
Mexico	5 653	7 335	7 335	7 335	..	..	..	..
Venezuela	1 637	2 400	2 400	2 000	2 000	1 000	1 000	1 000
Subtotal	67 235	65 311	58 731	46 087	29 489	16 193	12 863	7 534
Grand total	93 511	80 315	71 731	55 484	38 039	19 612	15 882	10 553

a) 2002-05: actual data; 2006-09: estimations.
Source: UNEP, Ozone Secretariat Technical and Economic Assessment Panel (TEAP).

used to phase out ODS through plant closures, technological innovation and the introduction of chemical alternatives in a *project-by-project* approach. Through 1994, over 100 enterprises had received assistance, with a phase-out of some 30 000 tonnes of ODS (Box 9.6).

Beginning in 1997, a *sector approach* was introduced which, by 2004, had been applied to seven sectors: mobile air conditioning, domestic refrigeration, aerosols, polyurethane foam, halon production and consumption, process agents, and CFC production. Grants from the Multilateral Fund amounting to almost USD 500 million were allocated to the sector-based component of the China-World Bank ODS programme which is expected to phase out an estimated 184 000 ODP tonnes upon completion, and at a *significantly lower price per kg ODP than the project-by-project approach*. Under the sector approach, China has closed down and dismantled 32 CFC production plants, along with 12 halon production facilities, and is in the process of closing or converting 80 fire equipment manufacturing companies. This is expected to

Box 9.6 **China and the Montreal Protocol Multilateral Fund**

Parties to the Montreal Protocol on ozone depletion agreed that a *financial mechanism* was needed to help countries meet their obligations to *phase out ozone-depleting substances*. In 1990, the Multilateral Fund was established for this purpose, with UNDP, UNIDO, UNEP and the World Bank to serve as its implementing agencies. The Multilateral Fund has been a key element of the success of the Montreal Protocol through its assistance to countries that otherwise lacked the technical and financial means to phase out ODS. As of 2003, nearly USD 1.5 billion had been allocated from the Fund to projects in 131 countries designed to phase out 245 000 ODP (ozone-depleting potential) tonnes. Some 45% of the total funding was provided by the World Bank under its designated Montreal Protocol programme, exclusively to developing countries.

In 1993, the World Bank and China agreed to build a partnership to help meet China's obligations as a party to the Montreal Protocol. Subsequently, China has become *the largest beneficiary of the Multilateral Fund*, through the World Bank's Montreal Protocol programme. As of 2004, the Fund had committed USD 740 million to China, including funding for approved multi-year projects up to 2010, when final phase-out targets for principal ODS are to be reached. A total phase-out by China of 285 000 ODP tonnes is envisioned through this partnership programme.

SEPA, which is heading a *multi-agency group* of 18 national ministries and agencies, is the focal point for projects carried out under the Multilateral Fund in China,. The Ministry of Finance serves as the financial window.

Source: World Bank, 2004.

enable China to meet its target for eliminating all halon 1211 production by the end of 2006 (four years ahead of the Protocol schedule), and to halt all remaining halon production by 2009.

In 1999, SEPA *promulgated a policy and regulations on import and export manufacturing measures for ozone-depleting substances*, setting out an initial list of ODS substances to be controlled in trade. Eight ODS substances were identified, with carbon tetrachloride imports prohibited and most major CFCs and halons placed under a quota system for both import and export. China's *Air Pollution Prevention and Control Act was amended* in 2000 to, inter alia , encourage ODS substitution; promote a gradual reduction of ODS consumption; and introduce a system of *financial penalties for companies* which violate the regulations (up to USD 20 000 per violation). SEPA and the Ministry of Foreign Trade and Economic Co-operation (MOFTEC) are the competent authorities for administering China's ODS programme.

A broad spectrum of *further actions have been taken* to achieve ODS emission reduction rapidly and efficiently. These include: introducing additional regulations (e.g. banning CFC-based mobile air-conditioning systems in new vehicles by January 2002); providing technical training for government officials and firms on standards development; establishing testing facilities and certification systems; closing additional production lines and individual plants; accelerating research and development on chemical substitutes; and establishing an industrial park in North China's Hebei province to bring together enterprises that promote ozone-friendly technologies.

4.3   Prospects

Specific next steps under the China-World Bank Partnership Programme include introducing an ODS import licensing system with permits and quotas; strengthening quality control for ODS substitutes; and improving the auditing and inspection of enterprises. Much of this work is proceeding, with additional bilateral support from a number of OECD countries (e.g. funding and technical support from Italy for methyl bromide substitution under the Sino-Italian programme for environmental co-operation).

While China is on schedule to meet its Protocol commitments along with the ambitious goals of its national CFC phase-out strategy, there is *considerable work yet to be done* if China is to achieve its final objectives. Despite the progress over 15 years, *China remains the world's largest producer and consumer of ODS*, accounting for an estimated 30% of global production in 2005[7] (Table 9.5). And, taking the final steps toward the Montreal ODS elimination targets will be *challenging for all countries*, particularly since the higher cost of CFC and halon substitutes increases the *risk of illegal production and trafficking*. Thus, the emphasis that China is attaching to improved auditing, inspection and enforcement is well placed. It is essential that sufficient funding and manpower be brought to bear on these critical tasks, and that high priority be given to investment and international co-operation to find and introduce cost-competitive and effective ODS alternatives.

5.   Global Issues: Climate Change

5.1   China and the UNFCCC

As a *non-Annex I (developing) country under the United Nations Framework Convention on Climate Change (UNFCCC), China* is not committed to the greenhouse gas emission reduction targets of the convention's Kyoto Protocol, which

China ratified in 2002. The ultimate success of the convention is, however, heavily dependent on China's ability to control its GHG contributions, since China is the *world's second largest emitter of carbon dioxide and other GHGs* (after the United States) (Figure 9.3). China's share is expected to increase substantially over the next few decades.

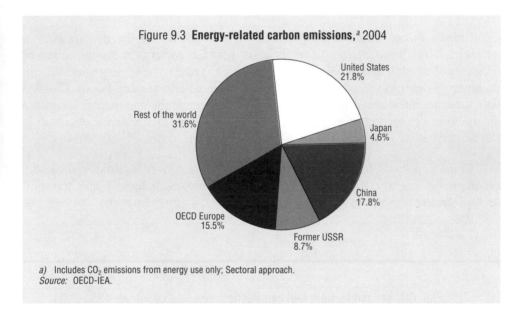

Figure 9.3 **Energy-related carbon emissions,**[a] **2004**

a) Includes CO_2 emissions from energy use only; Sectoral approach.
Source: OECD-IEA.

Since the early 1990s, Chinese officials have increasingly voiced concern about the *possible negative impacts of climate change* on China's plans for economic growth and social development. Today, some 20 ministries and agencies are engaged in climate-related activities, with lead responsibility for oversight and co-ordination vested in the *NDRC*, linked to its central role in the energy field. No overall national plan or integrated strategy on climate change has yet been developed.

In 2004, China produced its *Initial National Communication on Climate Change*, an inventory of anthropogenic emissions from all sources and removals by sinks of greenhouse gases, as required under the Framework Convention of all signatories. Covering the energy sector, industrial processes, agriculture, land use changes, forestry and waste, it was based on decade-old data. The assessment revealed that

China's net emissions of *carbon dioxide in 1994* were 2 666 mmt[8] (based on emissions of 2 795 mmt tonnes from the energy sector plus 278 mmt from industrial processes minus 407 mmt of CO_2 removals from land use modification and forest sinks). Methane emissions amounted to a net 34 mmt while nitrogen oxide contributed 850 000 tonnes of GHGs. China's total GHG emissions in 1994 were thus 3 650 mmt of CO_2 equivalent, with CO_2, methane and nitrogen oxide contributing 73.05%, 19.73% and 7.22%, respectively. Over the 1990-2004 period, *China's CO_2 contributions increased 110%,* from 2 256 mmt to 4 732 mmt (Table 9.6, Figure 9.4).

China's *Initial National Communication* also pointed to the *potential for a number of significant disruptive impacts* on Chinese society, including increased incidents of drought and flooding in a country that is already grappling with difficult problems of desertification, shortages of fresh water, and devastating floods. Coupled with other recent analyses by China's scientific and academic communities, concern about climate change is *raising the priority for preventive and adaptive measures in the national planning process.*

China's approach has, to date, been based on a *"no regrets" approach,* focussing on improved energy efficiency in the power generation, industrial and transport sectors, greater use of renewable and nuclear energy, improved forestry and land use

Table 9.6 **CO_2 emissions from fuel combustion,** selected countries, 1980-2004

(million tonnes of CO_2)

	1980	1990	1995	2000	2004	Change 1990-2004 (%)
China	1 390	2 256	2 976	2 978	4 732	*109.8*
Canada	428	429	461	530	551	*28.5*
United States	4 668	4 842	5 109	5 701	5 800	*19.8*
Japan	869	1 058	1 140	1 185	1 215	*14.8*
Korea (Rep. of)	122	226	361	428	462	*104.6*
France	465	355	357	379	387	*8.9*
Germany	1 061	966	878	827	849	*−12.2*
Italy	363	398	411	426	462	*16.1*
United Kingdom	573	558	528	525	537	*−3.7*
India	295	588	785	971	1 103	*87.5*
Former USSR	3 063	3 345	2 424	2 210	2 313	*−30.9*

Source: OECD-IEA, "CO_2 emissions from fuel combustion, 1971-2004".

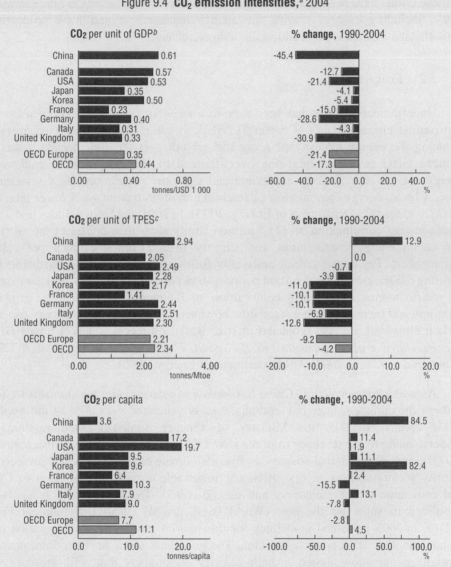

Figure 9.4 **CO$_2$ emission intensities,^{a} 2004**

CO$_2$ per unit of GDP^b

China	0.61
Canada	0.57
USA	0.53
Japan	0.35
Korea	0.50
France	0.23
Germany	0.40
Italy	0.31
United Kingdom	0.33
OECD Europe	0.35
OECD	0.44

tonnes/USD 1 000

% change, 1990-2004

-45.4	
-12.7	
-21.4	
-4.1	
-5.4	
-15.0	
-28.6	
-4.3	
-30.9	
-21.4	
-17.3	

CO$_2$ per unit of TPES^c

China	2.94
Canada	2.05
USA	2.49
Japan	2.28
Korea	2.17
France	1.41
Germany	2.44
Italy	2.51
United Kingdom	2.30
OECD Europe	2.21
OECD	2.34

tonnes/Mtoe

% change, 1990-2004

12.9	
0.0	
-0.7	
-3.9	
-11.0	
-10.1	
-10.1	
-6.9	
-12.6	
-9.2	
-4.2	

CO$_2$ per capita

China	3.6
Canada	17.2
USA	19.7
Japan	9.5
Korea	9.6
France	6.4
Germany	10.3
Italy	7.9
United Kingdom	9.0
OECD Europe	7.7
OECD	11.1

tonnes/capita

% change, 1990-2004

84.5	
11.4	
1.9	
11.1	
82.4	
2.4	
-15.5	
13.1	
-7.8	
-2.8	
4.5	

a) Includes CO$_2$ emissions from energy use only; excludes international marine and aviation bunkers; sectoral approach.
b) At 2000 prices and purchasing power parities.
c) Total primary energy supply.
Source: OECD-IEA; World Bank.

practices, and adaptation-to-change strategies (e.g. planning for sea level rise). Most of these efforts have been justified on the basis of *their contributions to other national goals*, including energy security, air quality management and food production, notwithstanding their roles in addressing climate change.

5.2   Prospects

Understandably, China has been *targeting its fossil fuel-based energy sector* in confronting climate change (Chapter 3). While it has made significant progress in reducing the energy intensity of economic growth over the past decade, gains in reduced GHG emissions are being overwhelmed by the expansion of coal-based energy use to serve a rapidly expanding and wealthier population. On a *per capita basis*, China's energy consumption and carbon emissions *remain much lower than the world average* (US Department of Energy, 2003). In 2001, the United States had a per capita energy consumption of 341.8 million Btu, greater than 5.2 times the world's per capita energy consumption and slightly over 11 times China's per capita consumption. Per capita carbon emissions follow this same pattern, with the US emitting 5.5 metric tonnes of carbon per person in comparison to the world average of 1.1 metric tonnes and China's contribution of 0.6 metric tons. With its growing economy and increasing living standards, however, China's per capita energy use and carbon emissions are also projected to rise. Based on recent trends, China's *GHG energy-related carbon emissions would reach 17.8% of the world total by 2025 (compared to 13% in 2001)* (US Department of Energy, 2003).

As with ozone depletion, China has *looked to international mechanisms* to help address the climate change risk and has already generated *over 30% of the total of CDM credits*. In 2005, the Ministry of Finance signed a Memorandum of Understanding to use revenues from the *UNFCCC's "clean development mechanism" (CDM)*, and other external sources, to finance climate change mitigation projects in priority sectors such as energy efficiency, renewable energy, sequestration of CO_2, and coal mine methane recovery and use (Box 9.7). The NDRC, which has been appointed to supervise the new (World Bank driven) Clean Development Fund (CDF), in 2005 published guidelines for the interim mechanism for operation and management of CDM projects in China. The guidelines allow for both domestic and foreign participation; assign a high priority to projects that will promote the introduction of environmentally-sound technologies into China; and indicate that funding from developed countries should be in addition both to official development assistance (ODA) and to other financial obligations they assume under the Climate Change Convention. A second fund, the (largely ADB funded) Clean Development Mechanism Fund (CDMF) is being established.

Box 9.7 **Clean Development Mechanism (CDM)**

Since 2001, China has set up an impressive institutional framework concerning CDM. The main authority is the *National CDM Board (NCDMB)*, led by NDRC and including the Ministry of Science and Technology, the Ministry of Foreign Affairs, the Ministry of Finance, the Ministry of Agriculture and SEPA.

According to the *"permanent guidelines for project approval"* (2005), a policy priority is given to: new and renewable energy development; energy efficiency improvements; and recovery and use of methane from landfills and coal beds. China also encourages fuel switching, re-forestation and afforestation. Only projects whose participants are wholly or mostly owned by Chinese companies qualify for approval. CDM projects need a broad set of approvals at different governmental levels (e.g. the local Land Bureau, the local subsidiary of the NDRC, and the local subsidiary of the Ministry of Finance). The average time for national approval is one month.

To date, *164 CDM projects* have been approved by the NCDMB, including 32 projects approved internationally under the Kyoto Protocol (i.e. two HFC23 projects, three landfill methane recovery projects, and three renewable energy projects). The number of approved projects is growing rapidly.

By 2010, it is estimated that China could represent *35% to 45% of the global CDM market* (measured in GHG emission reductions), or 100 to 200 million tonnes per year of GHG emission reductions (40% of which come from the power sector). These emission reductions represented 3 to 6% of China's net emissions in 1994.

By 2020, this potential would reach 777 million tonnes per year of GHG emission reductions (of which 545 million tonnes from *improving energy efficiency*, 138 million tonnes from usage of renewable sources, 67 million tonnes from coal-methane capture and usage, and 27 million tonnes from fuel switching and new power generation technologies). These figures do not include potential afforestation in China (e.g. 63 million ha).

The *revenue* from the projects will belong to the Chinese government and the enterprises that implement the projects. The government revenue will be shared as follows: 2% for energy efficiency, renewable energy, forestry, and small-scale projects; 30% for N_2O reduction projects; 65% for projects concerning HFC23, SF6 or PFCs. The revenue would also be used for climate change-related capacity building, preparing project design documents, and other preparatory expenses of CDM projects for poor regions.

China is the largest recipient of *FDI* among developing countries (USD 60 billion in 2004). The Chinese government is encouraging FDI in certain industries (e.g. new agricultural technology, energy, transportation and key raw materials industries) to be combined with CDM projects. This concerns: i) areas designed to attract FDI (special zones), and ii) central and western regions with large CDM potential, such as for afforestation, coal-bed methane utilisation, and high energy efficiency projects. It is estimated that CDM will contribute 0.5% per year of the Chinese GDP in 2030, mainly due to the advanced technology transfer.

Box 9.7 **Clean Development Mechanism (CDM)** *(cont.)*

Improving the awareness and capacity of stakeholders (e.g. governmental officials, banks and entrepreneurs) is key to expanding CDM use in China. Currently, considerable CDM know-how is available at some Beijing think-tanks (e.g. Tsinghua University, the Energy Research Institute). The next step will be to disseminate such know-how at the local level.

China is the principal generator of credits under the World Bank's *"Umbrella Carbon Facility"*. Under contracts valued at USD 930 million, two private Chinese chemical firms signed emission purchase agreements under which the firms will capture and destroy 19 million tonnes of CO_2 equivalent of HFC-23 (trifluoromethane), one of the most powerful GHGs, and a waste gas otherwise released in some manufacturing processes. The Chinese government will retain 65% of the HFC-reduction revenues for investing in climate change projects through the new CDF.

In July 2005, China joined with Australia, Japan, India, the United States and the Republic of Korea in an *Asia-Pacific Partnership on Clean Development* to promote, inter alia , technology development and deployment to reduce GHG emissions in eight industry sectors, including power generation, iron and steel, and aluminium. China is also a participant in the recently-established *"carbon sequestration leadership forum"*, involving both developed and developing countries that account for over 75% of CO_2 emissions. One of the projects selected is examining the potential for storing captured CO_2 in geologic formation and unminable coal seams in China. Carbon sequestration is also the focus of a USD 2.3-million grant from the Canadian Development Agency, which is enabling the University of Toronto to work with the Chinese government and university scientists on modelling and remote sensing to understand how *land use changes affect China's carbon cycle*, including the role of forests in capturing and retaining carbon. Climate change is also a focus of the *Sino-Italian Co-operation Programme for Environmental Protection*, which includes activities on the development and dissemination of renewable energy sources, energy efficiency, hydrogen technology, and the development of methodologies to reduce GHG emissions in the steel and building industries.

Domestically, China has taken in 2005 a number of actions relating to GHG emissions control: automobile efficiency standards (tighter than the new California standard and only slightly lower than proposed new EU standards), increased tax on fuels, increased tax on large cars, new standards for the building industry. As set out in the 11th FYP (2006-10), China will seek to reduce its energy intensity by 20% during that period, and raise the percentage of renewable sources in the overall energy mix to 10% from its current 5% level. China's climate change plans also look to large increases in nuclear energy to meet future energy demand, while intensifying technological innovation and deployment to control CO_2 emissions from the burning of coal. Even if these targets are successfully met, however, projections are that China will be the largest contributor to a still-growing amount of GHG release into the world's atmosphere by 2025 (US Department of Energy, 2003).

China's efforts on climate change would benefit, in terms of both effectiveness and efficiency, from the *preparation of a coherent national policy and implementing strategy* that would pull together the many significant, but to date dispersed, government-supported programmes and projects that contribute to the pursuit of goals under the UNFCCC. Such a national policy and strategy, which would ideally set out domestic *GHG-reduction targets* (without being commitments under the Kyoto Protocol), would symbolise for the public and the international community China's commitment to: i) address the problem; ii) provide an improved mechanism to set priorities and assess results; iii) promote improved inter-ministerial co-operation and co-ordination; and iv) define more focussed contributions by bilateral and multilateral donors, and by Chinese private sector institutions.

6. Global Issues: Trade and Investment

China is the world's *second largest trading country*, with export revenues of USD 762 billion and imports reaching USD 660 billion in 2005 (increases of 28% and 18% respectively above 2004 levels). Further, the growth of *foreign direct investment (FDI) has been one of China's major success stories of the past ten years*, as it has become the leader among all developing countries. The *environmental consequences* of this for China, and other countries, are myriad and complex. First and overall, the accelerated economic growth generated by trade and investment expansion will help move China more rapidly *along the historic path of OECD countries*, which have been able to afford increasing levels of expenditure for environmental protection as they became wealthier. Evidence of this happening in China is already being seen.

Secondly, China's membership in the WTO should *contribute significantly to improved environmental performance* (International Institute for Sustainable Development, 2004). The CCICED particularly concluded that:

– the need to meet *stringent health and safety standards* established internationally and in advanced industrialised countries (e.g. for automobile emission levels and agricultural products) will reduce environmental pollution in China;

- expanded foreign investment by OECD countries will introduce *advanced environmental management systems and technologies* to China, and will stimulate domestic industries to follow suit due to the increased competition;

- increased imports of raw materials will *relieve pressures on China's environment* (e.g. on China's limited forest lands); and

- *production shifts will occur* in China toward less-polluting activities in certain sectors (e.g. in agriculture, where there is a move away from crops and systems which use the greatest amount of fertilisers and pesticides).

Thirdly, the *environmental impact of China's liberalised trade and investment policies on the rest of the international community will be large and potentially negative*, particularly given China's demand for raw materials (e.g. oil and gas, logs, pulp and paper, cotton, copper, steel, aluminium, soya). The challenge will be to mitigate such effects through close consultation between China and its trading partners, expanded analysis and information exchange on trade-investment-environment relationships, and sound environmental management policies by China's trading partners (e.g. on sustainable forest and fisheries management, and energy development). *Chinese officials are aware of this*, voicing concern about the "reputational aspects" of China's worldwide demand for resources, and asking its firms to display good "enterprise citizenship" abroad.

Fourthly, China has been *constructively engaging other countries* in ongoing discussion and negotiation of rules and procedures for integrating trade and environment policies, within the framework of the WTO as well as in ASEAN and Asia-Pacific Economic Co-operation (APEC). It has also joined some existing agreements relating to trade and environmental policies. For instance, it became the ninth country to adhere (in 2000) to a UN/ECE agreement on global vehicle regulations designed to establish internationally-recognized safety, energy-efficiency, anti-theft and environmental performance criteria for motor vehicles.

6.1 Free trade agreements (FTA)

The important role that regional trade agreements (RTAs) and free trade agreements (FTAs) play in trade liberalisation and, hence, in promoting economic growth, *has been recognized by APEC Economic Leaders and Ministers*. To strengthen this contribution and ensure high standard agreements, in 2005 trade ministers from China and other APEC member states endorsed a set of *"best practices" for RTA/FTAs in APEC*. These voluntary guidelines included recognition that all such agreements should "reflect the interdependent and mutually supportive linkages between the three pillars of sustainable development, economic development, social development and environmental protection, of which trade is an integral component".

At present, however, *none of China's FTAs under negotiation* (e.g. with Chile and New Zealand) *contains explicit environmental references*, nor does the government have a formal policy that would require the introduction of environmental concerns into future agreements. The recent discussions within APEC, and the best practices guidelines, could serve as an impetus *for China to prepare a domestic policy* requiring trade and environment officials to assess jointly the environmental implications of proposed RTAs and FTAs, and to recommend any appropriate modifications. A *systematic review mechanism* could be an important step in helping to ensure that any potentially adverse environmental impacts are eliminated or mitigated, and that environmental benefits are maximized. Such prior assessments in East Asia would follow the pattern which has emerged in Europe and North America over the last decade.

6.2   Foreign direct investment (FDI)

FDI in China *has grown rapidly since 2000, rising 50% to reach USD 60.3 billion in 2005*. It is projected to continue to grow substantially as the country further liberalises in a period of rapid globalisation. While much of the FDI has heretofore originated in developing Asia (some 62% in 2005, and of that, one-half from Hong Kong), a growing amount has been coming from OECD member countries in Europe and North America.

Good prospects thus exist for China to *utilise FDI to support its quest for sound environmental management and sustainable development* by insisting on high environmental standards and best practices by foreign firms in, for example, management of hazardous waste, air and water pollution control, and energy efficiency. At the same time, foreign enterprises can bring with them advanced environmental management skills and clean technology, and can provide environmental training for local employees.

To date, China has *not articulated a specific policy* on the environmental component of FDI, nor does MOFTEC or the Ministry of Commerce have an explicit mandate to pursue this. However, some important first steps have been undertaken. A 2002 guide for foreign investment industries lists industries that are "encouraged", "restricted" or "banned" based on *environmental criteria*, with the latter class reserved for firms that pollute the environment, destroy natural resources, impair human health or occupy large amounts of arable land. "Restricted" firms include those that use dated technology and are wasteful of resources. Further, new regulations on general FDI, adopted in 2002, reference WTO guidelines for attracting FDI through, inter alia, curtailing local protectionism and corruption, improving workplace safety, promoting S&T innovation, and protecting the environment.

6.3   Overseas investments

Globalisation and national liberalisation policies have encouraged Chinese firms to increasingly invest abroad to open new markets and to help expand and diversify China's import needs, especially for food and raw materials. Since 2000, major investments have been made in oil and gas exploration (e.g. Indonesia, Algeria), power generation (e.g. Georgia, Algeria, Cambodia, Nigeria), telecommunications (c.g. Unitcd States), steel (e.g. Australia), airlines (e.g. Cambodia), and forest products (e.g. Gabon, Malaysia, Indonesia, Russia).

In recent years, Chinese authorities have voiced with increasing frequency the need both to expand overseas investments and to ensure that they remain stable and reliable over the long term, *based, inter alia , on "good behaviour" by Chinese firms abroad*, including being responsible and involved custodians of the local environment. This no doubt is partially reactive to international criticism that many Chinese firms, especially in the natural resources extraction area, are failing to live up to such goals, and that greater efforts are needed to ensure that China's "ecological footprint" abroad is as small and benign as possible.

Currently, however, there are *no explicit policies or regulations* in place to reward or penalize firms based on their environmental performance. While China has put in place an environmental impact assessment (EIA) requirement and procedure, it does not apply to the overseas activities of its enterprises. China has not adhered to the OECD Guidelines on Multinational Enterprises (MNEs), and is therefore not committed to their obligations, notably that of establishing national contact points to promote and monitor guidelines applications. China is, however, committed to following the environmental guidelines of the World Bank and other development institutions when utilizing loan financing for international investment purposes.

A *high-level examination of China's overall policy and strategy* in this area would be timely, and potentially very beneficial, politically, economically, and environmentally.

6.4   Forest products

Compared to OECD countries, *China is "forest poor"*, with only 21% of the country in forest land (197 million hectares) in contrast to the OECD average of 34%. Until the mid-1980s, China's policy was to look to domestic forests to fill its requirements for wood and wood products. During most of the 1990s, as China accelerated the liberalisation of its economy and sought to expand its markets for furniture and other finished wood products, the country's forests rapidly degraded. In 1998, devastating floods in the Yangtze river basin, associated in part with deforestation, triggered an abrupt change in national focus from maximum timber harvest and the opening up of unutilised forest areas to concern about forest conservation and sustainable management. *New policies and regulations were introduced* to promote rehabilitation of deforested and degraded lands, afforestation in areas prone to desertification, and the banning of logging in national forests. Long-range goals were set to increase China's forest cover over the next three decades to 26% of the territory, and to more than double the forest cover to 45% in the upper Yangtze river basin and to 27% in the Yellow river valley.

These measures have resulted in a *significant expansion of China's forest cover*, with attendant positive contributions to flood control, anti-desertification programmes, water quality and habitat protection. As China has sought to place its domestic forests under sound management, it has turned to overseas suppliers to meet a rapidly expanding demand for wood and wood products for housing construction, newsprint and other industrial uses, including raw material for an expanding furniture export industry. Consequently, its *imports of timber and other forest products have surged*.

Even before joining the WTO in 2001, China had *made changes in its trade policies* which impacted on the forestry sector. In 1997 it joined ASEAN members in an agreement to pursue trade liberalisation in nine sectors, including forestry. Subsequently, quotas on imports were eliminated or reduced, and by 2002 (apart from a continuing zero tariff policy applied to round logs, sawn wood and wood chip imports), there was an overall reduction which averaged 33.5% for all other forest products (including veneer panels, furniture, wood pulp, resin, and pulp and paper products).

As the result of market liberalisation in combination with its domestic forest conservation policy, China's *annual imports of all wood and wood products*, which amounted to less than 10 million cubic meters of roundwood equivalent (RWE) in 1981 and 20 million RWE in 1995, rose rapidly to reach some 75 million cubic meters RWE by 2003. Projections indicate that this will increase by one-third, to 100 million cubic meters RWE, by 2010 (Figure 9.5). By the latter date, China's forests and plantations will be providing less than half of the country's industrial wood needs, putting greater pressure on foreign sources.

International environmental organisations have expressed *concern in recent years about the environmental impact* that China's demand for wood and wood products is having on the world's forests. This is due, in particular, to the fact that much of China's imports come from developing countries in Asia and Africa with poor records in forest stewardship. Major sources of hardwood and softwood logs and lumber are: Malaysia, Gabon, Papua New Guinea, Indonesia, Thailand and Myanmar, along with New Zealand, Australia, Russia and the United States. Processed timber

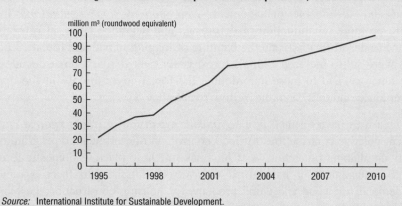

Figure 9.5 **Chinese imports of forest products**, 1995-2010

million m³ (roundwood equivalent)

Source: International Institute for Sustainable Development.

products such as wood pulp come mainly from Indonesia, Canada, Russia, Chile and the United States, while paper and paper products are imported principally from Chinese Taiwan, Korea, Indonesia, United States and Japan. China is today the world's biggest importer of waste paper, and the largest supplier of re-exported forest products, mainly furniture.

It is noteworthy that some major suppliers of China's forest products, which have sound forest management in place, are *not experiencing damaging impacts on their forests* (e.g. New Zealand, Germany, Canada and the United States). Extending good management practices more broadly will be key if China's growing demand on the world's forests, particularly in the tropics, is not to trigger a growing international outcry, and particularly if China is to secure reliable import markets over the long term based on purchases and imports from well-managed forests.

The problem of *illegal timber trade* also requires China's attention. Data from the International Tropical Timber Organization (ITTO) indicates a wide gap between what China's customs officials are reporting as import volumes and what many of China's timber *suppliers* are reporting as exports (Chapter 6). In 2000, for example, China's record of the volume of imported logs from Indonesia was 102 times lower than Indonesia's report of exports, and for Myanmar it was 27 times lower (International Institute for Sustainable Development, 2004). The discrepancies[9] exceed possible statistical error, and are indicative of the level of illegal traffic in forest products. While the responsibility and solution lie principally with the exporting countries, *China can play a significant role* by engaging its suppliers in discussions of the situation, by strengthening its own monitoring of forest product imports, and by working within the international community on certification programmes for wood from sustainably managed forests. China has, in fact, been engaged in discussions on certification with Japan and Korea, and also within the framework of the ITTO and WTO.

6.5   Hazardous waste and toxic substances

International NGOs that *track countries' performance in controlling trade in toxic substances* recently (2005) *ranked China in the "very good" category* for its ratification of three of four key international treaties in this area. These include the 1989 Basel Convention on the transboundary movement of hazardous waste, together with the 1995 Basel Ban Amendment that bans exports of such waste from OECD members and Liechtenstein to all other countries; the 1998 Rotterdam Convention requiring prior informed consent for trade in certain dangerous chemicals; and the 2004 Stockholm Convention which promotes the phase-out of persistent organic compounds. China has yet to ratify the fourth instrument, the 1996 London Dumping Convention protocol on most forms of ocean dumping.

Waste

China *ratified the Basel Convention in December 1991* after having enacted a provisional domestic ban on the import of non-recyclable hazardous waste earlier that year. Following concerns about the amount and character of hazardous waste being shipped to China by industrialised countries, *China was a leading proponent* at the Second Meeting of the Conference of Parties to the Basel Convention, in 1994, *of a total ban on the movement of all such waste between OECD and developing countries* (which led directly to the Basel Ban Amendment one year later).

To provide a *legal basis for controlling the importation of unwanted hazardous waste*, in 1995 China published the "Emergency Announcement on Strictly Controlling Trans-Boundary Movement of Waste to China". The following year, the Law on Prevention and Control of Environmental Pollution by Solid Waste was enacted, followed by a series of regulations on the import and export of hazardous waste which complied with and enforced the Basel Convention (e.g. the Provisional Regulations on Environmental Protection and Management of Waste Importation).

Under current regulations, *exports of hazardous waste* are permitted *only when China has no appropriate disposal/recovery facilities*, and then only when Basel requirements are followed for notification of the receiving country and any transit countries. *Imports* are also required to comply with Basel and its amendments, with imports of waste that is unusable as raw material prohibited and others with recovery potential restricted. Controls are based on an official list of waste to be banned or restricted which has been updated periodically over the last decade.

China's total *imports and exports in hazardous waste* are not documented or reported to the Basel Convention Secretariat or to the OECD. Regarding *waste imports overall* (hazardous and non-hazardous), in 2004, SEPA, which has administrative responsibility for the environmental management of waste trade, approved 400 enterprises in 29 provinces as designated facilities for the processing and utilisation of imported waste (e.g. waste hardware, old electrical appliances, waste wires and cables, and used electric motors). SEPA also issued a circular on strengthening the approval and management of the import of restricted waste, as a means to further restrict the approved volume of imported waste and related products, and also to intensify on-the-spot inspection of enterprises that treat waste with high environmental risks. In 2004, SEPA, the Ministry of Commerce, the General Administration of Commerce and the General Administration of Quality Supervision, Inspection and Quarantine of the People's Republic of China jointly issued an updated list of "waste as raw materials" to be restricted as imports. The same year, SEPA approved the import of 23.4 million tonnes of restricted waste as raw material, an increase of 63% over 2003, of which 10.5 million tonnes were actually imported. An additional 22.5 million tonnes were imported under automatic permission provisions of the regulations.

A recent concern has been the control of *electronic waste* (e.g. computers, printers and circuit boards) imported to support an e-waste processing industry which emerged in China in the late 1990s to reclaim precious metals, copper, plastic and non-ferrous metals. Chinese authorities cracked down on the industry and the waste traffic that supported it, after international attention was attracted in early 2002 by reports of *health problems associated with e-waste recycling in the town of Guiyu* in Guangdong province. In May 2002, a new list of banned waste imports was issued which included most electronic waste; and in September of that year, Chinese customs officials seized 450 tonnes of junked computers and other e-waste from the United States as a first step toward a proclaimed rigorous enforcement of the ban. The following year, SEPA issued a notification on enhancing environmental management of discarded electronic products and, together with the General Administration of Customs, began a *first-of-its-kind campaign with counterpart institutions in Hong Kong* to crack down on illegal trafficking in e-waste as part of a broader joint anti-smuggling effort (SEPA, 2005a).

Despite these measures, Chinese government officials and international institutions, both public and private, *remain concerned about illegal shipments of* hazardous waste into China. This is based on the very large amount of cargo which enters China's numerous ports in relation to the limited number of customs agents and inspectors, and continued reports of bribery of importers and local officials.

Chemicals

China has also moved to control trade in *potentially toxic chemicals*. In 1994, the government adopted the first comprehensive set of rules addressing toxic chemicals, the "Regulations on Environmental Management Registration of the First Import and Export of Toxic Chemicals". A detailed regulatory scheme followed, along with the establishment of a State Committee for Evaluation of Toxic Chemicals. The regulations have two major components: an official classification of toxic substances; and requirements for proper notification and registration by both Chinese importers and foreign exporters. In 2003, 89 environmental management registration licences were approved and issued by SEPA, along with 4 200 specific approval notices for toxic chemical imports and 1 200 notices of export (resulting in 500 000 tonnes of imports and 51 000 tonnes of exported chemicals).

China *ratified the Rotterdam Convention* requiring prior informed consent for exports and imports of toxic chemicals in 2005, and followed up with implementing legislation. It also participates in *UNEP's Registry of Potentially Toxic Substances* as well as the *Intergovernmental Forum on Chemical Safety (IFCS)*.

China ratified the *Stockholm Convention on Persistent Organic Pollutants (POPs)* in 2004 and has begun work on a National Implementation Plan (NIP) as the convention requires (Box 9.4). Experts from 11 ministries, under the leadership of SEPA, are preparing the NIP with major support from the Sino-Italian co-operation programme for environmental protection.

China also attaches considerable importance to the *Cartagena Protocol on Biosafety*, under the UN Convention on Biological Diversity, which is engaging SEPA, the Ministry of Agriculture, the Ministry of Commerce and the General Administration of Quality Supervision, Inspection and Quarantine in international co-operative activities in the fields of risk evaluation and management of transgenic organisms, product labelling, compensation for damages, and informed consent for transboundary movements.

Chinese officials and experts from SEPA and the Ministries of Health and Agriculture have been consulting with the OECD on China's possible adherence to the *OECD system on the harmonisation of tools and instruments* used by governments in the registration and notification of chemicals that enter international trade. The goal is to avoid and remove technical barriers to trade in potentially dangerous chemicals, while reducing costs of laboratory testing.

6.6   Endangered species

China is a party to the 1973 *Washington Convention on International Trade in Endangered and Threatened Species of Wild Fauna and Flora (CITES)*, and its Bonn and Gaborone Amendments (1979 and 1983), and has established a good record of submitting timely reports to the Convention Secretariat. It has established a National Office of the People's Republic of China on the Management of Import and Export of Endangered Species with 19 branch offices in the provinces, and also a science committee on endangered species. Regulations and circulars are periodically issued to try to enhance the scope and levels of protection afforded to threatened and endangered species through, for example, public education, increased monitoring, incentive programmes for species protection, and stiffer fines and penalties. Some of these are *species specific*, for example, a circular by the State Council forbidding the trade of rhinoceros horns and tiger bones.

China has worked *bilaterally and regionally* to strengthen its ability to promote CITES goals and its domestic laws. Under a long-standing partnership programme with the United States, Chinese law enforcement and customs agents visit US ports of entry to observe operations and discuss control of endangered species trade. In 2002, China and the NGO International Fund for Animal Welfare signed a Memorandum of Co-operation on enhancing public awareness of wildlife protection, enforcing implementation of CITES regulations, and strengthening the investigation of wildlife trade. Assistance for species protection is also being provided from the GEF and under China's environmental co-operation programme with the European Union.

China is a member of the International Whaling Commission and is a *non-whaling country*.

Despite efforts by China to control illegal wildlife trade, reports persist of *major illegal trafficking* by Chinese enterprises and individuals (Table 6.2). There is thus a *critical need for enhanced surveillance* by more and better-trained wildlife officers and customs agents, supported by improved technologies, more stringent penalties for infractions and crimes, coupled with a public education programme. The government has, in fact, been taking steps domestically to upgrade its enforcement capabilities, and also to co-operate internationally. It supported the *establishment within ASEAN, in 2005, of a Wildlife Enforcement Network*, involving national CITES authorities, customs officials and police to help combat transboundary criminal activities of wildlife trafficking and trade, and to otherwise help enforce CITES. This initiative was undertaken due to the fact that the ASEAN region has been a hot spot for illegal wildlife trading involving tigers, elephants, rare orchids, indigenous herbal medicines, rare marine species, endemic reptiles and songbirds. China also intends to participate in a new *international network on environmental compliance and enforcement (INECE)*, formally launched in 2005 by the Asian Development Bank, the US Agency for International Development (USAID) and the Philippines Department of Environment and Natural Resources. The organising meeting, attended by Chinese experts, included environmental officials, judges, lawyers and leaders from civil society from 13 Asian countries; it was followed by a training workshop held by INECE in co-operation with USAID and the OECD on the development of compliance and enforcement indicators.

7. Environment and Development

7.1 Environmental assistance

Environmental assistance to China from *developed countries and international organisations* has played an increasingly important role over the last decade as China has elevated the priority it attaches to the protection and management of environmental values. In 2004, China received a *total ODA* (in the form of commitments for grants, loans and some other official flows) of USD 5 billion, comprised of contributions from Japan (USD 1.3 billion), the ADB (USD 1.3), World Bank (USD 1.2), EU countries (USD 1.0) and the United States (USD 0.043). The environmental share of the total ODA was over 10% (over 20% for Japan). Adding other officials flows commitments (under the Global Environment Facility and the Montreal Fund), this leads to a total environmental assistance to China of some USD 750 million in 2004. Foreign capital for environmental activities consists in large part of international convention implementation funding from the World Bank, ADB, UNDP, UNEP, GEF and the Montreal Protocol Multilateral Fund (Boxes 9.3, 9.6), and bilateral government assistance funds. Grants are used mainly for implementation of international agreements, especially for capacity building, while loan financing is used mainly for construction, technology acquisition and management associated with pollution control projects.

The *World Bank has been the leading external supporter* of China's environmental efforts. For example, it has provided some 70% of all GEF funding in China (over USD 300 million from the GEF trust account since 1991, plus another USD 1.65 billion in "co-financing"). The Bank has also been the largest contributor to China's efforts to phase out ozone-depleting substances by providing grants totalling USD 445 million through its Montreal Protocol programme. The UNDP has been the second largest source of external funding, largely for grant-supported capacity building and technical assistance; while the ADB has steadily increased its environmental funding over the past five years.

Beginning in the early 1990s, the *developed countries began to readjust their foreign assistance* programmes *to place greater emphasis on environmental protection*. This emphasis continues today, even though overall levels of bilateral assistance to China are being reduced by many OECD countries as China's economic status improves. *Japan* has had a long, successful partnership with China in the environmental field, providing more than USD 260 million in environmental ODA in 2004, and more than USD 100 million since 1996 for the construction and operation of the Sino-Japan Friendship Centre for Environmental Protection. *Germany* has had a longstanding co-operation with China concerning environmental policy, environmental protection and the sustainable use of natural resources. Since the mid 1980s to 2005, it has provided some USD 800 million as financial co-operation and USD 135 million as technical co-operation. In 2006, new funding has been provided for environmental co-operation (i.e. USD 142 million in financial co-operation and USD 17 million in technical co-operation). The Sino-Italian co-operation programme for environmental protection has also been providing valuable assistance to China environmental management efforts. Over 2001-05, *Italy*, through its Ministry for the Environment and Territory, funded the programme with some USD 110 million through direct contributions and through Trust Funds established at the World Bank and in other multilateral institutions. Co-funding was provided by Chinese institutions (USD 30 million) and Italian companies (USD 28 million) in addition to contributions from the United Nations Foundation, assorted United Nations agencies (UNEP, UNDP, UNIDO), the GEF, the World Bank, and the Multilateral Fund for the Montreal Protocol (with an aggregate USD 19 million). Since 2000, the European Commission has provided grants of EUR 150 million to China for environmental co-operation programmes covering areas such as energy and environment (energy efficiency, renewable energies), cleaner industrial production, climate change, air pollution, water pollution, natural forest management, and overall environmental management. In 2006, major programmes on biodiversity protection and integrated river basin management were initiated.

It is *vital to the future success of China's environmental agenda* for the external financing to continue. *OECD countries should maintain their support* in this sector in their own long-term interests, as well as China's, as overall bilateral ODA to China shrinks.

As a developing country, *China has historically been a recipient of external assistance* and not an *aid-donor*. That *situation is beginning to change* as China grows wealthier and sees that it is in its interest to support the environmental efforts of other countries. The technical assistance it is now providing to Mongolia for anti-desertification efforts, and its co-operative agreements with African countries that will involve China-led training and technical services, are likely the beginnings of a growing trend. While no *formal environmental assistance policy and strategy* has been promulgated, it would be timely for the government to prepare one, with emphasis on assisting those developing countries which are China's principal sources of raw materials and food with sustainable natural resource management and rehabilitation.

7.2   Follow-up to UNCED and the WSSD

The *1992 UN Conference on Environment and Development (UNCED) was a defining event* in China's approach to international environmental protection. Previously, China's posture had been low-key and defensive in international forums. Against the backdrop of deteriorating environmental conditions in China through the 1980s, China positioned itself as a leader for developing countries in the run-up to UNCED, hosting a major ministerial-level preparatory meeting which issued the "Beijing Declaration" on developing country perspectives and goals.

Following the Conference, China integrated *"Agenda 21"* programme proposals (a major product of UNCED) into its 9th FYP, issued a National Agenda 21, and adopted "sustainable development" as a national policy objective. An administrative centre (ACCA 21) was established in 1993 to implement Agenda 21, and the CCICED was created "to further strengthen co-operation and exchange between China and the international community in the field of environment and development" (Heggelund et al., 2005).

A decade later, China *played a major role in the World Summit on Sustainable Development* (WSSD) in Johannesburg, sending an active, high-level delegation with broad representation, including from the private sector. While the impact of the conference on China's environmental agenda is difficult to establish, Chinese officials have endorsed the "programme of action" that emerged from the conference as influential in the development of domestic strategies and projects. Most important, China is a participant in a number of the multi-country *partnership programmes* that were endorsed at WSSD (e.g. the methane-to-market partnership and the Asia-Pacific partnership on reduction of air pollution and greenhouse gas emissions).

Notes

1. See also OECD Environmental Performance Reviews of Japan and Korea.

2. And more broadly in the framework of the OECD-China co-operation, including the 2005 OECD reviews of China on economic policy, agriculture and governance.

3. Concurrently, the China National Research and Development Centre for Combating Desertification and the Chinese National Desertification Monitoring Centre were established. Counterpart institutions have also been established by key provinces and autonomous regions.

4. In 2002 UNEP projected that sulphur dioxide emissions from Asia coal burning could triple in the following 12 years if the trend continued, noting that emission levels in the region already exceeded those of the US and Europe combined (UNEP, 2000).

5. This was substantially behind India, which scrapped 2 245 vessels (45% of world tonnage), and Bangladesh, which scrapped 529 ships (22% of world tonnage).

6. Production plus imports minus exports.

7. China and India together account for 54% of worldwide production.

8. I.e. million metric tonnes.

9. Such discrepancies exist with other wood-sourcing countries as well.

Selected Sources

The government documents, OECD documents and other documents used as sources for this chapter included the following. Also see list of websites at the end of this report.

APEC (Asia-Pacific Economic Co-operation) (2004), "Best Practices for RTAs/FTAs in APEC", Agenda Paper for 16th APEC Ministerial Meeting in Santiago, Chile, 2004.

APEC Information Office, China National Coordinating Committee on Climate Change (2004), *The Peoples Republic Of China Initial National Communication on Climate Change*, Beijing.

Chen, Jiyong, Wei, Liu, Yi, Hu (陈继勇，刘威，胡艺) 论中国对外贸易，环境保护与经济的可持续增长 (On Chinese Foreign Trade，Environmental Protection and Sustainable Economic Growth), 亚太经济, 2005年4月 (Zsia-Pacific Economic Review, April 2005).

Dai, Guilin and Xue Wang (戴桂林，王雪) (2005), 如何实现环境保护与贸易自由化的协调发展 (The Coordinated Development Between Environmental Protection and Free Trade), 中国海洋大学经济学院，山东青岛 (Ocean University of China School of Economics, Qingdao, Shandong Province), 商业研究, 2005/17, 总第325期 (Commercial Research, 2005/17, Total Vol. #325).

Danish Environmental Protection Agency (2004), *Shipbreaking in OECD*, Danish EPA, Copenhagen.

Guo, Peiyuan (郭沛源) (2004), 环境保护的国际合作：以SARS为例 (International Co-operation on Environmental Protection: Using SARS as Case Study), 清华大学21世纪发展研究院 (Tsinghua University School of 21st Century Development Studies), 生态经济, 2004年第8期 (Ecological Economy, 2004 Issue #8).

Heggelund, Andresen and Sun (2005), *Performance of the Global Environment Facility (GEF) in China: Achievements and Challenges as Seen by the Chinese*, The Fridjof Nansen Institute, Norway.

Huang, Miao (黄淼) (2004), 深入研究环境援助，加强环境国际合作 (Lucubrating on Environmental Assistance, Strengthening International Co-operation of Environmental Protection), 环境经济杂志, 2004年12月，总第12期 (Environmental Economy, Total Issue #12).

IEA (International Energy Agency) (2006a), CO_2 *Emissions from Fuel Combustion*, OECD-IEA, Paris.

IEA (2006b), *Energy Balances of OECD and Non-OECD Countries 2003-2004*, OECD-IEA, Paris.

Institute for Global Environmental Strategies, Chinese Renewable Energy Industries Association (2005), *CDM Country Guide for China*, 1st Edition, Beijing.

International Institute for Sustainable Development (2004), *An Environmental Impact Assessment of China's WTO Accession*, Report by Task Force on WTO and Environment of the China Council for International Co-operation on Environment and Development (CCICED), Winnipeg, Canada.

Korea National Institute of Environmental Research (2004), *International Joint Research on Long-Range Transboundary Air Pollution in Northeast Asia*, Incheon, Seoul.

Liu, Tuo (2001), *Influence of the Convention to Combat Desertification on Forestry in China*, Bureau to Combat Desertification, State Forestry Administration of China, Beijing.

Nautilus Institute for Security and Sustainable Development (2001), *Energy, Environment and Security in NE Asia: Defining a US – Japan Partnership for Regional Comprehensive Security*, ESENA Project Final Report 1996-99, Berkeley, California.

NBCD (National Bureau to Combat Desertification of the State Forestry Administration of China) (2005a), Zhu, Lieke, Deputy-Director General, *Conference presentation*, June 2005, Beijing.

NBCD (2005b), Liu, Tuo, Director General, *Influence of the Convention to combat desertification on forestry in China*, December 2005, Beijing; Appeared in FAO Unasylva publication 2006.

NBSC (National Bureau of Statistics of China) (2006), China Statistical Yearbook on Environment, China Statistics Press, Beijing.

Rosenberg, David and Dire Straits, Competing Security Priorities in the South China Sea, Japan Focus, 2005.

SEPA (State Environmental Protection Administration) (2006a), State of the Environment Report, SEPA, Beijing.

SEPA (2006b), China Environment Yearbook, China Environment Yearbook Press, Beijing.

SEPA (2005a), State of the Environment Report, SEPA, Beijing.

SEPA (2005b), China Environment Yearbooks (1995-2004), China Environment Yearbook Press, Beijing.

SEPA (2005c), *Tripartite Environment Ministers Meetings Among Korea, China and Japan, 1999-2004: Summary Record*, SEPA, Beijing.

SEPA (2003), State Bulletin on Offshore Environmental Quality of China, 2003, SEPA, Beijing.

State Council of the Peoples Republic of China (1998), *The Development of China's Marine Programs*, State Council Information Office, Beijing.

Tang, Dingding (唐丁丁) (2005), 环境保护国际合作的优先领域 (Priority Fields for International Co-operation of Environmental Protection), 国家环境保护总局, 北 (SEPA), 中国环保产业, 2005.9 (China Environmental Protection Industry, 2005.9), Beijing.

The Chinese Customs (2005), *The Statistical Yearbooks of China's Customs, 1999-2004*, China Customs Office, Beijing.

UNEP (United Nations Environment Programme) (2005), *Production and Consumption of Ozone Depleting Substances under the Montreal Protocol 1986-2004*, Ozone Secretariat, UNEP, Montreal.

UNEP (2000), *Global Environmental Outlook 2000*, UNEP, Nairobi.

US Department of Energy (2003), *China: Environmental Issues, Country Analysis Briefing Paper*, Energy Information Agency, Washington DC.

Wang, Mingyuan (王明远) (2000), 我国的环境立法与国际环境合作 (China's Environmental Legislation and International Environmental Co-operation), 清华大学法学院 （Tsinghua University Law School), 能源工程，2000年第4期 (Energy Engineering, No. 4）.

World Bank (2005), *World Development Indicators 2005*, World Bank, Washington DC.

World Bank (2004a), *Montreal Protocol: Successful Partnerships for Ozone Protection – The Case of China*, World Bank, Washington DC.

World Bank (2004b), Press Review, *China's energy crisis sends pollution across Asia*, World Bank, Washington DC.

World Bank (2004c), *Clean Development Mechanisms in China: Taking a proactive and sustainable approach*, World Bank, Washington DC.

Worldwatch Institute (2006), *State of the World 2006: Special Focus on China and India*, W.W. Norton and Company, New York NY.

Xie, Zhenhua (解振华) (1999), 努力把环境保护国际合作全面推向21世纪（上） [Actively Push Forward International Co-operation of Environmental Protection Towards the 21st Century (1)], 国家环境保护总局局长 (Head of SEPA), 环境保护, 1999.10, SEPA, Beijing.

Zhang, Kunmin and Can Wang (张坤民，王灿) (1999), 中国的可持续发展战略与国际环境合作 (China's Sustainable Developmental Strategy and International Environmental Co-operation), 国家环保总局，清华大学 （中国环境管理，第2期，1999年4月) (SEPA, Tsinghua University, China Environmental Management, Issue #2, April 1999), SEPA, Beijing.

Zhang, Liping (张莉萍) (2001), 经济全球化与国际环境合作 (Economy Globalization and International Environmental Cooperation), 国际论坛，2001年4月，第3卷第2期 (International Forum, April 2001, Vol. 3, Issue 2).

Zhao, Shouliang (赵守梁) (2005), 论海洋环境保护的国际合作（International Cooperation of Marine Environmental Protection), 齐鲁渔业，2005年第22卷第12期 (Shandong Fisheries, 2005, Vol. 22, Issue 12).

REFERENCES

I.A: SELECTED ENVIRONMENTAL DATA (1)

		CHN	CAN	MEX	USA	JPN	KOR	AUS	NZL	AUT	BEL	CZE	D
LAND													
Total area (1000 km^2)		9600	9971	1958	9629	378	100	7713	270	84	31	79	
Major protected areas (% of total area)	2	16.0	8.7	9.2	25.1	17.0	9.6	18.5	32.4	28.0	3.4	15.8	1
Nitrogenous fertiliser use (t/km^2 of agricultural land)		6.0	2.7	1.1	2.6	8.8	18.9	0.2	2.1	3.5	10.8	6.8	
Pesticide use (t/km^2 of agricultural land)		0.31	0.06	0.04	0.08	1.24	1.20	-	0.02	0.09	0.69	0.10	0
Livestock densities (head of sheep eq./km^2 of agr. land)		437	192	256	191	1011	1560	62	685	492	1790	287	
FOREST													
Forest area (% of land area)		32.3	45.3	33.9	32.6	68.9	63.8	21.4	34.7	41.6	22.4	34.1	1
Use of forest resources (harvest/growth)		..	0.4	0.2	0.6	0.4	0.1	0.6	..	0.7	0.9	0.7	
Tropical wood imports (USD/cap.)	3	..	1.6	0.2	2.2	10.7	6.1	4.0	3.4	0.4	24.2	0.3	
THREATENED SPECIES													
Mammals (% of species known)		13.8	31.6	34.0	18.8	24.0	17.9	24.7	18.0	22.0	30.5	18.9	2
Birds (% of species known)		6.6	12.9	17.0	11.6	12.9	13.3	12.5	21.0	27.3	28.1	49.5	1
Fish (% of species known)		2.1	7.3	34.4	14.4	25.3	9.2	0.8	10.0	41.7	23.8	40.0	
WATER													
Water withdrawal (% of gross annual availability)		19.6	1.5	15.5	19.2	20.3	35.6	6.4	1.7	4.2	32.5	11.9	
Public waste water treatment (% of population served)		33	72	25	71	64	78	..	80	86	46	70	
Fish catches (% of world catches)		0.2	1.2	1.6	5.4	5.0	1.9	0.2	0.6	-	-	-	
AIR													
Emissions of sulphur oxides (kg/cap.)		17.3	76.3	12.2	49.4	6.7	10.4	126.9	18.6	4.4	14.6	22.2	
(kg/1000 USD GDP)	4	2.9	2.6	1.4	1.4	0.3	0.6	4.6	0.8	0.2	0.5	1.4	
change (1990-early 2000s)		16	-27	..	-31	-14	-46	59	39	-55	-58	-88	
Emissions of nitrogen oxides (kg/cap.)		9.3	78.4	12.0	63.9	15.8	24.4	84.2	39.0	24.7	26.3	32.3	3
(kg/1000 USD GDP)	4	1.7	2.7	1.4	1.8	0.6	1.3	3.0	1.7	0.9	1.0	2.0	
change (1990-early 2000s)		41	-6	18	-19	-2	47	29	16	-3	-24	-40	
Emissions of carbon dioxide (t./cap.)	5	3.6	17.2	3.6	19.7	9.5	9.6	17.6	8.1	9.2	11.2	11.6	
(t./1000 USD GDP)	4	0.61	0.57	0.39	0.53	0.35	0.50	0.63	0.36	0.31	0.41	0.71	C
% change (1990-2004)		110	29	27	20	15	105	36	49	31	7	-23	
WASTE GENERATED													
Industrial waste (kg/1000 USD GDP)	4, 6	160	..	..	..	40	40	..	10	..	50	50	
Municipal waste (kg/cap.)	7	120	380	320	740	410	390	450	400	550	470	280	
Nuclear waste (t./Mtoe of TPES)	8	..	4.0	0.1	1.1	1.6	3.0	-	-	-	1.9	1.4	

.. not available.   - nil or negligible.

1) Data refer to the latest available year. They include provisional figures and Secretariat estimates.
   Partial totals are underlined. Varying definitions can limit comparability across countries.
2) IUCN management categories I-VI and protected areas without IUCN category assignment; national classifications may differ.
3) Total imports of cork and wood from non-OECD tropical countries.
4) GDP at 2000 prices and purchasing power parities.
Source: OECD Environmental Data Compendium.

FIN	FRA	DEU	GRC	HUN	ISL	IRL	ITA	LUX	NLD	NOR	POL	PRT	SLO	ESP	SWE	CHE	TUR	UKD*	OECD*
438	549	357	132	93	103	70	301	3	42	324	313	92	49	506	450	41	779	245	35042
9.1	13.3	31.5	5.2	8.9	9.5	1.2	19.0	17.1	18.9	6.4	29.0	8.5	25.2	9.5	9.5	28.7	4.3	30.1	16.4
6.0	7.5	10.5	3.0	6.2	0.5	8.1	6.0	-	14.6	9.6	4.5	2.6	3.6	3.6	6.0	3.5	3.1	6.8	2.2
.06	0.27	0.17	0.14	0.14	-	0.05	0.58	0.33	0.41	0.08	0.06	0.40	0.16	0.14	0.05	0.10	0.06	0.21	0.07
90	514	689	245	207	65	1139	488	4351	2142	845	315	498	226	339	409	794	290	674	208
5.5	31.6	30.2	22.8	19.5	1.3	9.4	23.3	34.5	9.5	39.2	30.0	36.9	41.6	33.3	73.5	30.8	27.0	11.6	34.4
0.7	0.6	0.5	0.6	0.5	-	0.7	0.5	0.5	0.6	0.5	0.6	0.8	0.5	0.5	0.7	0.8	0.5	0.6	0.6
1.4	6.8	1.8	2.7	0.1	2.8	11.2	7.1	-	15.6	3.6	0.3	17.6	0.1	6.2	2.2	0.6	0.5	2.7	4.0
1.9	19.0	41.8	37.8	71.1	-	1.8	40.7	51.6	18.6	3.4	14.1	17.7	22.2	26.3	22.4	32.9	22.2	6.3	..
3.3	19.2	27.3	1.9	18.8	44.0	5.4	18.4	50.0	21.5	7.7	8.6	13.7	14.4	25.5	19.1	36.4	30.8	15.4	..
1.8	31.9	68.2	26.2	32.1	-	23.1	29.0	27.9	48.9	-	7.0	22.9	24.1	52.9	16.4	38.9	9.9	11.1	..
2.1	17.5	20.2	12.1	4.7	0.1	2.3	24.0	3.7	9.9	0.7	18.6	15.1	1.4	34.7	1.5	4.7	17.0	20.8	11.4
81	79	92	56	32	50	73	69	95	98	73	55	41	53	55	86	97	17	95	64
0.2	0.7	0.3	0.1	-	2.2	0.3	0.3	-	0.5	2.9	0.2	0.2	-	1.0	0.3	-	0.6	0.7	27.9
6.4	9.0	7.4	46.3	35.3	35.0	24.5	11.5	6.7	5.3	4.9	38.1	28.4	19.0	37.3	6.5	2.3	31.3	16.9	27.8
0.6	0.3	0.3	2.6	2.7	1.3	0.8	0.5	0.1	0.2	0.1	3.6	1.6	1.6	1.8	0.2	0.1	4.6	0.6	1.1
-64	-60	-89	4	-64	22	-48	-63	-80	-58	-58	-55	-9	-81	-29	-45	-60	33	-73	-40
.5	22.7	17.2	28.9	17.7	90.5	31.0	21.8	38.1	26.6	46.9	20.8	27.8	19.0	34.7	27.1	11.4	14.1	26.8	34.3
.5	0.8	0.7	1.7	1.4	3.3	1.0	0.9	0.8	1.0	1.3	2.0	1.6	1.6	1.7	1.0	0.4	2.1	1.0	1.4
32	-29	-48	11	-24	-2	5	-34	-27	-28	-5	-38	13	-53	14	-25	-46	48	-43	-17
3.2	6.4	10.3	8.5	5.6	7.7	10.2	7.9	24.9	11.4	7.9	7.8	5.7	7.0	7.7	5.8	6.0	2.9	9.0	11.1
.47	0.23	0.40	0.45	0.40	0.26	0.31	0.31	0.47	0.42	0.21	0.66	0.33	0.54	0.36	0.20	0.20	0.40	0.33	0.44
25	9	-12	33	-19	19	37	16	7	18	26	-15	52	-34	59	-	8	63	-4	17
30	70	20	..	20	-	60	20	..	40	20	140	50	40	20	90	10	30	30	60
50	540	640	430	460	730	750	520	660	600	700	260	450	300	640	470	660	360	620	550
.9	4.1	1.4	-	1.8	-	-	-	-	0.1	-	-	-	3.1	1.5	3.8	2.1	-	4.0	1.5

UKD: pesticides and threatened species: Great Britain; water withdrawal and public waste water treatment plants: England and Wales.

5) CO_2 from energy use only; sectoral approach; international marine and aviation bunkers are excluded.

6) Waste from manufacturing industries.

7) CAN, NZL: household waste only.

8) Waste from spent fuel arising in nuclear power plants, in tonnes of heavy metal, per million tonnes of oil equivalent
   of total primary energy supply.

I.B: SELECTED ECONOMIC DATA (1)

		CHN	CAN	MEX	USA	JPN	KOR	AUS	NZL	AUT	BEL	C
GROSS DOMESTIC PRODUCT												
GDP, 2004 (billion USD at 2000 prices and PPPs)		**7767**	964	957	10842	3447	921	561	91	241	285	1
% change (1990-2004)		**284.2**	47.2	49.7	52.4	19.7	116.2	59.5	53.7	35.5	31.0	1
per capita, 2004 (1000 USD/cap.)		**6.0**	30.2	9.2	36.9	27.0	19.2	27.9	22.4	29.5	27.4	1
Exports, 2004 (% of GDP)		**40.2**	38.2	30.1	10.0	13.1	44.1	18.2	29.2	50.6	83.5	7
INDUSTRY	2											
Value added in industry (% of GDP)		**41**	32	27	23	31	43	26	25	32	27	
Industrial production: % change (1990-2003)		**394.8**	42.3	42.2	43.4	-3.2	164.7	29.8	26.9	53.2	17.6	
AGRICULTURE												
Value added in agriculture (% of GDP)	3	**13**	3	4	2	1	4	4	7	2	1	
Agricultural production: % change (1990-2005)		**96.7**	25.6	41.5	27.6	-12.3	19.3	25.4	47.9	9.9	13.0	
Livestock population, 2005 (million head of sheep eq.)		**1845**	118	275	787	53	30	283	99	17	25	
ENERGY												
Total supply, 2004 (Mtoe)		**1609**	269	165	2326	533	213	116	18	33	58	
% change (1990-2004)		**85.7**	28.5	33.1	20.7	19.6	129.9	32.2	28.2	32.6	17.5	-
Energy intensity, 2004 (toe/1000 USD GDP)		**0.21**	0.28	0.17	0.21	0.15	0.23	0.21	0.19	0.14	0.20	0
% change (1990-2004)		**-51.7**	-12.7	-11.1	-20.8	-0.1	6.3	-17.1	-16.6	-2.1	-10.3	-1
Structure of energy supply, 2004 (%)	4											
Solid fuels		**61.7**	10.6	4.3	23.5	21.8	23.5	42.7	10.7	12.0	10.2	4
Oil		**19.3**	36.4	58.1	40.8	47.8	47.6	32.0	39.9	43.3	40.4	2
Gas		**2.6**	28.9	26.4	22.1	13.2	11.9	19.6	19.6	23.1	25.5	1
Nuclear		**0.8**	8.7	1.4	9.1	13.8	16.0	-	-	-	21.6	1
Hydro, etc.		**15.6**	15.3	9.8	4.5	3.4	1.0	5.6	29.9	21.5	2.3	
ROAD TRANSPORT	5											
Road traffic volumes per capita, 2002 (1000 veh.-km/cap.)		**0.8**	10.1	0.7	15.9	6.2	2.3	9.8	11.2	8.3	8.8	
Road vehicle stock, 2003 (10 000 vehicles)		**2383**	1850	2051	23139	7254	1454	1279	255	490	544	
% change (1990-2003)		**332.2**	11.7	107.5	22.6	28.4	328.4	30.8	38.1	32.7	27.7	5
per capita (veh./100 inh.)		**2**	58	20	80	57	30	64	64	60	52	

.. not available. - nil or negligible.

1) Data may include provisional figures and Secretariat estimates. Partial totals are underlined.

2) Value added: includes mining and quarrying, manufacturing, gas, electricity and water and construction;
production: excludes construction.

Source: OECD Environmental Data Compendium.

OECD EPR / SECOND CYCLE

NK	FIN	FRA	DEU	GRC	HUN	ISL	IRL	ITA	LUX	NLD	NOR	POL	PRT	SLO	ESP	SWE	CHE	TUR	UKD	OECD
60	145	1650	2109	210	141	9	132	1495	24	446	175	446	181	70	910	258	225	529	1649	29441
2.1	31.6	28.1	23.0	48.7	24.4	43.6	144.4	21.3	91.1	36.6	54.6	60.2	34.0	30.7	44.2	31.6	14.0	63.6	39.4	40.9
).6	27.8	27.4	25.6	19.0	14.0	29.9	32.7	25.7	53.3	27.4	38.1	11.7	17.2	12.9	21.3	28.7	30.4	7.4	27.6	25.3
3.5	37.1	25.9	38.2	20.2	64.9	36.8	80.2	26.7	146.2	65.4	43.7	39.1	30.9	76.8	27.0	46.2	45.9	28.9	24.7	23.3
27	32	25	30	23	31	27	42	29	20	26	38	30	29	32	30	28	27	31	26	29
6.1	71.3	13.4	9.7	11.7	67.4	..	302.4	11.8	39.3	16.5	33.6	81.5	17.7	10.9	24.1	45.0	19.4	65.9	9.0	26.2
3	4	3	1	7	4	9	3	3	1	3	2	3	4	5	3	2	1	12	1	3
).7	-3.9	0.9	-4.7	10.1	-10.5	5.4	2.6	10.7	13	-9.2	-9.4	-15.8	1.1	..	7.4	-10.2	-4.3	18.2	-8.0	..
24	8	156	117	21	12	1	50	64	6	42	9	58	19	6	100	13	12	111	113	2639
20	38	275	348	30	26	3	15	184	5	82	28	92	27	18	142	54	27	82	234	5508
2.2	30.6	21.1	-2.3	37.4	-7.7	61.0	46.1	24.6	33.0	23.1	28.7	-8.1	49.6	-14.0	56.1	13.4	8.6	54.6	10.1	21.7
13	0.26	0.17	0.17	0.15	0.19	0.40	0.12	0.12	0.20	0.18	0.16	0.21	0.15	0.26	0.16	0.21	0.12	0.15	0.14	0.19
5.1	-0.7	-5.5	-20.6	-7.6	-25.8	12.1	-40.2	2.8	-30.4	-9.9	-16.8	-42.6	11.6	-34.2	8.3	-13.9	-4.7	-5.5	-21.0	-13.6
4.5	20.0	5.0	24.6	30.1	13.6	2.9	15.1	9.2	2.1	10.8	3.5	58.6	13.0	24.5	14.8	5.5	0.5	27.3	16.1	20.5
4.3	29.8	32.8	36.0	57.2	24.9	25.0	58.5	46.2	69.4	39.6	39.7	23.4	59.3	17.4	49.7	28.8	46.1	36.7	35.9	40.7
2.8	10.5	14.3	22.6	7.4	45.5	-	24.2	36.6	26.9	45.5	16.7	12.8	12.7	29.7	17.7	1.6	10.0	22.8	37.5	21.7
-	15.7	41.6	12.5	-	12.1	-	-	-	-	1.2	-	-	-	24.3	11.6	37.5	25.9	-	8.9	11.0
4.4	24.0	6.2	4.3	5.3	3.8	72.0	2.2	8.0	1.6	2.9	40.2	5.2	15.0	4.2	6.3	26.5	17.5	13.2	1.6	6.1
3.9	9.4	8.7	7.2	7.5	2.3	8.2	8.8	8.3	7.9	7.7	7.3	3.6	6.3	2.4	4.4	8.5	7.8	0.8	8.1	8.2
32	263	3563	4736	500	320	19	179	3848	33	787	240	1364	542	154	2311	451	406	645	3296	62611
2.7	17.6	25.2	27.0	98.1	..	41.5	88.2	28.6	64.7	37.3	23.8	113.2	146.5	56.9	60.0	15.0	25.0	173.2	30.7	33.3
43	50	59	57	45	32	66	45	66	73	48	53	36	52	29	55	50	55	9	55	54

3) Agriculture, forestry, hunting, fishery, etc.

4) Breakdown excludes electricity trade.

5) Refers to motor vehicles with four or more wheels, except for Italy, which include three-wheeled goods vehicles.

I.C: SELECTED SOCIAL DATA (1)

		CHN	CAN	MEX	USA	JPN	KOR	AUS	NZL	AUT	BEL	C
POPULATION												
Total population, 2004 (100 000 inh.)		**12999**	319	1040	2939	1277	481	201	41	82	104	
% change (1990-2004)		**13.7**	15.3	28.0	17.6	3.4	12.2	17.8	20.8	5.9	4.3	
Population density, 2004 (inh./km^2)		**135.4**	3.2	53.1	30.5	338.0	482.8	2.6	15.0	97.5	340.6	12
Ageing index, 2004 (over 64/under 15)		**44.4**	72.3	18.6	59.7	140.3	44.4	65.4	54.9	97.1	97.2	9
HEALTH												
Women life expectancy at birth, 2003 (years)		**73.3**	82.1	77.4	79.9	85.3	80.8	82.8	81.1	81.6	81.1	7
Infant mortality, 2003 (deaths /1 000 live births)		**34.7**	5.4	20.1	7.0	3.0	6.2	4.8	5.6	4.5	4.3	
Expenditure, 2003 (% of GDP)		**5.6**	9.9	6.2	15.0	7.9	5.6	9.3	8.1	7.6	9.6	
INCOME AND POVERTY												
GDP per capita, 2004 (1000 USD/cap.)		**6.0**	30.2	9.2	36.9	27.0	19.2	27.9	22.4	29.5	27.4	1
Poverty (% pop. < 50% median income)		..	10.3	20.3	17.0	15.3	..	11.2	10.4	9.3	7.8	
Inequality (Gini levels)	2	**44.7**	30.1	48.0	35.7	31.4	..	30.5	33.7	26.0	26.0	2
Minimum to median wages, 2000	3	x	42.5	21.1	36.4	32.7	25.2	57.7	46.3	x	49.2	3
EMPLOYMENT												
Unemployment rate, 2004 (% of civilian labour force)	4	**4.2**	7.2	3.0	5.5	4.7	3.7	5.5	3.9	4.9	8.4	
Labour force participation rate, 2004 (% 15-64 years)		**81.9**	79.6	59.9	74.9	77.5	67.8	76.1	78.0	79.7	66.2	7
Employment in agriculture, 2004 (%)	5	**46.9**	2.6	15.9	1.6	4.5	8.1	3.7	7.5	5.0	2.0	
EDUCATION												
Education, 2003 (% 25-64 years)	6	**14.9**	83.6	21.5	87.5	83.9	73.2	62.5	77.5	78.6	62.0	8
Expenditure, 2002 (% of GDP)	7	..	6.1	6.3	7.2	4.7	7.1	6.0	6.8	5.7	6.4	
OFFICIAL DEVELOPMENT ASSISTANCE	8											
ODA, 2005 (% of GNI)		..	0.34	..	0.22	0.28	..	0.25	0.27	0.52	0.53	
ODA, 2005 (USD/cap.)		..	116	..	93	103	..	82	67	189	189	

.. not available.   - nil or negligible.   x not applicable.

1) Data may include provisional figures and Secretariat estimates. Partial totals are underlined.

2) Ranging from 0 (equal) to 100 (inequal) income distribution; figures relate to total disposable income (including all incomes, taxes and benefits) for the entire population.

3) Minimum wage as a percentage of median earnings including overtime pay and bonuses.

Source: OECD.

OECD EPR / SECOND CYCLE

	FIN	FRA	DEU	GRC	HUN	ISL	IRL	ITA	LUX	NLD	NOR	POL	PRT	SLO	ESP	SWE	CHE	TUR	UKD	OECD
4	52	603	825	111	101	3	40	582	5	163	46	382	105	54	427	90	74	718	598	11617
1	4.9	6.3	4.0	9.6	-2.6	14.8	15.4	2.6	17.6	8.9	8.3	0.2	6.4	1.6	9.9	5.1	10.1	27.8	4.4	11.6
3	15.5	109.8	231.1	83.8	108.6	2.8	57.5	193.1	174.8	391.9	14.2	122.1	114.2	109.8	84.4	20.0	179.0	92.1	244.1	33.2
5	89.6	88.5	134.5	121.5	98.7	52.2	53.5	133.1	75.3	74.2	74.3	76.9	107.8	66.8	116.0	97.3	100.8	19.4	87.1	70.2
5	81.8	82.9	81.3	80.7	76.5	82.5	80.3	82.9	81.5	80.9	81.9	78.9	80.6	77.8	83.7	82.4	83.0	71.0	80.7	..
4	3.1	3.9	4.2	4.8	7.3	2.4	5.1	4.3	4.9	4.8	3.4	7.0	4.1	7.9	4.1	3.1	4.3	29.0	5.3	..
0	7.5	10.1	11.1	9.9	7.8	10.5	7.3	8.4	6.1	9.8	10.3	6.1	9.6	5.9	7.7	9.2	11.5	6.6	7.7	..
6	27.8	27.4	25.6	19.0	14.0	29.9	32.7	25.7	53.3	27.4	38.1	11.7	17.2	12.9	21.3	28.7	30.4	7.4	27.6	25.3
3	6.4	7.0	9.8	13.5	8.2	..	15.4	12.9	5.5	6.0	6.3	9.8	13.7	..	11.5	5.3	6.7	15.9	11.4	10.2
0	25.0	28.0	28.0	33.0	27.0	35.0	32.0	33.0	26.0	27.0	25.0	31.0	38.0	33.0	31.0	23.0	26.7	45.0	34.0	30.7
x	x	60.8	x	51.3	37.2	x	55.8	x	48.9	47.1	x	35.5	38.2	..	31.8	x	x	..	41.7	..
5	8.9	9.6	9.5	10.5	6.1	3.1	4.5	8.0	4.8	4.6	4.4	19.0	6.7	18.2	10.6	6.4	4.4	10.2	4.7	6.9
3	74.1	69.6	77.0	64.8	59.2	83.6	71.0	62.2	66.6	78.6	79.4	63.8	76.9	69.4	69.2	77.8	87.3	51.7	75.9	70.9
,1	4.9	3.5	2.4	12.6	5.3	6.3	6.4	4.5	1.3	3.0	3.5	18.0	12.1	5.1	5.5	2.1	3.7	34.0	1.3	6.1
,5	75.9	64.9	83.4	51.1	74.1	59.0	61.6	44.4	59.1	66.5	87.4	48.3	22.6	86.7	42.8	82.2	69.7	26.5	65.1	65.6
,1	6.0	6.1	5.3	4.1	5.6	7.4	4.4	4.9	3.6	5.1	6.9	6.1	5.8	4.2	4.9	6.9	6.2	3.8	5.9	5.7
31	0.47	0.47	0.35	0.24	..	..	0.41	0.29	0.87	0.82	0.93	..	0.21	..	0.29	0.92	0.44	..	0.48	0.33
39	171	165	120	48	..	..	168	86	580	314	600	..	35	..	72	363	238	..	179	121

4) Standardised unemployment rates; MEX, ISL, TUR: commonly used definitions.

5) Civil employment in agriculture, forestry and fishing.

6) Upper secondary or higher education; OECD: average of rates.

7) Public and private expenditure on educational institutions; OECD: average of rates.

8) Official Development Assistance by Member countries of the OECD Development Assistance Committee.

II.A: SELECTED MULTILATERAL AGREEMENTS (WORLDWIDE)

Y = in force S = signed R = ratified D = denounced

		CHN	CAN	MEX	USA
1946 Washington	Conv. - Regulation of whaling	Y R	D	R	R
1956 Washington	Protocol	Y R	D	R	R
1949 Geneva	Conv. - Road traffic	Y	R		R
1957 Brussels	Conv. - Limitation of the liability of owners of sea-going ships	Y	S		
1979 Brussels	Protocol	Y			
1958 Geneva	Conv. - Fishing and conservation of the living resources of the high seas	Y	S	R	R
1959 Washington	Treaty - Antarctic	Y R	R		R
1991 Madrid	Protocol to the Antarctic treaty (environmental protection)	Y R	R		R
1960 Geneva	Conv. - Protection of workers against ionising radiations (ILO 115)	Y		R	
1962 Brussels	Conv. - Liability of operators of nuclear ships				
1963 Vienna	Conv. - Civil liability for nuclear damage	Y		R	
1988 Vienna	Joint protocol relating to the application of the Vienna Convention and the Paris Convention	Y			
1997 Vienna	Protocol to amend the Vienna convention	Y			
1963 Moscow	Treaty - Banning nuclear weapon tests in the atmosphere, in outer space and under water	Y	R	R	R
1964 Copenhagen	Conv. - International council for the exploration of the sea	Y	R		R
1970 Copenhagen	Protocol	Y	R		R
1969 Brussels	Conv. - Intervention on the high seas in cases of oil pollution casualties (INTERVENTION)	Y R		R	R
1973 London	Protocol (pollution by substances other than oil)	Y R		R	R
1969 Brussels	Conv. - Civil liability for oil pollution damage (CLC)	Y D	D	D	S
1976 London	Protocol	Y D	R	R	
1992 London	Protocol	Y R	R	R	
1970 Bern	Conv. - Transport of goods by rail (CIM)	Y			
1971 Brussels	Conv. - International fund for compensation for oil pollution damage (FUND)	Y D	D	D	S
1976 London	Protocol	Y D	R	R	
1992 London	Protocol (replaces the 1971 Convention)	Y R	R	R	
2000 London	Amendment to protocol (limits of compensation)	Y	R	R	
2003 London	Protocol (supplementary fund)				
1971 Brussels	Conv. - Civil liability in maritime carriage of nuclear material	Y			
1971 London, Moscow, Washington	Conv. - Prohib. emplacement of nuclear and mass destruct. weapons on sea-bed, ocean floor and subsoil	Y R	R	R	R
1971 Ramsar	Conv. - Wetlands of international importance especially as waterfowl habitat	Y R	R	R	R
1982 Paris	Protocol	Y R	R	R	R
1987 Regina	Regina amendment	Y	R	R	
1971 Geneva	Conv. - Protection against hazards of poisoning arising from benzene (ILO 136)	Y			
1972 London, Mexico, Moscow, Washington	Conv. - Prevention of marine pollution by dumping of wastes and other matter (LC)	Y R	R	R	R
1996 London	Protocol to the Conv. - Prevention of marine poll. by dumping of wastes and other matter	S	R		S
1972 Geneva	Conv. - Protection of new varieties of plants (revised)	Y R	R	R	R

OECD EPR / SECOND CYCLE

Y = in force S = signed R = ratified D = denounced

	AUS	NZL	AUT	BEL	CZE	DNK	FIN	FRA	DEU	GRC	HUN	ISL	IRL	ITA	LUX	NLD	NOF	POL	PRT	SVK	ESP	SWE	CHE	TUR	UKD	EU
R	R	R	R	R	R	R	R	R	R		R	R	R		R	R		R	R	R		R	R		R	
R	R	R	R	R	R	R	R	R	R		R	R	R	R	R	R		R	R	R		R	R		R	
R	R	R	R	R	R	R	R	R		R	R	R	R	R	R	R	R	R	R	R	R	R	S	R		R
D			D		D	D	D	D		R		S		D	D	R	R		R		D	R		D		
R			R			S		S					R			R	R		R		R		D			
R	S		R		R	R	R			S	S		R			R		R		R		R		R		
R	R	R	R	R	R	R	R	R	R			R		R	R	R		R	R	R		R	R	R	R	
R	R	S	R	R	S	R	R	R	R	S		R		R	R	R		S	R	R	S		R			
		R	R	R	R	R	R	R	R			R		R	R	R	R	R	R	R	R	R	R	R	R	
		S					S				S		R			R										
			R					R				R		R	S							S				
		S	R	R	R	S	R	R	R			R		R	R	R	S	R	S	R	S	S	S	S		
			S					S			S			S												
R	R	R	R	R	R		R	R	R	R	R	R	R	R	R	S	R	R	R	R	R	R			
			R		R	R	R	R		R	R		R	R	R	R		R	R		R	R			R	
			R		R	R	R	R		R	R		R	R	R	R		R	R		R	R			R	
R	R		R		R	R	R	R	S		R	R	R		R	R	R	R		R	R	R			R	
R	S		R		R	R	R	R			R	R		R	R	R	R		R	R	R			R		
D	D		D		D	D	D	D	D		D	D	D	R	D	D	D	D		D	D	D				
R			R		R	R	R	R	R		R	D	R	R	R	R	R	R		R	R	R				
R	R		R		R	R	R	R	R		R	R	R	R	R	R	R	R		R	R	R	R	R	R	
	R	R	R	R	R	R	R	R	R	R		R	R	R	R	R	R	R	R	R	R	R	R	R	R	
D	D		R		D	D	D	D	D		D	D	D		D	D	D	R		D	D	D			D	
R	R		R		R	R	R	R	R		R	D	R	R	R	R	R		R	R	R			D	
R	R		R		R	R	R	R	R		R	R	R	R	R	R	R		R	R	R	R	R	R	
R	R		R		R	R	R	R	R		R	R	R		R	R	R		R	R	R		R	R	
	R		R		R	R	R	R				R	R		R	R		R		R	R		R			
	R		R		R	R	R	R				R		R	R		S		R	R			S			
R	R	R	R	R	R	R		R	R	R	R	R	R	R	R	R	R	R	R	R		R	R	R	R
R	R	R	R	R	R	R	R	R	R	R	R	R	R	R	R	R	R	R	R	R	R	R	R	R	R
R	R	R	R	R	R	R	R	R	R	R	R	R	R	R	R	R	R	R	R	R	R	R	R	R	R
R	R	R	R		R	R	R	R	R	R	R		R	R	R		R	R	R			R	R	R	R
		R			R	R	R	R	R			R						R	R				R			
R	R		R		R	R	R	R	R	R	R	R	R	R	R	R	R	R			R	R		R	
R	R		R		R	S	R	R			R	R			S	R				R		R		R	
R	R	R	R	R	R	R	R	R		R		R	R		R	R	R	R	R	R	R	R		R	

II.A: SELECTED MULTILATERAL AGREEMENTS (WORLDWIDE) (cont.)

Y = in force S = signed R = ratified D = denounced

			CHN	CAN	MEX	USA
1978 Geneva	Amendments	Y	R	R	R	R
1991 Geneva	Amendments	Y				R
1972 Geneva	Conv. - Safe container (CSC)	Y	R	R	R	R
1972 London, Moscow, Washington	Conv. - International liability for damage caused by space objects	Y	R	R	R	R
1972 Paris	Conv. - Protection of the world cultural and natural heritage	Y	R	R	R	R
1973 Washington	Conv. - International trade in endangered species of wild fauna and flora (CITES)	Y	R	R	R	R
1974 Geneva	Conv. - Prev. and control of occup. hazards caused by carcinog. subst. and agents (ILO 139)	Y				
1976 London	Conv. - Limitation of liability for maritime claims (LLMC)	Y		R		
1996 London	Amendment to convention	Y		S		
1977 Geneva	Conv. - Protection of workers against occupational hazards in the working environment due to air pollution, noise and vibration (ILO 148)	Y				
1978 London	Protocol - Prevention of pollution from ships (MARPOL PROT)	Y	R	R	R	R
1978 London	Annex III	Y	R	R		R
1978 London	Annex IV	Y				
1978 London	Annex V	Y	R		R	R
1997 London	Annex VI	Y				S
1979 Bonn	Conv. - Conservation of migratory species of wild animals	Y				
1991 London	Agreem. - Conservation of bats in Europe	Y				
1992 New York	Agreem. - Conservation of small cetaceans of the Baltic and the North Seas (ASCOBANS)	Y				
1996 Monaco	Agreem. - Conservation of cetaceans of the Black Sea, Mediterranean Sea and Contiguous Atlantic Area	Y				
1996 The Hague	Agreem. - Conservation of African-Eurasian migratory waterbirds	Y				
2001 Canberra	Agreem. - Conservation of albatrosses and petrels (ACAP)	Y				
1982 Montego Bay	Conv. - Law of the sea	Y	R	R	R	
1994 New York	Agreem. - relating to the implementation of part XI of the convention	Y	R	R	R	S
1995 New York	Agreem. - Implementation of the provisions of the convention relating to the conservation and management of straddling fish stocks and highly migratory fish stocks	Y	S	R		R
1983 Geneva	Agreem. - Tropical timber	Y	R	R		R
1994 New York	Revised agreem. - Tropical timber	Y	R	R	R	R
1985 Vienna	Conv. - Protection of the ozone layer	Y	R	R	R	R
1987 Montreal	Protocol (substances that deplete the ozone layer)	Y	R	R	R	R
1990 London	Amendment to protocol	Y	R	R	R	R
1992 Copenhagen	Amendment to protocol	Y	R	R	R	R
1997 Montreal	Amendment to protocol	Y	R			R
1999 Beijing	Amendment to protocol	Y	R			R
1986 Vienna	Conv. - Early notification of a nuclear accident	Y	R	R	R	R
1986 Vienna	Conv. - Assistance in the case of a nuclear accident or radiological emergency	Y	R	R	R	R
1989 Basel	Conv. - Control of transboundary movements of hazardous wastes and their disposal	Y	R	R	R	S

OECD EPR / SECOND CYCLE

Y = in force   S = signed   R = ratified   D = denounced

	AUS	NZL	AUT	BEL	CZE	DNK	FIN	FRA	DEU	GRC	HUN	ISL	IRL	ITA	LUX	NLD	NOF	POL	PRT	SVK	ESP	SWE	CHE	TUR	UKD	EU
R	R	R		R	R	R	R	R		R			R	R	R	R	R			R	R			R		
R		R		R	R	R		R					R		R		R	R	R	R				R		
R	R	R	R	R	R	R	R	R	R	R			R	R	R	R	R	R	R	R	R	R	S	R		
R	R	R	R	R	R	R	R	R	R	R	S	R	R	R	R	R		R	R	R	R		R		
R	R	R	R	R	R	R	R	R	R	R	R	R	R	R	R	R	R	R	R	R	R	R	R	R	
R	R	R	R	R	R	R	R	R	R	R	R	R	R	R	R	R	R	R	R	R	R	R	R	R	
		R	R	R	R	R	R		R	R	R	R		R		R	R		R	R					
R	R		R		D	D	R	D	R			R		R	R	R		R	R	R	R	R			
R				R	R	S	R				R	S	R		R	R		R	R						
		R	R	R	R	R	R		R		R		R		R	R	R	R		R					
R	R	R	R	R	R	R	R	R	R	R	R	R	R	R	R	R	R	R	R	R	R	R	R	R	
R	R	R	R	R	R	R	R	R	R	R	R	R	R	R	R	R	R	R	R	R	R	R	R	R	
R		R	R	R	R	R	R	R		R	R	R	R	R	R	R	R		R	R				R	
R	R	R	R	R	R	R	R	R	R	R	R	R	R	R	R	R	R	R	R	R	R	R	R	R	
	R		R	R	R	R		R	R	R	R		R	R		R	R	R		R					
R	R	R	R	R	R	R	R	R	R	R	R	R	R	R	R	R	R	R	R	R	R	R		R	R
	R	R	R	R	R	R	R		R	R		R	R	R	R	R	R		R	R				R	
	R		R	R	R	R				R		R			R		R		R	R				R	S
		R		R		R			R			R		R											
	R		R	R	R	R	S	R		R	R	R	R			R	R	R		R	R			R	R
R	R					S												R						R	
R	R	R	R	R	R	R	R	R	R	R	R	R	R	R	R	R	R	R	R	R	R	S		R	R
R	R	R	R	R	R	R	R	R	R	R	R	R	R	R	R	R	R	R	R	R	R	S		R	R
R	R	R	R		R	R	R	R	R		R	R	R	R	R	R	R		R	R				R	R
R	R	R	R		R	R	R	R	R		R	R	R	R		R			R	R			R	R	
R	R	R	R		R	R	R	R	R		R	R	R	R		R			R	R			R	R	
R	R	R	R	R	R	R	R	R	R	R	R	R	R	R	R	R	R	R	R	R	R	R	R	R	R
R	R	R	R	R	R	R	R	R	R	R	R	R	R	R	R	R	R	R	R	R	R	R	R	R	R
R	R	R	R	R	R	R	R	R	R	R	R	R	R	R	R	R	R	R	R	R	R	R	R	R	R
R	R	R	R	R	R	R	R	R	R	R	R	R	R	R	R	R	R	R	R	R	R	R	R	R	R
R	R	R	R	R	R	R	R	R	R	R	R	R	R	R	R	R	R	R	R	R	R	R	R	R	R
R	R	R	R	R	R	R	R	R	R	R	R	R	R	R	R	R	R	R	R	R	R	R	R	R	R
R	R	R	R	R	R	R	R	R	R	R	R	R	R	R	R	R	R	R	R	R	R	R	R	R	
R	R	R	R	R	S	R	R	R	R	R	R	S	R	R	R	R	R	R	R	R	R	R	R	R	R
R	R	R	R	R	R	R	R	R	R	R	R	R	R	R	R	R	R	R	R	R	R	R	R	R	R

II.A: SELECTED MULTILATERAL AGREEMENTS (WORLDWIDE) (cont.)

Y = in force S = signed R = ratified D = denounced

		CHN	CAN	MEX	USA	
1995 Geneva	Amendment	R				
1999 Basel	Prot. - Liability and compensation for damage					
1989 London	Conv. - Salvage	Y	R	R	R	R
1990 Geneva	Conv. - Safety in the use of chemicals at work (ILO 170)	Y	R		R	
1990 London	Conv. - Oil pollution preparedness, response and co-operation (OPRC)	Y	R	R	R	R
2000 London	Protocol - Pollution incidents by hazardous and noxious substances (OPRC-HNS)					
1992 Rio de Janeiro	Conv. - Biological diversity	Y	R	R	R	S
2000 Montreal	Prot. - Biosafety (Cartagena)	Y	R	S	R	
1992 New York	Conv. - Framework convention on climate change	Y	R	R	R	R
1997 Kyoto	Protocol	Y	R	R	R	S
1993 Paris	Conv. - Prohibition of the development, production, stockpiling and use of chemical weapons and their destruction	Y	R	R	R	R
1993 Geneva	Conv. - Prevention of major industrial accidents (ILO 174)	Y				
1993	Agreem. - Promote compliance with international conservation and management measures by fishing vessels on the high seas	Y		R	R	R
1994 Vienna	Conv. - Nuclear safety	Y	R	R	R	R
1994 Paris	Conv. - Combat desertification in those countries experiencing serious drought and/or desertification, particularly in Africa	Y	R	R	R	R
1996 London	Conv. - Liability and compensation for damage in connection with the carriage of hazardous and noxious substances by sea (HNS)		S			
2000 London	Protocol - Pollution incidents by hazardous and noxious substances (OPRC-HNS)					
1997 Vienna	Conv. - Supplementary compensation for nuclear damage				S	
1997 Vienna	Conv. - Joint convention on the safety of spent fuel management and on the safety of radioactive waste management	Y		R		R
1997 New York	Conv. - Law of the non-navigational uses of international watercourses					
1998 Rotterdam	Conv. - Prior informed consent procedure for hazardous chemicals and pesticides (PIC)	Y	R	R	R	S
2001 London	Conv. - Civil liability for bunker oil pollution damage					
2001 London	Conv. - Control of harmful anti-fouling systems on ships				S	
2001 Stockholm	Conv. - Persistent organic pollutants	Y	R	R	R	S

Source: IUCN; OECD.

OECD EPR / SECOND CYCLE

Y = in force   S = signed   R = ratified   D = denounced

	AUS	NZL	AUT	BEL	CZE	DNK	FIN	FRA	DEU	GRC	HUN	ISL	IRL	ITA	LUX	NLD	NOF	POL	PRT	SVK	ESP	SWE	CHE	TUR	UKD	EU
		R	R	R	R	R	R	R	R		R				R	R	R	R	R	R		R	R	R	R	R
			S	S	S				S						S							S	S		S	
R	R		R		R	S	R	R	R		R	R	R		R	R	S		R	R	R		R		
											R				R	R				R					
R	R				R	R	R	R	R		R	R	R		R	R	R	R		R	R	R	R		
			S	S	S	S	R								R		R			R					
R	R	R	R	R	R	R	R	R	R	R	R	R	R	R	R	R	R	R	R	R	R	R	R	R	R
R	R	R	R	R	R	R	R	R	R	R	S	R	R	R	R	R	R	R	R	R	R	R	R	R	R
R	R	R	R	R	R	R	R	R	R	R	R	R	R	R	R	R	R	R	R	R	R	R	R	R	R
S	R	R	R	R	R	R	R	R	R	R	R	R	R	R	R	R	R	R	R	R	R	R		R	R
R	R	R	R	R	R	R	R	R	R	R	R	R	R	R	R	R	R	R	R	R	R	R	R		
		R									R				R				R						
R											R				R				R					R	
R		R	R	R	R	R	R	R	R	S	R	R	R	R	R	R	R	R	R	R	R	R	R	R	R
R	R	R	R	R	R	R	R	R	R	R	R	R	R	R	R	R	R	R	R	R	R	R	R	R	R
			S	S		S						S	S						S			S			
			S	S	S	S	R					R		R					R						
S		S						S																	
R		R	R	R	R	R	R	R	R		R	R	R	R	R	R	R	R		R	R		R		
			R		S		R				S	R	R		R		R								
R	R	R	R	R	R	R	R	R	R	R	R	R	R	R	R	R	R	R		R	R	R	S	R	R
					R						S	R			R		S								
S			R	S		R					R		R	R		R	R								
R	R	R	R	R	R	R	R	R	R	S	R	S	S	R	R	R	S	R	R	R	R	S	R	R	

II.B: SELECTED MULTILATERAL AGREEMENTS (REGIONAL)

Y = in force S = signed R = ratified D = denounced

		CHN	CAN	MEX	
1948 Baguio	Agreem. - Establishment of the Asia-Pacific fishery commission	Y R			
1956 Rome	Agreem. - Plant protection for the Asia and Pacific region	Y R			
1957 Washington	Conv. - Conservation of North Pacific fur seals	Y	R		F
1969 Washington	Extension		R		F
1964 Brussels	Agreem. - Measures for the conservation of Antarctic Fauna and Flora	Y R			F
1966 Rio de Janeiro	Conv. - International convention for the conservation of Atlantic tunas (ICCAT)	Y R	R	R	F
1969 Rome	Conv. - Conservation of the living resources of the Southeast Atlantic	Y			
1972 London	Conv. - Conservation of Antarctic seals	Y	R		F
1980 Canberra	Conv. - Conservation of Antarctic marine living resources	Y	R		F
1993 Apia	Agreem. - South Pacific Regional Environment Programme (SPREP)	Y			S
1988 Bangkok	Agreem. - Network of aquaculture centres in Asia and the Pacific	Y R			
1990	Conv. - establishing a marine scientific organization for the North Pacific Region (PICES)	Y R	R		F
1992 Moscow	Conv. - Conservation of anadromous stocks (North Pacific Ocean)	Y	R		F
1993 Tokyo	Memorandum of understanding on port state control in the Asia-Pacific region	Y R	R		
1993 Canberra	Conv. - Conservation of Southern Pacific bluefin tuna	Y			
1993 Rome	Agreem. - Establishment of the Indian Ocean Tuna Commission	Y R			
1994 Lisbon	Treaty - Energy Charter	Y			
1994 Lisbon	Protocol (energy efficiency and related environmental aspects)	Y			

Source: IUCN; OECD.

OECD EPR / SECOND CYCLE

Y = in force S = signed R = ratified D = denounced

KOR	AUS	NZL	AUT	BEL	CZE	DNK	FIN	FRA	DEU	GRC	HUN	ISL	IRL	ITA	LUX	NLD	NOF	POL	PRT	SVK	ESP	SWE	CHE	TUR	UKD	EU
R	R	R						R														R				
R	R	R						R								R		R				R				
	R	R	R					R						R			R	R				R				
R								R					R	S			R		S		S		R	R	R	
R			R					R	R					R				R	R	R						
	R	S	R					R	R					R			R	R				R				
R	R	R	R			R	R	R	R					R		R	R	R		R	R	R	R			
	R	R						R																		
	R																									
R																										
R	R	R																								
R	R	R																								
R	R							R																	R	R
	S		R	R	R	R	R	R	R	R	R	S	R	R	R	R	R	S	R	R	R	R	R	R	R	R
	S		R	R	R	R	R	R	R	R	R	S	R	R	R	R	R	S	R	R	R	R	R	R	R	R

Reference III
ABBREVIATIONS

ADB	Asian Development Bank
APEC	Asia-Pacific Economic Co-operation
ASEAN	Association of Southeast Asian Nations
BOD	Biochemical oxygen demand
BRT	Bus rapid transit
CCICED	China Council for International Co-operation on Environment and Development
CDF	Clean Development Fund
CDM	Clean Development Mechanism
CFC	Chlorofluorocarbon
CITES	Washington Convention on International Trade in Endangered Species of Wild Fauna and Flora
COD	Chemical oxygen demand
CNY	Chinese yen
CPC	Communist Party of China
DRC	Development Research Center of the State Council
EANET	East Asia Acid Deposition Monitoring Network
EIA	Environmental impact assessment
EMS	Environmental Management System
ESCAP	United Nations Economic and Social Commission for Asia and the Pacific
EPB	Environmental Protection Bureau
ERPC	Environmental and Resources Protection Committee
ESCAP	United Nations Economic and Social Commission for Asia and the Pacific
EU	European Union
FAO	United Nations Food and Agriculture Organisation
FDI	Foreign direct investment
FGD	Flue gas desulphurisation
FTA	Free trade agreement
FYP	Five-Year Plan for National Economic and Social Development
FYEP	Five-Year Plan for Environmental Protection
GDP	Gross domestic product

GEF	Global Environment Facility
GHG	Greenhouse gas
GWt	Gigawatt
IFCS	Intergovernmental Forum on Chemical Safety
ILO	International Labour Organisation
IMO	International Maritime Organisation
IPPC	Integrated pollution prevention and control
ITTO	International Tropical Timber Organisation
IUCN	International Union for the Conservation of Nature and Natural Resources
LNG	Liquefied natural gas
LPG	Liquified petroleum gas
MEA	Multilateral environmental agreement
MOA	Ministry of Agriculture
MOC	Ministry of Construction
MOF	Ministry of Finance
MOFTEC	Ministry of Foreign Trade and Co-operation
MOWR	Ministry of Water Resources
Mtoe	Million tonnes of oil equivalent
MWt	Megawatt (a million watts)
NBS	National Bureau of Statistics of China
NDRC	National Development and Reform Commission
NEAC	Northeast Asian Conference on Environmental Co-operation
NEASPEC	Northeast Asian Subregional Programme of Environmental Co-operation
NGO	Non governmental organisation
NIP	National Implementation Plan
NPC	National People's Congress
NOWPAP	Northwest Pacific Action Plan
ODA	Official development assistance
ODP	Ozone-depleting potential
ODS	Ozone-depleting substance(s)
PAC	Pollution abatement and control
PM	Particulate matter
POP	Persistent organic pollutant
PPPs	Purchasing power parities
PRC	People's Republic of China
PRTR	Pollutant release and transfer register
R&D	Research and development
RTA	Regional trade agreement

SEPA	State Environmental Protection Administration
SFA	State Forestry Administration
SICP	Sino-Italian Co-operation Programme
SME	Small and medium-sized enterprise
TEMM	Tripartite Environment Ministers Meeting
TFC	Total final consumption of energy
TPES	Total primary energy supply
TPLM	Total pollution load management
UNCLOS	UN Convention on the Law of the Sea
UNCSD	UN Commission on Sustainable Development
UNCCD	UN Convention to Combat Desertification
UNDP	UN Development Programme
UNEP	UN Environment Programme
UNFCCC	UN Framework Convention on Climate Change
UNESCO	UN Educational, Scientific and Cultural Organisation
VAT	Value added tax
VOC	Volatile organic compound
WHO	World Health Organisation
WSSD	World Summit on Sustainable Development
WTO	World Trade Organization
WWF	World Wide Fund for Nature

Reference IV
SELECTED ENVIRONMENTAL WEBSITES

Website	Host institution

Government

Website	Host institution
www.zhb.gov.cn	State Environmental Protection Administration of China
www.china.org.cn	China Internet Information Center
www.gov.cn/ Official	Portal of the Government of People's Republic of China
www.ndrc.gov.cn/	National Development and Reform Commission (NDRC)
www.cciced.org/	China Council for International Cooperation on Environment and Development
www.mlr.gov.cn	Ministry of Land and Resources
www.forestry.gov.cn	State Forestry Administration
www.stats.gov.cn	National Bureau of Statistics of China
www.mwr.gov.cn	Ministry of Water Resources
www.cnemc.cn	Environmental Monitoring of China
www.agri.gov.cn	Ministry of Agriculture
www.soa.org.cn	State Oceanic Administration
www.chinacp.com/	Cleaner Production in China
www.hrc.gov.cn/	The HuaiHe river commission of the Ministry of Water Resources
www.yrwr.com.cn/	Yellow river water resources protection bureau
www.ywrp.gov.cn/	ChangJiang water resources protection bureau
www.cjw.gov.cn/	ChangJiang water conservancy net

Website	Host institution

Other

www.chinaenvironmental.com	China Environmental Net
www.sdino.net.cn	Sustainable Development in China
www.sdinfo.net.cn/hjinfo/hjinfo/	Environment Information Network of China
http://cbis.brim.ac.cn/cbise/index.html	Chinese Biodiversity Information System (CBIS)
www.plant.csdb.cn/	Chinese Plant Database
www.zoology.csdb.cn/index.asp	Chinese Zoology Database
www.chinabiodiversity.com/index.htm	Conserving China's Biodiversity
www.caf.ac.cn/db/slzy/forest.cfm	Chinese Forest Resource Database
www.china.org.cn/chinese/zhuanti/ 2004cxfz.htm	Report of Strategy for China's Sustainable Development 2004
www.efchina.org/home.cfm	Energy Foundation
www.cn-greenlights.gov.cn/ english/2B.htm	China Green Lights project
www.fon.org.cn/	Friends of Nature, China (NGO)
www.chinagev.org/	Green Earth Volunteers (NGO)
www.gvbchina.org/	Global Village of Beijing (NGO)

OECD PUBLICATIONS, 2, rue André-Pascal, 75775 PARIS CEDEX 16
PRINTED IN FRANCE
(97 2007 05 1 P) ISBN 978-92-64-03116-6 – No. 55419 2007